QUANTUM PHYSICS AND MODERN APPLICATIONS

Problems and Solutions

QUANTUM PHYSICS AND MODERN APPLICATIONS

Problems and Solutions

Seng Ghee Tan
Chinese Culture University, Taiwan

Ching Hua Lee
National University of Singapore, Singapore

Mansoor B A Jalil
National University of Singapore, Singapore

World Scientific

NEW JERSEY • LONDON • SINGAPORE • BEIJING • SHANGHAI • HONG KONG • TAIPEI • CHENNAI • TOKYO

Published by

World Scientific Publishing Co. Pte. Ltd.
5 Toh Tuck Link, Singapore 596224
USA office: 27 Warren Street, Suite 401-402, Hackensack, NJ 07601
UK office: 57 Shelton Street, Covent Garden, London WC2H 9HE

British Library Cataloguing-in-Publication Data
A catalogue record for this book is available from the British Library.

QUANTUM PHYSICS AND MODERN APPLICATIONS
Problems and Solutions

ISBN 978-981-127-039-0 (hardcover)
ISBN 978-981-127-101-4 (paperback)
ISBN 978-981-127-040-6 (ebook for institutions)
ISBN 978-981-127-041-3 (ebook for individuals)

For any available supplementary material, please visit
https://www.worldscientific.com/worldscibooks/10.1142/13250#t=suppl

Typeset by Stallion Press
Email: enquiries@stallionpress.com

Printed in Singapore

Preface

Quantum mechanics is an extension of classical mechanics in the absence of general relativity. Today, quantum mechanics permeates all walks of physics: nanoscience, high energy, condensed matter, bio-physics, quantum computation, optical physics, and quantum communication. This book is written with the view of providing learners with a fast track into the modern applications of quantum physics. This is a book of Problems and Solutions with about 125 long problems and their explicit solutions. As each problem is further divided into smaller parts, this book will be equivalent in strength to one with about 400 solved problems.

Modern research topics are definitely on our mind while writing this book. Problems are designed to suit recent contexts like the 2D graphene, topological materials, spintronics, and quantum information. We categorize the problems into 8 chapters. Chapter 1 is primarily about quantum mechanics for undergraduates with an emphasis on the Dirac formalism and its representation in matrix and function spaces. Problems and discussions around the Dirac formalism may be useful for advanced understanding. Chapter 2 is dedicated as a separate chapter to the spin physics. This is because in modern research, spin or to be more precise, spinor formalism is present in nearly all emerging topics, e.g., spintronics, graphene, topological systems, Dirac, Weyl, and all branches of quantum information sciences. Chapter 3 deals with second quantization and its applications in nanoscience and condensed matter physics. This chapter is also suitable for postgraduate students taking courses in advanced quantum mechanics. Chapter 4 deals with the non-equilibrium Green's Function (NEGF) — a modern topic with problems designed to suit applications in nanoscale electronic and spintronic systems. The requisite knowledge of second-quantization in Chapter 3 and spinor formalism in Chapter 2 is needed here. Chapter 5 is about gauge and topology with modern emphasis on applications in new materials like graphene and topological systems. Chapter 6 consists of numerous sub-topics in condensed matter physics. It comprises conventional topics like crystal lattice and band structure as well as modern problems in entanglement entropy. In Chapter 7, we present general problems of cross-disciplinary nature. Problems in this chapter are not necessarily quantum, but many deal with issues and visualizations that have inspired the development of quantum mechanics. Chapter 8 would cater specifically to the field of quantum computation and information. This chapter begins with a preamble on the quantum mechanical foundation of projection, measurement, and density matrix that would lead to applications in quantum gates, teleportation, and entanglement. Requisite knowledge in the quantum concepts of Chapter 1 and the spinor formalism of Chapter 2 will be useful. Problems in this chapter could also be exercises for a standard quantum mechanics course as well as an introduction to research in quantum information science.

All in all, we consider this book to be a useful material for

(1) postgraduate and senior undergraduate students keen on tackling problems in modern research fields,
(2) professors in all walks of physics and engineering fields,
(3) engineers, chemists, or physics experimentalists,
(4) self-learners who desire a glimpse into the world of theoretical physics.

Due to the explicit nature of our presentation, readers will find the learning method in this book handy and efficient. There is one caveat though, a solid command of undergraduate mathematics is necessary to make learning truly effective.

Seng Ghee Tan, C.H. Lee, and Mansoor B.A. Jalil

Contents

Preface		v
Chapter 1. Quantum Mechanics and Concepts		1
Basic Quantum Concepts		2
Problem 1.01	The Schrodinger Equations	2
Problem 1.02	Energy Eigenstates and Expectation	3
Problem 1.03	Quantum and Classical Physics of Momentum	5
Problem 1.04	Particle Current and Flux	7
Problem 1.05	Completeness relation: Resolution of Identity	9
Problem 1.06	Non-commutative Physics	10
Problem 1.07	Quantum Mechanic Pictures of Hamiltonian	12
Problem 1.08	Quantum Mechanic Pictures of Operators	13
Problem 1.09	Hermiticity: Let's Be Real	15
Problem 1.10	Hermiticity: Symmetry 1	17
Problem 1.11	Hermiticity: Symmetry 11	17
Problem 1.12	Parity Operator	19
Harmonic Oscillators		21
Problem 1.13	Harmonic Oscillator: Energy spectrum	21
Problem 1.14	Harmonic Oscillator: Raising and Lowering Operation	24
Problem 1.15	Harmonic Oscillator: Position and Momentum Matrix	25
Problem 1.16	Harmonic Oscillator: Energy Matrix	26
Problem 1.17	Harmonic Oscillator: Expectation Matrix Element	28
Problem 1.18	Harmonic Oscillator: Uncertainty Principle	29
Dirac's Bra-Ket Formalism		31
Problem 1.19	Dirac's Bra-Ket: General and Function Spaces	31
Problem 1.20	Momentum Operator	33

Problem 1.21 Position Operator 35
Problem 1.22 Non-commutative Position and Momentum 37
Problem 1.23 Explicit: Momentum Operator in Coordinate Representation 38
Problem 1.24 Explicit: Position Operator in Momentum Representation 39
Problem 1.25 Explicit: Momentum Eigenfunctions in Coordinate Representation 41

Chapter 2. Spin Physics 43

Introduction to Spin Physics 44
Problem 2.01 Quantum Spin States 44
Problem 2.02 Spin Expectation 48
Problem 2.03 Spin Operators and Hermiticity 53
Problem 2.04 Spin Operators and Commutative Property 55
Problem 2.05 Spin Eigenvalues and Eigenstates 56
Problem 2.06 Spin and Magnetism 58
Spin Transformation 60
Problem 2.07 Spin Precession: Time Dependence 60
Problem 2.08 Spin Precession: Heisenberg 62
Problem 2.09 Spin Precession: Heisenberg and Compact Notation 63
Problem 2.10 Spinor Transformation 66
Problem 2.11 Frame Rotation 69

Chapter 3. Second Quantization and Applications 73

Second Quantization 74
Problem 3.01 Second Quantization 1 74
Problem 3.02 Second Quantization 11 76
Problem 3.03 Field Operators 77
Problem 3.04 Second-Quantized Kinetic Energy 80
Problem 3.05 Second-Quantized Kinetic Energy for Discrete System 83
Problem 3.06 Second-Quantized Electron Spin 85
Problem 3.07 Second-Quantized Spin–Orbit Coupling 1 87
Problem 3.08 Second-Quantized Spin–Orbit Coupling 11 90
Dirac Delta Calculus 92
Problem 3.09 Dirac Delta and Fourier Transform 92

Problem 3.10 Dirac Delta and Calculus 94
Problem 3.11 Dirac Delta Representations 95
Problem 3.12 Vector Calculus and Dirac Delta 97
Condensed Matter Applications 101
Problem 3.13 Electron Background Energy 101
Problem 3.14 Electron Interaction Formulation 103
Problem 3.15 Electron–Electron Energy: Direct 104
Problem 3.16 Electron–Electron Energy: Exchange 106
Problem 3.17 Scattering Strength of Electron Interaction 109
Problem 3.18 Quantifying the Exchange Energy 112
Problem 3.19 Kinetic Energy per Particle 115
Appendix 3A. Dirac Delta Identities 118
Appendix 3B. Vector Calculus Identities 119
Appendix 3C. Fourier Transform Identities 120

Chapter 4. Non-equilibrium Green's Function 121

Green's Function for Quantum Transport 122
Problem 4.01 Green's Function for Quantum Electronics 122
Problem 4.02 Retarded and Advanced Green's Function — Fourier Transform 124
Problem 4.03 Green's Function: Density of States 126
Problem 4.04 Green's Function in Electronic Charge Distribution 1 128
Problem 4.05 Green's Function in Electronic Charge Distribution 11 129
Problem 4.06 Mathematical Methods: Trigonometric Integral . . . 131
Non-equilibrium Green's Function (NEGF) 133
Problem 4.07 Non-equilibrium Green's Function (NEGF): Electronic Charge Current 133
Problem 4.08 Non-equilibrium Green's Function (NEGF): Kinetic Spin Current 136
Problem 4.09 Non-equilibrium Green's Function (NEGF): Magnetic Spin Current 138
Problem 4.10 Non-equilibrium Green's Function: Spin–Orbit Spin Current 141

Chapter 5. Gauge and Topology 149

Coordinate Transformation 150
Problem 5.01 Coordinate Change and Basis Vectors 150
Problem 5.02 Coordinate Change and Component Vectors 153

Problem 5.03 Pauli Matrices and Coordinate Change 155
Berry-Pancharatnam Gauge 158
Problem 5.04 Frame Rotation Gauge Field 158
Problem 5.05 Path Integral and Gauge Field 159
Problem 5.06 Magnetic Monopole 162
Problem 5.07 Berry–Pancharatnam Phase in 2D Systems 165
Problem 5.08 Graphene Gauge Field 168
Problem 5.09 Graphene Berry–Pancharatnam Curvature 170
Problem 5.10 Spin Hall Gauge in a 2D system 1 171
Problem 5.11 Spin Hall Gauge in a 2D System 11 175
Problem 5.12 Spin Hall Gauge: Integral Calculus 177
Non-Abelian Gauge 179
Problem 5.13 Non-Abelian Gauge Transformation 1 179
Problem 5.14 Non-Abelian Gauge Transformation 11 180
Problem 5.15 Non-Abelian Gauge Curvature 182
Topology and Gauge 183
Problem 5.16 Topological Models 183
Problem 5.17 Exact Solution to the Su–Schrieffer–Heeger (SSH) Model 184
Problem 5.18 Topological Characterization of the SSH Model . . . 186
Problem 5.19 Topological Polarization of the 2D Chern Insulator 187
Problem 5.20 Berry Curvature of the Dirac Model 190
Appendix 5A. Spherical Coordinates in Different Bases 191

Chapter 6. Advanced Condensed Matter Physics 193

Crystal Lattice and Band structure 194
Problem 6.01 Crystal Unit Cell 194
Problem 6.02 Condensed Matter: Rotational Symmetry 194
Problem 6.03 Wannier Functions 195
Problem 6.04 Heat Capacity of a 3D Electron Gas 196
Problem 6.05 Magnetic Susceptibility 196
Problem 6.06 Quantum Harmonic Oscillator 197
Problem 6.07 Heat Capacity: D-Dimensional Lattice of Phonons 198
Problem 6.08 Quantum Heisenberg Chain 200
Problem 6.09 Band structure of a 2D Lattice 201

Entanglement Entropy 204
Problem 6.10 Entanglement Entropy of the Free Fermions 204
Problem 6.11 Entanglement Entropy of a Critical System 206
Problem 6.12 Entanglement Entropy and Conformal Mapping . . . 208

Chapter 7. General Physics 211

Minimization Physics 212
Problem 7.01 Minimum Principles in Physics 212
Problem 7.02 Shortest Time Trajectory 213
Problem 7.03 The Hamilton Principle 217
Electron Physics and Visualizations 220
Problem 7.04 What is the Size and Shape of an Electron? 220
Problem 7.05 What is Spin? 222
Problem 7.06 Single Electronic Motion: Master Equation 223
Problem 7.07 Semi-classical Electronic Transport 228
Mathematical Methods 232
Problem 7.08 Integral Method 232
Problem 7.09 Divergence Theorem 232
Problem 7.10 Barycentric Coordinates 235

Chapter 8. Quantum Computation and Information 239

Quantum States of Qubits 240
Problem 8.01 Mixed State: Statistical Ensemble 240
Problem 8.02 Density Matrix and Expectation 1 241
Problem 8.03 Density Matrix and Expectation 11 243
Problem 8.04 Density Matrix of a 2-Qubit System 245
Problem 8.05 Quantum Measurement 1 247
Problem 8.06 Quantum Measurement II 250
Problem 8.07 Projection Operator 255
Quantum Gates and Circuits 257
Problem 8.08 Quantum Logic Gates: NOT Gate 257
Problem 8.09 Quantum Logic Gates: Hadamard Gate 259
Problem 8.10 CNOT Gate 261
Problem 8.11 Toffoli Gate 263
Problem 8.12 Parallelism: Concept 266
Problem 8.13 Parallelism: Quantum Circuit and Bell States 268
Problem 8.14 No-Cloning Principle 271

Problem 8.15 Quantum Circuit: No-Cloning 272
Problem 8.16 Quantum Circuit: Swapping 273
Problem 8.17 Mathematical Methods: Boolean Algebra 276
Problem 8.18 Quantum Communication: Teleportation 278
Problem 8.19 Deutsch Algorithm 281

Chapter 1

Quantum Mechanics and Concepts

This chapter contains primarily problems in basic quantum mechanics. It covers the numerous aspects of quantum mechanics as follows: Schrodinger equations, eigenfunctions and eigenvalues, completeness, and so on. The concepts of non-commutativity and Hermiticity in physics are also introduced. There are problems of harmonic oscillations which also introduce the matrix formulation of quantum mechanics. In the Dirac formalism, discussion hovers around its representation in general, matrix and function spaces. This chapter is suitable as a supplementary material for students learning quantum mechanics at the level of undergraduate. However, problems and discussions around the Dirac formalism may also be useful for deeper thought and advanced understanding.

Basic Quantum Concepts

Problem 1.01 The Schrodinger Equations
Problem 1.02 Energy Eigenstates and Expectation
Problem 1.03 Quantum and Classical Physics of Momentum
Problem 1.04 Particle Current and Flux
Problem 1.05 Completeness Relation: Resolution of Identity
Problem 1.06 Non-commutative Physics
Problem 1.07 Quantum Mechanic Pictures of Hamiltonian
Problem 1.08 Quantum Mechanic Pictures of Operators
Problem 1.09 Hermiticity: Let's Be Real
Problem 1.10 Hermiticity: Symmetry 1
Problem 1.11 Hermiticity: Symmetry 11
Problem 1.12 Parity Operator

Harmonic Oscillators

Problem 1.13 Harmonic Oscillator: Energy Spectrum
Problem 1.14 Harmonic Oscillator: Raising and Lowering Operation
Problem 1.15 Harmonic Oscillator: Position and Momentum Matrix
Problem 1.16 Harmonic Oscillator: Energy Matrix
Problem 1.17 Harmonic Oscillator: Expectation Matrix Element
Problem 1.18 Harmonic Oscillator: Uncertainty Principle

Dirac's Bra-Ket Formalism

Problem 1.19 Dirac's Bra-Ket: General and Function Spaces
Problem 1.20 Momentum Operator
Problem 1.21 Position Operator
Problem 1.22 Non-commutative Position and Momentum
Problem 1.23 Explicit: Momentum Operator in Coordinate Representation
Problem 1.24 Explicit: Position Operator in Momentum Representation
Problem 1.25 Explicit: Eigenfunctions in Coordinate Representation

Basic Quantum Concepts

Problem 1.01 The Schrodinger Equations

In quantum mechanics, the time-dependent Schrodinger equation is

$$i\hbar \frac{\partial \psi(x,t)}{\partial t} = -\hbar^2/2m \frac{\partial^2 \psi(x,t)}{\partial x^2} + V\psi(x,t)$$

where V is the potential energy of the system. Show that a time-independent Schrodinger equation can be derived from the above.

Solution

The time-dependent Schrodinger equation consists of the kinetic energy and the potential energy on the RHS. It's important that the potential energy V is not time-dependent, i.e., V $\neq V(x,t)$. Now, one writes for the solution of the system a function which is separable in time and space.

$$\psi(x,t) = R(x)D(t)$$

Substituting into the time-dependent Schrodinger equation,

$$i\hbar \frac{1}{D(t)} \frac{\partial D(t)}{\partial t} = -\hbar^2/2m \frac{1}{R(x)} \frac{\partial^2 R(x)}{\partial x^2} + VR(x)$$

Since variables t and x are now separated, both sides must be a constant, one has

$$i\hbar \frac{1}{D(t)} \frac{\partial D(t)}{\partial t} = E \to D(t) = Ce^{-\frac{iEt}{\hbar}}$$

and

$$-\hbar^2/2m \frac{1}{R(x)} \frac{\partial^2 R(x)}{\partial x^2} + VR(x) = E \to -\hbar^2/2m \frac{\partial^2 R(x)}{\partial x^2} + VR(x) = ER(x)$$

The time-independent Schrodinger equation is

$$H = -\hbar^2/2m \frac{\partial^2}{\partial x^2} + V$$
$$HR(x) = ER(x)$$

H is the operator aka Hamiltonian. $R(x)$ is the spatial part of the eigenfunction. E is the eigenvalue.

Problem 1.02 Energy Eigenstates and Expectation

A particle in an infinite square well is prepared in an initial state of $|\psi\rangle = \frac{\sqrt{3}}{2}|\phi_1\rangle + \frac{1}{2}|\phi_2\rangle$. States $|\phi_1\rangle$ and $|\phi_2\rangle$ are the two lowest energy eigenstates of the square well, with eigenvalues ε_1 and ε_2, respectively, where $\varepsilon_1 \neq \varepsilon_2$. The initial state is clearly not the eigenstate of the system. Which one of the following is true as the state $|\psi\rangle$ evolves with time?

(A) The expectation value of the energy varies with a beat-like pattern in time consisting of two frequencies of $\omega_1 = \frac{\varepsilon_1}{\hbar}$ and $\omega_2 = \frac{\varepsilon_2}{\hbar}$.
(B) The expectation value of energy is constant at $\frac{3}{4}\varepsilon_1 + \frac{1}{4}\varepsilon_2$ at all times.
(C) The expectation value of energy varies sinusoidally in time between ε_1 and ε_2.
(D) The expectation value of energy varies sinusoidally with an amplitude $\frac{3}{4}\varepsilon_1 + \frac{1}{4}\varepsilon_2$.

Note: Bra-Ket notations, e.g., $|\psi\rangle$ are used in this problem, but a detailed understanding of its formalism is not required at this stage. Further elaboration of Dirac's Bra-Ket formalism will be covered. At this stage, just take note that Ket $|\psi\rangle$ is an alternative to the function of $\psi(x)$ when taken in coordinate representation.

Solution

Initial state: $|\psi\rangle = \frac{\sqrt{3}}{2}|\phi_1\rangle + \frac{1}{2}|\phi_2\rangle$ is not the eigenstate of the system. Like its function form, the initial state can be written as

$$|\psi\rangle = \frac{\sqrt{3}}{2}|\phi_1\rangle e^{-i\frac{\varepsilon_1 t}{\hbar}} + \frac{1}{2}|\phi_2\rangle e^{-i\frac{\varepsilon_2 t}{\hbar}}$$

The Hamiltonian of the system is H, thus

$$\begin{aligned} H|\psi\rangle &= \frac{\sqrt{3}}{2}H|\phi_1\rangle e^{-i\frac{\varepsilon_1 t}{\hbar}} + \frac{1}{2}H|\phi_2\rangle e^{-i\frac{\varepsilon_2 t}{\hbar}} \\ &= \frac{\sqrt{3}}{2}\varepsilon_1|\phi_1\rangle e^{-i\frac{\varepsilon_1 t}{\hbar}} + \frac{1}{2}\varepsilon_2|\phi_2\rangle e^{-i\frac{\varepsilon_2 t}{\hbar}} \end{aligned}$$

due to the fact that the system has $|\phi_1\rangle$ and $|\phi_2\rangle$ as the two lowest energy eigenstates, with eigenvalues ε_1 and ε_2. Now, let us write down the transpose conjugate of $|\psi\rangle$,

$$\langle\psi| = \frac{\sqrt{3}}{2}\varepsilon_1\langle\phi_1|e^{i\frac{\varepsilon_1 t}{\hbar}} + \frac{1}{2}\varepsilon_2\langle\phi_2|e^{i\frac{\varepsilon_2 t}{\hbar}}$$

For a clearer picture, note is to be taken of the following correspondence:

$$\begin{aligned} |\psi\rangle &\rightarrow \psi(x) \\ \langle\psi| &\rightarrow \psi^*(x) \end{aligned}$$

The expectation value for the energy is

$$\begin{aligned} \langle\psi|H|\psi\rangle &= \left(\frac{\sqrt{3}}{2}\langle\phi_1|e^{i\frac{\varepsilon_1 t}{\hbar}} + \frac{1}{2}\langle\phi_2|e^{i\frac{\varepsilon_2 t}{\hbar}}\right)\left(\frac{\sqrt{3}}{2}\varepsilon_1|\phi_1\rangle e^{-i\frac{\varepsilon_1 t}{\hbar}} + \frac{1}{2}\varepsilon_2|\phi_2\rangle e^{-i\frac{\varepsilon_2 t}{\hbar}}\right) \\ &= \frac{3}{4}\varepsilon_1 + \frac{1}{4}\varepsilon_2 \end{aligned}$$

Conclusion: the expectation value of energy is constant at $\frac{3}{4}\varepsilon_1 + \frac{1}{4}\varepsilon_2$ at all times.

Remarks and Reflections

Note there is a correspondence between a state in function space and one in the general Bra-Ket space.

$$\langle \phi_{\rm m} | \phi_{\rm n} \rangle = \delta_{\rm mn} \qquad \text{General Bra-Ket space}$$

$$\int_{-\infty}^{+\infty} \phi_{\rm m}^*(x)\phi_{\rm n}(x)dx = \delta_{\rm mn} \qquad \text{Function space}$$

In the solution, the general Bra-Ket space is used.

Problem 1.03 Quantum and Classical Physics of Momentum

In classical physics, momentum is given by $p = mv = m\frac{dx}{dt}$. It makes sense to speculate that momentum in quantum physics can be written as $\langle p \rangle = m\frac{d\langle x \rangle}{dt}$, where p is the momentum operator. In the following,

(a) show that a valid expression for the momentum operator is

$$p = \frac{\hbar}{i}\frac{\partial}{\partial x}$$

(b) show that the expectation of p with this operator expression is real so that classical and quantum physics is consistent.

Solution

(a) The expectation for an operator A in a quantum mechanical system with wavefunction $\psi(x,t)$ is given by

$$\langle A \rangle = \int_{-\infty}^{+\infty} dx \psi^*(x,t) A \psi(x,t)$$

Therefore, the classical expression for momentum is related to its quantum mechanical expression as follows:

$$\langle p \rangle = m\frac{d\langle x \rangle}{dt} \rightarrow \langle p \rangle = m\frac{d}{dt}\int_{-\infty}^{+\infty} dx \psi^*(x,t) x \psi(x,t)$$

Note that the Schrodinger equation is

$$i\hbar\frac{\partial \psi(x,t)}{\partial t} = -\frac{\hbar^2}{2m}\frac{\partial^2 \psi(x,t)}{\partial x^2}$$

The momentum expectation is therefore

$$\langle p \rangle = \frac{\hbar}{2i}\int_{-\infty}^{+\infty} dx \left(\frac{\partial^2 \psi^*}{\partial x^2} x \psi - \psi^* x \frac{\partial^2 \psi}{\partial x^2} \right)$$

The following pair of expressions are written to simplify the process of integration above

$$\frac{\partial}{\partial x}\left[\frac{\partial \psi^*}{\partial x} x\psi\right] = \frac{\partial^2 \psi^*}{\partial x^2} x\psi + \frac{\partial \psi^*}{\partial x}\psi + \frac{\partial \psi^*}{\partial x} x \frac{\partial \psi}{\partial x}$$

$$\frac{\partial}{\partial x}\left[\frac{\partial \psi}{\partial x} x\psi^*\right] = \frac{\partial^2 \psi}{\partial x^2} x\psi^* + \frac{\partial \psi}{\partial x}\psi^* + \frac{\partial \psi}{\partial x} x \frac{\partial \psi^*}{\partial x}$$

One can now derive the following:

$$\langle p \rangle = \frac{\hbar}{2i}\int_{-\infty}^{+\infty} dx \frac{\partial}{\partial x}\left(\frac{\partial \psi^*}{\partial x} x\psi - \frac{\partial \psi}{\partial x} x\psi^* - \psi^*\psi\right) + 2\psi^* \frac{\partial \psi}{\partial x}$$

As the wavefunction ψ is square-integrable and vanishes much more rapidly than x tending to infinity, the integrand

$$\frac{\partial}{\partial x}\left(\frac{\partial \psi^*}{\partial x} x\,\psi - \frac{\partial \psi}{\partial x} x\,\psi^* - \psi^*\psi\right)$$

is eliminated upon integration. One is left with

$$\langle p \rangle = \int_{-\infty}^{+\infty} dx\, \psi^* \frac{\hbar}{i}\frac{\partial \psi}{\partial x}$$

And the operator expression

$$p = \frac{\hbar}{i}\frac{\partial}{\partial x}$$

is deduced.

(b) The expectation of p must be real if quantum and classical physics are to be consistent. Let's examine the following:

$$\langle p \rangle^* = -\int_{-\infty}^{+\infty} dx\, \psi \frac{\hbar}{i}\left(\frac{\partial \psi}{\partial x}\right)^* = -\int_{-\infty}^{+\infty} dx\, \psi \frac{\hbar}{i}\frac{\partial \psi^*}{\partial x}$$

The above leads to

$$\langle p \rangle - \langle p \rangle^* = \int_{-\infty}^{+\infty} dx\, \psi^* \frac{\hbar}{i}\frac{\partial \psi}{\partial x} + \psi \frac{\hbar}{i}\frac{\partial \psi^*}{\partial x}$$

$$\langle p \rangle - \langle p \rangle^* = \frac{\hbar}{i}\int_{-\infty}^{+\infty} dx \frac{\partial(\psi^*\psi)}{\partial x} = [\psi^*\psi]_{-\infty}^{+\infty} = 0$$

The above follows from the square integrability of the wavefunctions, i.e., they necessarily vanish at the infinite limits. The expectation of p is therefore real regardless of the state of the system.

Problem 1.04 Particle Current and Flux

In quantum mechanics, flux (J) or the probability current is an important physical quantity that obeys the conservation law of

$$\frac{\partial P(x,t)}{\partial t} + \frac{\partial J(x,t)}{\partial x} = 0$$

where $P = \psi^* \psi$ is the probability density of the particle in space.

(a) Show that flux or probability current in 1D is given by

$$J = \frac{-i\hbar}{2m}\left(\psi^\dagger(x,t)\frac{\partial}{\partial x}\psi(x,t) - \psi(x,t)\frac{\partial}{\partial x}\psi^\dagger(x,t)\right)$$

(b) Show that the energy term of $e\boldsymbol{J}.\boldsymbol{A}$ can be derived from

$$H = \int \frac{1}{2m}\psi^\dagger(p + eA)^2\psi\, d\boldsymbol{x}$$

Solution

(a) In quantum physics, current is given by $\psi^\dagger \frac{\partial H}{\partial p}\psi$, where $\frac{\partial H}{\partial p}$ has the physical meaning of velocity where H is the Hamilton in classical physics. To make sure the expression is Hermitian, we perform a symmetrization process as follows:

$$J = \frac{1}{2}\int\left(\psi^\dagger\frac{\partial H}{\partial p}\psi + \left(\frac{\partial H}{\partial p}\psi\right)^\dagger\psi\right)$$

Now,

$$H = \frac{p^2}{2m} \to \frac{\partial H}{\partial p} = \frac{p}{m}$$

It thus follows that

$$J = \frac{1}{2m}(\psi^\dagger p\,\psi + (p\,\psi)^\dagger\psi) = \frac{-i\hbar}{2m}\left(\psi^\dagger\frac{\partial}{\partial x}\psi - \left(\frac{\partial}{\partial x}\psi^\dagger\right)\psi\right)$$

(b) To find the interaction energy, we expand the kinetic energy,

$$H = \int \frac{1}{2m}\psi^\dagger(p + e\boldsymbol{A})^2\psi\, d\boldsymbol{x}$$

Note that integration limit is taken from $-\infty$ to $+\infty$.

$$H = \frac{1}{2m}\int \psi^\dagger(\boldsymbol{p}^2 + e\,\boldsymbol{p}\cdot\boldsymbol{A} + e\boldsymbol{A}\cdot\boldsymbol{p})\psi + e^2\psi^\dagger\boldsymbol{A}^2\psi\,d\boldsymbol{x}$$

This leads to

$$H = \frac{1}{2m}\int \psi^\dagger(-\hbar^2\nabla^2\psi) + [-i\hbar\boldsymbol{\nabla}.(\psi^\dagger e\boldsymbol{A}\psi) - (-i\hbar\boldsymbol{\nabla}\psi^\dagger)\cdot(e\boldsymbol{A}\psi)]$$
$$- i\hbar\psi^\dagger e\boldsymbol{A}\cdot\boldsymbol{\nabla}\psi + e^2\psi^\dagger\boldsymbol{A}^2\psi\,d\boldsymbol{x}$$

The first term (left) in the square bracket is the surface term that goes to zero upon integration.

$$\int -i\hbar\boldsymbol{\nabla}.(\psi^\dagger e\boldsymbol{A}\,\psi)\,d\boldsymbol{x} \to 0$$

The expression $e^2\psi^\dagger\boldsymbol{A}^2\psi$ is a high-order term that can be neglected. Thus, what follows is

$$H = \frac{1}{2m}\int -\hbar^2\psi^\dagger\nabla^2\psi - i\hbar e\boldsymbol{A}\cdot(\psi^\dagger\boldsymbol{\nabla}\psi) + i\hbar\,\psi(\boldsymbol{\nabla}\psi^\dagger)\cdot e\boldsymbol{A}\,d\boldsymbol{x}$$

Dropping the kinetic energy term and considering only the interaction energy term, one has

$$H = \frac{-i\hbar e}{2m}\int \boldsymbol{A}\cdot(\psi^\dagger\boldsymbol{\nabla}\psi - \psi\boldsymbol{\nabla}\psi^\dagger)\,d\boldsymbol{x} = e\,\boldsymbol{J}\cdot\boldsymbol{A}$$

Remarks and Reflections

Note that the expression for current $J = \frac{-i\hbar}{2m}\left(\psi^\dagger(x,t)\frac{\partial}{\partial x}\psi(x,t) - \psi(x,t)\frac{\partial}{\partial x}\psi^\dagger(x,t)\right)$ is only true for $H = \frac{p^2}{2m}$ in the non-relativistic limit.

The expression $J = \frac{-i\hbar}{2m}\left(\psi^\dagger(x,t)\frac{\partial}{\partial x}\psi(x,t) - \psi(x,t)\frac{\partial}{\partial x}\psi^\dagger(x,t)\right)$ can also be derived from

$$\frac{\partial P(x,t)}{\partial t} + \frac{\partial J(x,t)}{\partial x} = 0$$

where $P(x,t)$ is the probability density, and

$$\frac{\partial P(x,t)}{\partial t} = \frac{\partial\psi^\dagger}{\partial t}\psi + \psi^\dagger\frac{\partial\psi}{\partial t}$$

Problem 1.05 Completeness relation: Resolution of Identity

Resolution of identity follows from the completeness relation. It is a familiar technique in the mathematics of quantum physics. Show that in discrete form, one can write

$$\sum_i |u_i\rangle\langle u_i| = 1$$

For illustration, we will use the completeness relation for a continuous system. The objective is to relate quantum mechanics in general vector form to function form. Let's choose the position of eigenstates as our continuous system:

$$\sum_i |u_i\rangle\langle u_i| = 1 \rightarrow \int |x\rangle\langle x|dx = 1$$

Note: Bra-Ket notations, e.g., $|\psi\rangle$ *are used in this problem, but a detailed understanding of its formalism is not required at this stage. Further elaboration of Dirac's Bra-Ket formalism will be covered. At this stage, just take note that Ket* $|\psi\rangle$ *is an alternative to the function of* $\psi(x)$ *when taken in coordinate representation.*

Solution

The state vector is written on the basis of discrete eigenkets $|u_i|$ as follows:

$$|\psi\rangle = \sum a_i|u_i\rangle$$

Focusing on one particular component,

$$\langle u_j|\psi\rangle = \langle u_j|\sum a_i|u_i\rangle$$

With the use of the Kronecker delta function (the discrete version of the Dirac delta), the above is reduced to one component as shown in the following:

$$\langle u_j|\psi\rangle = \sum_i \delta_{ij}a_i = a_j$$

Using the above, one can now write

$$|\psi\rangle = \sum a_i|u_i\rangle = \sum \langle u_i|\psi\rangle|u_i\rangle$$

As $\langle u_i|\psi\rangle$ is a scalar, one moves the eigenket to the left, resulting in

$$|\psi\rangle = \sum |u_i\rangle\langle u_i|\psi\rangle$$

Comparing both sides, one deduces that

$$\sum_i |u_i\rangle\langle u_i| = 1$$

Problem 1.06 Non-commutative Physics

In quantum mechanics, physical quantities are represented by operators. Therefore, a lot of times, they do not commute in sequence. The following are examples of operators that do not commute:

(a) Show that $[p_x, x] = -i\hbar \neq 0$.
(b) Show that when $[p, r] = a$ where a is a scalar, $[p^2, r] = 2p[p, r]$.
(c) Prove the identity $[AB, C] = A[B, C] + [A, C]B$.
(d) Hermitian physics has it that $H = x + ip_x$ corresponds to $H^* = x - ip_x$, show that $[H,\ H^*] = 2\hbar$.
(e) Show that $\left(\frac{1}{2\hbar}\right)[H^*H, H] = -H$ and $\left(\frac{1}{2\hbar}\right)[H^*H, H^*] = H^*$.

Note: The commutator bracket is defined as $[A, B] = AB - BA$.

Solution

(a) Note that in coordinate representation, the momentum and position operators are, respectively, $p_x = -\frac{i\hbar\partial}{\partial x}$, $x = x$. This leads to

$$\begin{aligned}\int \psi^\dagger(x)[p_x, x]\psi(x)\,dx &= \int \psi^\dagger(x)\left(-\frac{i\hbar\partial}{\partial x}x + x\frac{i\hbar\partial}{\partial x}\right)\psi(x)\,dx \\ &= \int \psi^\dagger\left(-i\hbar\psi - x\frac{i\hbar\partial}{\partial x}\psi + x\frac{i\hbar\partial}{\partial x}\psi\right)dx \\ &= \int \psi^\dagger(-i\hbar)\psi\,dx\end{aligned}$$

Comparing integrands on both sides, one has

$$[p_x, x] = -i\hbar \neq 0$$

(b) Since $[p, r] = a$,

$$pr - rp = a$$
$$\rightarrow pr = a + rp, \quad rp = pr - a$$

Now, as $[p^2, r] = p^2r - rp^2$, one substitutes the above into $p^2r - rp^2$ and obtains

$$\begin{aligned}[p^2, r] &= p(a + rp) - (pr - a)p \\ &= pa + ap = 2pa\end{aligned}$$

Recalling that $[p, r] = a$, one has

$$[p^2, r] = 2p\,[p, r]$$

(c) To show $[AB, C] = A[B, C] + [A, C]B$, one examines the RHS and the LHS separately:

$$LHS: \quad [AB, C] = ABC - CAB$$
$$\begin{aligned}RHS: \quad A\,[B, C] + [A, C]B &= ABC - ACB + ACB - CAB \\ &= ABC - CAB = [AB, C]\end{aligned}$$

(d) With $H = x + ip_{\mathrm{x}}$ and $H^* = x - ip_x$,

$$\begin{aligned}[H, H^*] &= (x + ip_x)(x - ip_x) - (x - ip_x)(x + ip_x) \\ &= -2ixp_x + 2ip_x x\end{aligned}$$

Recalling that $[p_x, x] = -i\hbar$, one has $[H, H^*] = 2\hbar$

(e) Let's begin from the LHS

$$\begin{aligned}\left(\frac{1}{2\hbar}\right)[H^*H, H] &= \left(\frac{1}{2\hbar}\right)(H^*HH - HH^*H) \\ &= -\left(\frac{1}{2\hbar}\right)[H, H^*]H \\ &\Longrightarrow \left(\frac{1}{2\hbar}\right)[H^*H, H] = -H\end{aligned}$$

$$\left(\frac{1}{2\hbar}\right)[H^*H, H^*] = \left(\frac{1}{2\hbar}\right)(H^*HH^* - H^*H^*H)$$

$$= \left(\frac{1}{2\hbar}\right) H^*[H, H^*]$$

$$\Longrightarrow \left(\frac{1}{2\hbar}\right) [H^*H, H^*] = H^*$$

Remarks and Reflections

Commutative relations are widely used in non-Abelian gauge which has an origin in the high-energy physics. Today, the application of non-Abelian physics is widespread, from condensed matter physics to the more specialized fields of quantum spintronics and photonics.

Exercise: Derive a similar identity involving $[A, BC]$ and prove the Jacobi identity $[A, [B, C]] + [B, [C, A]] + [C, [A, B]] = 0$.

Problem 1.07 Quantum Mechanic Pictures of Hamiltonian

The Schrodinger picture Hamiltonian is given by $H_S = H_0 + V_S$. The different pictures of the operators and the Hamiltonian are summarized in Table 1. Show that the Hamiltonians of H_H (Heisenberg) and H_S (Schrodinger) are time-independent. Find the Hamiltonian of H_I (interaction).

Solution

In the Schrodinger picture,

$$H_S = H_0 + V_S$$

where V_S is the potential energy. In the Heisenberg picture, an operator is time-dependent and is defined as

$$A_H(t) = e^{iH_S t} A_S e^{-iH_S t}$$

Table 1.

	Operators	**Hamiltonian**
Schrodinger Picture	A_S	$H_S = H_0 + V_S$
Heisenberg picture	$A_H(t) = e^{iH_S t} A_S e^{-iH_S t}$	$H_H = H_S$
Interaction picture	$A_I(t) = e^{iH_0 t} A_S e^{-iH_0 t}$	$H_I = H_0 + V_I(t)$

Applying this to the Hamiltonian would simply result in

$$H_H = e^{iH_S t} H_S e^{-iH_S t} = H_S$$

since H_S commute. Therefore, both the Hamiltonians of H_H (Heisenberg) and H_S (Schrodinger) are time-independent. We, therefore, write

$$H = H_H = H_S$$

Now, one can set out to find H_I. In the Heisenberg picture, an operator is time-dependent and is defined as

$$A_I(t) = e^{iH_0 t} A_S e^{-iH_0 t}$$

Applying this to the Hamiltonian,

$$\begin{aligned} H_I(t) &= e^{iH_0 t}(H_S) e^{-iH_0 t} \\ &= e^{iH_0 t}(H_0 + V_S) e^{-iH_0 t} \\ &= H_0 + e^{iH_0 t}(V_S) e^{-iH_0 t} \end{aligned}$$

Therefore,

$$H_I(t) = H_0 + V_I(t) \quad \text{where } V_I(t) = e^{iH_0 t}(V_S) e^{-iH_0 t}$$

Problem 1.08 Quantum Mechanic Pictures of Operators

Show the relationship between the arbitrary operators A of different pictures, i.e., A_S, A_H, A_I. Show that when $V_s = 0$, the interaction and the Heisenberg pictures coincide.

Solution

Referring to this table

	Operators	**Hamiltonian**
Schrodinger Picture	A_S	$H_S = H_0 + V_S$
Heisenberg picture	$A_H(t) = e^{iH_S t} A_S e^{-iH_S t}$	$H_H = H_S$
Interaction picture	$A_I(t) = e^{iH_0 t} A_S e^{-iH_0 t}$	$H_I = H_0 + V_I(t)$

One can set out to derive the following relationships:

Interaction and Schrodinger relationship is obtained straight from the above table

$$A_I(t) = e^{iH_0 t} A_S e^{-iH_0 t}$$

where $H_s = H_o + V_S$, and H_o is the non-interacting part of the Hamiltonian.

Heisenberg and Schrodinger relationship is obtained straight from the above table

$$A_H(t) = e^{iH_S t} A_s e^{-iH_S t}$$

Heisenberg and interaction:

$$\begin{aligned} A_H(t) &= e^{iH_S t}(A_S) e^{-iH_S t} \\ &= e^{iH_S t}(e^{-iH_0 t} A_I e^{iH_0 t}) e^{-iH_S t} \end{aligned}$$

When $V_S = 0$, one has $H_S = H_o + V_S = H_o$. Therefore,

$$\begin{aligned} A_H(t) = e^{iH_S t}(e^{-iH_0 t} A_I e^{iH_0 t}) e^{-iH_S t} \\ \Longrightarrow A_H = e^{iH_0 t}(e^{-iH_0 t} A_I e^{iH_0 t}) e^{-iH_0 t} \\ \Longrightarrow A_H = A_I \end{aligned}$$

Remarks and Reflections

Summary: Schrodinger, Heisenberg, and interaction pictures

	Operators	**State Vectors**
Schrodinger Picture	A_S Hamiltonian: $H_S = H_0 + V_S$	$\vert\psi_S(t)\rangle = e^{-iH_S t}\vert\psi_0\rangle$
Heisenberg picture	$A_H(t) = e^{iH_S t} A_S e^{-iH_S t}$ Hamiltonian: $H_H = H_S$	$\vert\psi_H\rangle = e^{iH_S t}\vert\psi_S(t)\rangle = \vert\psi_0\rangle$
Interaction picture	$A_I(t) = e^{iH_0 t} A_S e^{-iH_0 t}$ Hamiltonian: $H_I = H_0 + e^{iH_0 t}(V_S) e^{-iH_0 t}$ $= H_0 + V_I(t)$	$\vert\psi_I(t)\rangle = e^{iH_0 t}\vert\psi_S(t)\rangle$

$|\psi_0\rangle$ is the initial time-independent state vector. Expectation value of operator A under different pictures:

Schrodinger picture

$$\langle\psi_S(t)|A_S|\psi_S(t)\rangle$$

Heisenberg picture

$$\begin{aligned}\langle\psi_H|A_H(t)|\psi_H\rangle &= \langle\psi_S(t)|e^{-iH_St}e^{iH_St}A_Se^{-iH_St}e^{iH_St}|\psi_S(t)\rangle \\ &= \langle\psi_S(t)|A_S|\psi_S(t)\rangle\end{aligned}$$

Interaction picture

$$\begin{aligned}\langle\psi_I(t)|A_I(t)|\psi_I(t)\rangle &= \langle\psi_S(t)|e^{-iH_0t}e^{iH_0t}A_Se^{-iH_0t}e^{iH_0t}|\psi_S(t)\rangle \\ &= \langle\psi_S(t)|A_S|\psi_S(t)\rangle\end{aligned}$$

Problem 1.09 Hermiticity: Let's Be Real

A Hermitian operator is the physicists' term for what the mathematicians call a self-adjoint operator. Hermitian operators are particularly important in quantum mechanics because they signify real eigenvalues. In quantum mechanics, the momentum operator is written as follows:

$$p = \frac{\hbar}{i}\frac{\partial}{\partial x} = -i\hbar\partial_x$$

Referring to the expectation for O operator in different expressions (see Table 2), show that the momentum operator is a Hermitian operator.

Table 2.

	Form	Expressions
1.	Integral form	$\int \phi^*(O\psi)dx = \left(\int \psi^* O\phi dx\right)^* = \int \psi(O\phi)^* dx$
2.	Bra-ket form	$\langle\phi|O|\psi\rangle = \langle\psi|O|\phi\rangle^*$

Note: Notations ∂_x and $\frac{\partial}{\partial x}$ are interchangeable.

Solution

Referring to Table 2, for simplicity, operator O is Hermitian if its expectation satisfies

$$\int \psi^\dagger (O\psi)\, dx = \int (O\,\psi)^\dagger \psi\, dx$$

Note that $\int \psi^\dagger O\psi$ is the conjugate of $\int (O\psi)^\dagger \psi$, and symbol $\dagger$ is simply conjugate for scalar ψ. Now, the following is performed for the momentum operator:

$$\begin{aligned}\int \psi^\dagger p\psi &= \frac{\hbar}{i}\int \psi^\dagger \partial_x \psi\, dx \\ &= -i\hbar \int \psi^\dagger d\psi \\ &= -i\hbar\psi\psi^\dagger|_{-\infty}^{\infty} + i\hbar \int (d\psi^\dagger)\psi\end{aligned}$$

The surface term on the RHS vanishes, as wavefunctions vanish at the infinities. One is left with

$$\int \psi^\dagger p\psi = i\hbar \int (d\psi^\dagger)\psi = i\hbar \int (\partial_x \psi^\dagger)\psi dx$$

Now,

$$i\hbar \int (\partial_x \psi^\dagger)\psi dx = \int (-i\hbar\partial_x \psi)^\dagger \psi dx$$

Note in the above that use is made of

$$\left(-i\hbar\frac{d}{dx}\right)^* = \left(i\hbar\frac{d}{dx}\right)$$

Therefore,

$$\int \psi^\dagger p\psi = \int (p\psi)^\dagger \psi dx$$

and it is shown that p is Hermitian.

Problem 1.10 Hermiticity: Symmetry 1

Assuming that O is non-Hermitian. Show that one can perform a process of symmetrization as follows:

$$A = \frac{1}{2}\int (\psi^\dagger(O\psi) + (O\psi)^\dagger\psi)dx$$

and A will be Hermitian.

Solution

An operator O that is Hermitian would satisfy

$$\int \psi^\dagger O\psi = \int (O\psi)^\dagger \psi$$

The above implies that for a non-Hermitian O, the expression $\int (O\psi)^\dagger \psi$ would be the complex conjugate of $\int \psi^\dagger O\psi$. As a result,

$$A = \frac{1}{2}\int (\psi^\dagger O\psi + (O\psi)^\dagger\psi)$$

would be real as shown by

$$A = \int Re(\psi^\dagger O\psi)$$

In other words, A consists of only the real part of $\int \psi^\dagger O\psi$ and is, therefore, Hermitian.

Problem 1.11 Hermiticity: Symmetry 11

Hermiticity is a concept that applies equally to scalar, matrix, and operator. Hermiticy involving matrices is an important part of linear algebra. In the event of square matrices for S and P, one can show that if S and P are individually Hermitian, the expression

$$PS + SP$$

is always Hermitian even though $[S, P] \neq 0$.

Solution

To be sure that PS is Hermitian or self-adjoint, we need to show that $PS = (PS)^\dagger$, i.e., product matrix PS is its own adjoint. Let's take the adjoint of PS by writing

$$(PS)^\dagger = S^\dagger P^\dagger$$

If S and P are individually Hermitian,

$$S^\dagger = S; \quad P^\dagger = P$$

Therefore,

$$(PS)^\dagger = S^\dagger P^\dagger = SP$$

If $[S, P] = 0$, one would have

$$(PS)^\dagger = SP = PS$$

The above shows that PS is Hermitian only when $[S, P] = 0$. The fact that S, P are individually Hermitian is no promise that PS is Hermitian. Now, in the event that $[S, P] \neq 0$, or S doesn't commute with P, PS would not be Hermitian. However, by taking the conjugate of $(PS + SP)$,

$$\begin{aligned}(PS + SP)^\dagger &= (PS)^\dagger + (SP)^\dagger \\ &= (SP + PS)\end{aligned}$$

Therefore, $(PS + SP)$ is Hermitian as

$$(PS + SP)^\dagger = (PS + SP)$$

One concludes that $(PS + SP)$ is always Hermitian as long as S and P are individually Hermitian.

Remarks and Reflections

In summary, it can be proven as follows:

(1) When an invertible matrix A is Hermitian, A^{-1} will also be Hermitian.
(2) The sum of any two Hermitian matrices is Hermitian, i.e., is P and S are individually Hermitian, $(P + S)$ is definitely Hermitian.
(3) The entries on the main diagonal of a Hermitian matrix are real.

A fully real matrix is Hermitian if and only if it is symmetrical.

Problem 1.12 Parity Operator

The parity operator is defined as

$$P\psi(x) = \psi(-x)$$

(a) Prove that the parity operator is Hermitian.
(b) Show that the eigenvalues of P are ± 1.
(c) Find the eigenstates $\phi_e(x)$ and $\phi_o(x)$, corresponding to eigenvalues of $+1$ and -1, respectively.
(d) Show that the eigenstates $\phi_e(x)$ and $\phi_o(x)$ are orthogonal to one another.
(e) Show the orthogonality relation using the Bra-Ket formalism.

Note: The Bra-Ket formalism will be covered in greater detail from Problem 1.18 onwards. Part (e) can be read together with the Bra-Ket problems later if readers so prefer.

Solution

(a) Definition of Hermiticity:

$$\int_{-\infty}^{+\infty} \psi^*(x) A\phi(x) dx = \int_{-\infty}^{+\infty} (A\psi(x))^* \phi(x) dx$$

Apply the LHS to the parity operator,

$$\int_{-\infty}^{+\infty} \psi^*(x) P\phi(x) dx = \int_{-\infty}^{+\infty} \psi^*(x) \phi(-x) dx$$

Let $y = -x$

$$\int_{-\infty}^{+\infty} \psi^*(x) P\phi(x) dx = \int_{+\infty}^{-\infty} -\psi^*(-y) P\phi(y) dy = \int_{-\infty}^{+\infty} \psi^*(-y) \phi(y) dy$$

Since y is a dummy variable,

$$\int_{-\infty}^{+\infty} \psi^*(x) P\phi(x) dx = \int_{-\infty}^{+\infty} \psi^*(-x) \phi(x) dx = \int_{-\infty}^{+\infty} (P\psi(x))^* \phi(x) dx$$

The parity operator satisfies the definition of Hermiticity.

(b) To find eigenvalues, let $\phi_1(x)$ be the eigenvalue of P and note that

$$P\phi_1(x) = A_1\phi_1(x) \quad \textit{and} \quad P\phi_1(x) = \phi_1(-x)$$

Now,

$$\begin{aligned} P^2\phi_1(x) &= P(P\phi_1(x)) = P(A_1\phi_1(x)) = A_1 P\phi_1(x) \\ &= (A_1)^2\phi_1(x) \qquad \text{(A)} \\ P^2\phi_1(x) &= P(P\phi_1(x)) = P(\phi_1(-x)) \\ &= \phi_1(x) \qquad \text{(B)} \end{aligned}$$

From (A) and (B),

$$(A_1)^2\phi_1(x) = \phi_1(x) \rightarrow A^2 = 1, \quad A = \pm 1$$

(c) To find the eigenfunctions, note that the eigenvalues are $A = \pm 1$
For $A = +1$,

$$\begin{aligned} P\phi_e(x) &= (+1)\phi_e(x) \\ \phi_e(-x) = \phi_e(x) &\rightarrow \textit{any even function of } x \end{aligned}$$

For $A = -1$,

$$\begin{aligned} P\phi_e(x) &= (-1)\phi_o(x) \\ \phi_o(-x) = -\phi_o(x) &\rightarrow \textit{any odd function of } x \end{aligned}$$

(d) To show orthogonality,

$$I = \int_{-\infty}^{+\infty} \phi_e^*(x)\phi_o(x)dx$$

Let $y = -x$

$$I = \int_{\infty}^{-\infty} \phi_e^*(-y)\phi_o(-y)(-dy) = \int_{-\infty}^{+\infty} \phi_e^*(y)(-\phi_o(y))dy$$

y is a dummy variable,

$$I = -\int_{-\infty}^{+\infty} \phi_e^*(x)\phi_o(x)dx = -I \rightarrow I = 0$$

(e) Orthogonality can also be shown with the Bra-Ket formalism

$$\langle\phi_e|\phi_o\rangle = \langle\phi_e|I|\phi_o\rangle = \langle\phi_e|P^2|\phi_o\rangle$$
$$= \langle P^\dagger\phi_e|P\phi_o\rangle$$

Since $P = P^\dagger$ because of Hermiticity,

$$\langle\phi_e|\phi_o\rangle = \langle P\phi_e|P\phi_o\rangle = \langle +\phi_e| - \phi_o\rangle$$

As a result,

$$\langle\phi_e|\phi_o\rangle = -\langle\phi_e\phi_o\rangle \rightarrow \langle\phi_e|\phi_o\rangle = 0$$

Harmonic Oscillators

Problem 1.13 Harmonic Oscillator: Energy spectrum

(a) A simple harmonic oscillator energy is given by

$$H = \frac{p^2}{2m} + \frac{1}{2}m\omega^2 x^2$$

Show that this energy system can be constructed with a pair of operators a_-, a_+ which are Hermitian conjugate to one another, i.e., $(a_-)^\dagger = a_+$.

(b) Show that $[a_-, a_+] = 1$ and that the Hamiltonian can be represented by these operators as

$$H_{SHO} = \hbar\omega\left(a_+a_- + \frac{1}{2}\right)$$

(c) Find the eigenenergy of the SHO system.

Solution

(a) One can write

$$E_T = \frac{p^2}{2m} + \frac{1}{2}m\omega^2 x^2 = \omega\left(\sqrt{\frac{m\omega}{2}}x - \frac{ip}{\sqrt{2m\omega}}\right)\left(\sqrt{\frac{m\omega}{2}}x + \frac{ip}{\sqrt{2m\omega}}\right)$$

But in quantum mechanics, the momentum and the position operators do not commute as shown by $[p, x] = -i\hbar$. Expanding E_T and elevating the x and p to operators lead to

$$H_T = \omega\left(\sqrt{\frac{m\omega}{2}}x - \frac{ip}{\sqrt{2m\omega}}\right)\left(\sqrt{\frac{m\omega}{2}}x + \frac{ip}{\sqrt{2m\omega}}\right)$$
$$= \frac{p^2}{2m} + \frac{1}{2}m\omega^2x^2 - \frac{i\omega}{2}(px - xp)$$

Therefore, $E_T \Longrightarrow H_T$

$$H_T = H_{SHO} - \frac{1}{2}\hbar\omega$$

where H_{SHO} is the quantum version of the SHO energy system. By inspection of

$$H_T = \left(\sqrt{\frac{m\omega}{2}}x - \frac{ip}{\sqrt{2m\omega}}\right)\left(\sqrt{\frac{m\omega}{2}}x + \frac{ip}{\sqrt{2m\omega}}\right)$$

one can now introduce a pair of operators to represent the energy system as follows:

$$a_+ = \sqrt{\frac{m\omega}{2\hbar}}x - \frac{ip}{\sqrt{2\hbar m\omega}}, \quad a_- = \sqrt{\frac{m\omega}{2\hbar}}x + \frac{ip}{\sqrt{2\hbar m\omega}}$$

Check that the pair is Hermitian conjugate to one another, i.e.,

$$(a_+)^\dagger = \sqrt{\frac{m\omega}{2\hbar}}x^\dagger + \frac{ip^\dagger}{\sqrt{2\hbar m\omega}} = \sqrt{\frac{m\omega}{2\hbar}}x + \frac{ip}{\sqrt{2\hbar m\omega}} = a_-$$

(b) Let's start with the commutator as follows:

$$[a_-, a_+] = \left(\sqrt{\frac{m\omega}{2\hbar}}x + \frac{ip}{\sqrt{2\hbar m\omega}}\right)\left(\sqrt{\frac{m\omega}{2\hbar}}x - \frac{ip}{\sqrt{2\hbar m\omega}}\right)$$
$$- \left(\sqrt{\frac{m\omega}{2\hbar}}x - \frac{ip}{\sqrt{2\hbar m\omega}}\right)\left(\sqrt{\frac{m\omega}{2\hbar}}x + \frac{ip}{\sqrt{2\hbar m\omega}}\right)$$
$$= \frac{i}{2\hbar}px + \frac{i}{2\hbar}px - \frac{i}{2\hbar}xp - \frac{i}{2\hbar}xp = \frac{i}{\hbar}[p, x] = 1$$

From $H_T = H_{SHO} - \frac{1}{2}\hbar\omega$ one has

$$H_{SHO} = H_T + \frac{1}{2}\hbar\omega = \hbar\omega\left(a_+a_- + \frac{1}{2}\right)$$

(c) First of all, check that

$$[H_{SHO}, a_-] = \hbar\omega[a_+a_-, a_-] = -\hbar\omega a_-$$

$$[H_{SHO}, a_+] = \hbar\omega[a_+a_-, a_+] = \hbar\omega a_+$$

The eigenvalue equation should look like this

$$H_{SHO}|E_n\rangle = E_n|E_n\rangle$$

Let us now perform $H_{SHO}a_-|E_n\rangle$. Making use of $[H_{SHO}, a_-] = -\hbar\omega a_-$, one has

$$\begin{aligned} H_{SHO}a_-|E_n\rangle &= (a_-H_{SHO} - \hbar\omega a_-)|E_n\rangle \\ &= (E_n - \hbar\omega)a_-|E\rangle \end{aligned}$$

The above shows that $a_-|E_n\rangle$ is also an eigenstate of H_{SHO} with energy lower than E_n by $\hbar\omega$. This process can go on until the lowest energy state $|E_0\rangle$ is reached. The lowest energy state does not have to be zero. Since there is no more state below E_0, further lowering it is formally described by

$$a_-|E_0\rangle = 0$$

Recalling that $H_{SHO} = \hbar\omega\left(a_+a_- + \frac{1}{2}\right)$,

$$H_{SHO}|E_0\rangle = \frac{1}{2}\hbar\omega|E_0\rangle$$

The ground state energy is $E_0 = \frac{1}{2}\hbar\omega$. Let us now perform $H_{SHO}a_+|E_n\rangle$. Making use of $[H_{SHO}, a_+] = -\hbar\omega a_+$, one has

$$\begin{aligned} H_{SHO}a_+|E_n\rangle &= (a_+H_{SHO} + \hbar\omega a_+)|E\rangle \\ &= (E_n + \hbar\omega)a_+|E_n\rangle \end{aligned}$$

The above shows that $a_+|E_n\rangle$ is also an eigenstate of H_{SHO} with energy higher than E_n by $\hbar\omega$. Applying this to ground state E_0,

$$H_{SHO}a_+|E_0\rangle = (E_0 + \hbar\omega)a_+|E_n\rangle$$

This process will go one with each successive application of a_+ on the energy eigenstate. Therefore, one deduces that the eigenenergy of the SHO is

$$E_n = \left(n + \frac{1}{2}\right)\hbar\omega, \quad \langle\varphi_n|H_{SHO}|\varphi_n\rangle = \left(n + \frac{1}{2}\right)\hbar\omega$$

Problem 1.14 Harmonic Oscillator: Raising and Lowering Operation

In a simple harmonic oscillator (SHO) system, state $|\varphi_n$ represents its nth energy eigenstate, with eigenenergy of $E_n = \left(n + \frac{1}{2}\right)\hbar\omega$. The lowering operator acts on it as follows: $a_-|\varphi_n\rangle = k_n|\varphi_{n-1}\rangle$.

(a) Assuming that $|\varphi_n\rangle$ and $|\varphi_{n-1}\rangle$ are both normalized, show that $k_n = \sqrt{n}$.

(b) Evaluate the lowering operation $(a_-)^n|\varphi_n\rangle$ where $|\varphi_n\rangle$ is the normalized nth energy eigenstate of the SHO system, and n is a positive integer. Which one of the following is the answer?
(1) $\sqrt{n^n}|\varphi_1\rangle$, (2) $\sqrt{n!}|\varphi_0\rangle$, (3) $\sqrt{n!}|\varphi_1\rangle$, (4) $\sqrt{(n-1)!}|\varphi_0\rangle$,
(5) $\sqrt{(n-1)!}|\varphi_1\rangle$

Solution

(a) Let $a_-|\varphi_n\rangle = k_n|\varphi_{n-1}\rangle$, and thus

$$|\varphi_{n-1}\rangle = \frac{1}{k_n}a_-|\varphi_n\rangle$$

Take the transpose conjugate of the above and one has

$$\langle\varphi_{n-1}| = \frac{1}{k_n^*}\langle\varphi_n|(a_-)^\dagger = \frac{1}{k_n^*}\langle\varphi_n|a_+$$

The above leads to

$$\langle\varphi_{n-1}|\varphi_{n-1}\rangle = \frac{1}{k_n^2}\langle\varphi_n|a_+a_-|\varphi_n\rangle \Longrightarrow 1 = \frac{1}{k_n^2}\langle\varphi_n|a_+a_-|\varphi_n\rangle \qquad \text{(A)}$$

Now, we make use of the identity that $[a_-, a_+] = 1$, which quickly leads to $a_+a_- = a_-a_+ - 1$. The Hamiltonian of an SHO system is

$$H = \hbar\omega\left(a_-a_+ - \frac{1}{2}\right)$$

which once again leads quickly to

$$a_-a_+ = \frac{H}{\hbar\omega} + \frac{1}{2}, \quad a_+a_- = \frac{H}{\hbar\omega} - \frac{1}{2}$$

Substitute $a_+a_- = \frac{H}{\hbar\omega} - \frac{1}{2}$ into (A), and one has

$$1 = \frac{1}{k_n^2}\left(\frac{1}{\hbar\omega}\left(\langle\varphi_n|H|\varphi_n\rangle - \frac{1}{2}\hbar\omega\right)\right)$$

Since $\langle\varphi_n|H|\varphi_n\rangle = E_n = \left(n+\frac{1}{2}\right)\hbar\omega$, one has

$$1 = \frac{1}{k_n^2}\left(\frac{1}{\hbar\omega}\left(\left(n+\frac{1}{2}\right)\hbar\omega - \frac{1}{2}\hbar\omega\right)\right) = \frac{1}{k_n^2}n \Longrightarrow k_n = \sqrt{n}$$

(b) The lowering operation works as follows:

$$(a_-)|\varphi_n\rangle = \sqrt{n}|\varphi_{n-1}\rangle$$

Repeating the process,

$$\begin{aligned}(a_-)^2|\varphi_n\rangle &= (a_-)(a_-)|\varphi_n\rangle\\ &= \sqrt{n}(a_-)|\varphi_{n-1}\rangle\\ &= \sqrt{n(n-1)}|\varphi_{n-2}\rangle\end{aligned}$$

By inspection, repeating the above n times would lead to

$$\begin{aligned}(a_-)^n|\varphi_n\rangle &= \sqrt{n(n-1)(n-2)\cdots(n-(n-1))}|\varphi_{n-n}\rangle\\ &= \sqrt{n(n-1)(n-2)\cdots 1}|\varphi_0\rangle\\ &= \sqrt{n!}|\varphi_0\rangle \rightarrow \text{answer is (2)}\end{aligned}$$

Problem 1.15 Harmonic Oscillator: Position and Momentum Matrix

Express the momentum and the position operator as of an SHO particle in matrix representation up to matrix size 4×4 and in terms of $\hbar\omega$ and m.

Note: In previous problems, quantum mechanical states are presented mostly in function space. Dirac's Bra-Ket was sometimes used to illustrate some quantum mechanical concepts but depth was not required at the time. In this problem, the matrix formalism is introduced. Quantum states can thus be expressed in the general Bra-Ket space, as well as the function and the matrix spaces.

Solution

The raising and lowering operators have been defined as follows:

$$a_+ = \sqrt{\frac{m\omega}{2\hbar}}x - \frac{ip}{\sqrt{2\hbar m\omega}}, \quad a_- = \sqrt{\frac{m\omega}{2\hbar}}x + \frac{ip_x}{\sqrt{2\hbar m\omega}}$$

The position operator can be expressed in terms of these operators as follows:

$$a_- - a_+ = \frac{\sqrt{2}ip_x}{\sqrt{\hbar m\omega}}, \quad a_- + a_+ = \sqrt{\frac{2m\omega}{\hbar}}x$$

$$\Longrightarrow p_x = -i\sqrt{\frac{\hbar m\omega}{2}}(a_- - a_+), \quad x = \sqrt{\frac{\hbar}{2m\omega}}(a_- + a_+)$$

The matrix component for p_x is given by

$$[p_x]_{ij} = \langle\varphi_{i-1}|p_x|\varphi_{j-1}\rangle$$

e.g., $[p_x]_{11} = \langle\varphi_0|(a_- - a_+)|\varphi_0\rangle$. Recalling that

$$a_-|\varphi_n\rangle = \sqrt{n}|\varphi_{n-1}\rangle, \quad a_+|\varphi_{n-1}\rangle = \sqrt{n}|\varphi_n\rangle, \quad (a_-)|\varphi_0\rangle = 0$$

one can show by going up to matrix size 4×4 that

$$[p_x]_{ij} = i\sqrt{\frac{m\omega\hbar}{2}}\begin{bmatrix} 0 & -1 & 0 & 0 \\ 1 & 0 & -\sqrt{2} & 0 \\ 0 & \sqrt{2} & 0 & -\sqrt{3} \\ 0 & 0 & \sqrt{3} & 0 \end{bmatrix},$$

$$[x]_{ij} = \sqrt{\frac{\hbar}{2m\omega}}\begin{bmatrix} 0 & 1 & 0 & 0 \\ 1 & 0 & \sqrt{2} & 0 \\ 0 & \sqrt{2} & 0 & \sqrt{3} \\ 0 & 0 & \sqrt{3} & 0 \end{bmatrix}$$

Problem 1.16 Harmonic Oscillator: Energy Matrix

Express the Hamiltonian and the potential energy operator of an SHO particle in matrix representation (in the basis of the energy eigenstates), up to matrix size 4×4, and in terms of $\hbar\omega$ and m.

Note: In previous problems, quantum mechanical states are presented mostly in function space. Dirac's Bra-Ket was sometimes used to illustrate some quantum mechanical concepts but depth was not required at the time. In this problem, the matrix formalism is introduced. Quantum states can thus be expressed in the general Bra-Ket space, as well as the function and the matrix spaces.

Solution

(a) Consider the matrix component for the Hamiltonian:

$$[H]_{ij} = \langle \varphi_{i-1} | H | \varphi_{j-1} \rangle$$

Note the offset by -1. It follows that

$$\begin{aligned}
[H]_{ij} &= \left\langle \varphi_{i-1} \right| \left(j - 1 + \frac{1}{2} \right) \hbar\omega \left| \varphi_{j-1} \right\rangle \\
&= \left(j - \frac{1}{2} \right) \hbar\omega \langle \varphi_{i-1} | \varphi_{j-1} \rangle \\
&= \left(j - \frac{1}{2} \right) \hbar\omega \delta_{ij}
\end{aligned}$$

Therefore, the Hamiltonian of an SHO particle in matrix representation up to matrix size 4×4 is a diagonal matrix as shown in the following:

$$[H]_{ij} = \hbar\omega \begin{bmatrix} 1/2 & 0 & 0 & 0 \\ 0 & 3/2 & 0 & 0 \\ 0 & 0 & 5/2 & 0 \\ 0 & 0 & 0 & 7/2 \end{bmatrix}$$

Making use of

$$[x]_{ij} = [x]_{ij} = \sqrt{\frac{\hbar}{2m\omega}} \begin{bmatrix} 0 & 1 & 0 & 0 \\ 1 & 0 & \sqrt{2} & 0 \\ 0 & \sqrt{2} & 0 & \sqrt{3} \\ 0 & 0 & \sqrt{3} & 0 \end{bmatrix}$$

one has

$$\begin{aligned}
[x^2]_{ij} &= \frac{\hbar}{2m\omega} \begin{bmatrix} 0 & 1 & 0 & 0 \\ 1 & 0 & \sqrt{2} & 0 \\ 0 & \sqrt{2} & 0 & \sqrt{3} \\ 0 & 0 & \sqrt{3} & 0 \end{bmatrix} \cdot \begin{bmatrix} 0 & 1 & 0 & 0 \\ 1 & 0 & \sqrt{2} & 0 \\ 0 & \sqrt{2} & 0 & \sqrt{3} \\ 0 & 0 & \sqrt{3} & 0 \end{bmatrix} \\
&= \frac{\hbar}{2m\omega} \begin{bmatrix} 1 & 0 & \sqrt{2} & 0 \\ 0 & 3 & 0 & \sqrt{6} \\ \sqrt{2} & 0 & 5 & 0 \\ 0 & \sqrt{6} & 0 & 7 \end{bmatrix}
\end{aligned}$$

Therefore, the potential energy is

$$\frac{1}{2}m\omega^2[x^2]_{ij} = \frac{\hbar\omega}{4}\begin{bmatrix} 1 & 0 & \sqrt{2} & 0 \\ 0 & 3 & 0 & \sqrt{6} \\ \sqrt{2} & 0 & 5 & 0 \\ 0 & \sqrt{6} & 0 & 7 \end{bmatrix}$$

Remarks and Reflections

Note that the Hamiltonian is a diagonal matrix, i.e., all the off-diagonal components are zeros, so it is a constant of motion. However, this is not the case for the potential energy $\frac{1}{2}m\omega^2x^2$ — therefore, its expectation value varies with time t. This is consistent with the physics of a classical harmonic oscillator in which both the kinetic and the potential energies oscillate, while the total energy is a constant.

It's also worth noting that for any given energy eigenstate, the expectation value of the potential energy is $1/2$ that of the total energy. Once again, this is consistent with the classical harmonic oscillator.

Problem 1.17 Harmonic Oscillator: Expectation Matrix Element

A simple harmonic oscillator particle is in the $n = 2$ energy eigenstate. The ground energy eigenstate corresponds to $n = 0$ and has an eigenvalue of $\frac{\hbar}{2}\omega$. Find the expectation value of its momentum $\langle p \rangle$. In the following, $p_0 = \sqrt{\frac{m\omega\hbar}{2}}$.

Solution

Use is made of

$$[p_x]_{ij} = i\sqrt{\frac{m\omega\hbar}{2}}\begin{bmatrix} 0 & -1 & 0 & 0 \\ 1 & 0 & -\sqrt{2} & 0 \\ 0 & \sqrt{2} & 0 & -\sqrt{3} \\ 0 & 0 & \sqrt{3} & 0 \end{bmatrix}$$

Recall that

$$[H]_{ij} = \langle \varphi_{i-1}|H|\varphi_{j-1}\rangle$$

Therefore, the expectation value of the momentum in the $n = 2$ energy eigenstate is given by

$$\langle \varphi_2 | p_x | \varphi_2 \rangle = [p_x]_{33} = 0$$

as shown in the following:

$$[p_x]_{ij} = i\sqrt{\frac{m\omega\hbar}{2}} \begin{bmatrix} 0 & -1 & 0 & 0 \\ 1 & 0 & -\sqrt{2} & 0 \\ 0 & \sqrt{2} & \boxed{0} & -\sqrt{3} \\ 0 & 0 & \sqrt{3} & 0 \end{bmatrix}$$

Problem 1.18 Harmonic Oscillator: Uncertainty Principle

An SHO particle is in the state of

$$|\psi\rangle = \sqrt{\frac{2}{3}}|\varphi_0\rangle + \frac{1}{3}|\varphi_1\rangle$$

where $|\varphi_0\rangle$ and $|\varphi_1\rangle$ are, respectively, the ground state ($n = 0$) and the first excited state ($n = 1$). In other words, the SHO particle has twice the probability of being in the ground state compared to being in the first excited state. Evaluate the product of the uncertainty in position and momentum, i.e., $(\Delta x)(\Delta p_x)$, and show that the product is in accordance with Heisenberg's uncertainty principle.

Note: $\Delta x = \sqrt{\langle x^2 \rangle - \langle x \rangle^2}$ *and similarly for* Δp_x.

Solution

The following are listed down for references:

$$[x]_{ij} = \sqrt{\frac{\hbar}{2m\omega}} \begin{bmatrix} 0 & \sqrt{1} & 0 & 0 \\ \sqrt{1} & 0 & \sqrt{2} & 0 \\ 0 & \sqrt{2} & 0 & \sqrt{3} \\ 0 & 0 & \sqrt{3} & 0 \end{bmatrix},$$

$$[x^2]_{ij} = \frac{\hbar}{2m\omega} \begin{bmatrix} 1 & 0 & \sqrt{2} & 0 \\ 0 & 3 & 0 & 0 \\ \sqrt{2} & 0 & 5 & \sqrt{6} \\ 0 & \sqrt{6} & 0 & 7 \end{bmatrix}$$

$$[p_x]_{ij} = i\sqrt{\frac{m\omega\hbar}{2}} \begin{bmatrix} 0 & -1 & 0 & 0 \\ 1 & 0 & -\sqrt{2} & 0 \\ 0 & \sqrt{2} & 0 & -\sqrt{3} \\ 0 & 0 & \sqrt{3} & 0 \end{bmatrix},$$

$$[p_x^2]_{ij} = \frac{\hbar m\omega}{2} \begin{bmatrix} 1 & 0 & -\sqrt{2} & 0 \\ 0 & 3 & 0 & -\sqrt{6} \\ -\sqrt{2} & 0 & 5 & 0 \\ 0 & -\sqrt{6} & 0 & 3 \end{bmatrix}$$

The expectation value for the position is

$$\begin{aligned} \langle x \rangle &= \left(\sqrt{\frac{2}{3}} \langle \varphi_0 | + \frac{1}{\sqrt{3}} \langle \varphi_1 | \right) x \left(\sqrt{\frac{2}{3}} | \varphi_0 \rangle + \frac{1}{\sqrt{3}} | \varphi_1 \rangle \right) \\ &= \frac{2}{3} \langle \varphi_0 | x | \varphi_0 \rangle + \frac{\sqrt{2}}{3} \langle \varphi_0 | x | \varphi_1 \rangle + \frac{\sqrt{2}}{3} \langle \varphi_1 | x | \varphi_0 \rangle + \frac{1}{3} \langle \varphi_1 | x | \varphi_1 \rangle \\ &= \frac{2}{3} [x]_{11} + \frac{\sqrt{2}}{3} [x]_{12} + \frac{\sqrt{2}}{3} [x]_{21} + \frac{1}{3} [x]_{22} \end{aligned}$$

Referring to $[x]_{ij} = \sqrt{\frac{\hbar}{2m\omega}} \begin{bmatrix} 0 & \sqrt{1} & 0 & 0 \\ \sqrt{1} & 0 & \sqrt{2} & 0 \\ 0 & \sqrt{2} & 0 & \sqrt{3} \\ 0 & 0 & \sqrt{3} & 0 \end{bmatrix}$

$$\langle x \rangle = \frac{2\sqrt{2}}{3} \sqrt{\frac{\hbar}{2m\omega}}$$

Likewise,

$$\langle x^2 \rangle = \frac{2}{3} [x^2]_{11} + \frac{\sqrt{2}}{3} [x^2]_{12} + \frac{\sqrt{2}}{3} [x^2]_{21} + \frac{1}{3} [x^2]_{22} = \frac{5\hbar}{6m\omega}$$

Therefore,

$$\Delta x = \sqrt{\langle x^2 \rangle - \langle x \rangle^2} = \frac{\sqrt{7}}{3} \left(\frac{\hbar}{2m\omega} \right)^{\frac{1}{2}}$$

We will now turn our attention to the momentum operator.

$$\langle p_x \rangle = \frac{2}{3}[p_x]_{11} + \frac{\sqrt{2}}{3}[p_x]_{12} + \frac{\sqrt{2}}{3}[p_x]_{21} + \frac{1}{3}[p_x]_{22} = 0$$

$$\langle p^2 \rangle = \frac{2}{3}[p_x^2]_{11} + \frac{\sqrt{2}}{3}[p_x^2]_{12} + \frac{\sqrt{2}}{3}[p_x^2]_{21} + \frac{1}{3}[p_x^2]_{22} = \frac{5\hbar m\omega}{6}$$

$$\Delta p_x = \sqrt{\frac{5}{3}}\left(\frac{\hbar m\omega}{2}\right)^{\frac{1}{2}}$$

Now, let us check the uncertainty relation:

$$(\Delta x)(\Delta p_x) = \frac{\sqrt{7}}{3}\left(\frac{\hbar}{2m\omega}\right)^{\frac{1}{2}}\sqrt{\frac{5}{3}}\left(\frac{\hbar m\omega}{2}\right)^{\frac{1}{2}}$$
$$= 1.14\frac{\hbar}{2} > \frac{\hbar}{2}$$

The product is in accordance with Heisenberg's uncertainty principle.

Dirac's Bra-Ket Formalism

Problem 1.19 Dirac's Bra-Ket: General and Function Spaces

Dirac Bra-Ket is a general expression for operators and states that could take on different forms in their respective spaces. We will focus on the relations between the spaces, e.g., $A^v \leftrightarrow A^f$ (see definitions for "v" and "f" in the following). The physical state of a system can be described by a mathematical object (vector) known as the Ket. The Ket is thus known as a state vector. In the following, show that

$$\langle \phi | A^v | \psi \rangle = \int_{-\infty}^{+\infty} \phi^*(x) A^f \psi(x) dx$$

Note: For example, an electron with a well-defined momentum state is denoted by Ket $|p\rangle$, which takes on different forms in the spaces of matrices or functions. On the other hand, a spin up state is denoted by Ket $|\uparrow\rangle$.

Note: Superscript v denotes a general vector space. The operator can take on the matrix A^m or the function A^f forms in their respective spaces.

Solution

We need to first set up an equivalent relation as follows:

$$\langle\phi|\psi\rangle = \int_{-\infty}^{+\infty} \phi^*(x)\psi(x)dx \tag{A}$$

which leads by inspection to

$$\langle\psi|\phi\rangle = \int_{-\infty}^{+\infty} \psi^*(x)\phi(x)dx$$

It can thus be deduced that

$$\langle\phi|\psi\rangle = \langle\psi|\phi\rangle^*$$

In fact, recalling the resolution of identity (completeness relation), i.e., $\sum_i |x_i\rangle\langle x_i| = \int |x\rangle\langle x|dx = 1$, setting up (A) implies $\langle x|\psi\rangle = \psi(x)$ as follows:

$$\langle\phi|\psi\rangle = \int_{-\infty}^{+\infty} \langle\phi|x\rangle\langle x|\psi\rangle dx \rightarrow \langle x|\psi\rangle = \psi(x), \quad \langle\phi|x\rangle = \phi^*(x) \tag{B}$$

With this, can we now set out to prove $\langle\phi|A^v|\psi\rangle = \int_{-\infty}^{+\infty} \phi^*(x)A^f\psi(x)dx$? Not yet. We still need the following definitions in Bra-Ket, i.e.,

$$A^v|\psi\rangle = |\psi'\rangle = |A^v\psi\rangle$$

It then follows right from the above that

$$\langle\phi|A^v|\psi\rangle = \langle\phi|A^v\psi\rangle = \langle\phi|\psi'\rangle \tag{C}$$

Now, with (A), (B), and (C), one has

$$\langle\phi|A^v|\psi\rangle = \int_{-\infty}^{+\infty} \phi^*(x)\psi'(x)dx$$

But what is $\psi'(x)$? Recalling that $\langle x|\psi\rangle = \psi(x)$,

$$\psi'(x) = \langle x|A^v\psi\rangle \rightarrow \psi'(x) = \langle x|A^v|\psi\rangle$$

Note that $\langle x|A^v|\psi\rangle = A^f\psi(x)$. Therefore,

$$A^f\psi(x) = \psi'(x) \tag{D}$$

Finally, with (D),

$$\langle\phi|A^v|\psi\rangle = \int_{-\infty}^{+\infty} \phi^*(x)A^f\psi(x)dx$$

Remarks and Reflections

There are three important steps in the proof:

(1)	$\langle\phi\|\psi\rangle = \int_{-\infty}^{+\infty} \phi^*(x)\psi(x)dx$	which implies $\langle x\|\psi\rangle = \psi(x)$ via the completeness relation
(2)	$A^v\|\psi\rangle = \|\psi'\rangle = \|A^v\psi\rangle$	which implies $\langle\phi\|A^v\|\psi\rangle = \langle\phi\|A^v\psi\rangle = \langle\phi\|\psi'\rangle$
(3)	$\langle x\|A^v\|\psi\rangle = A^f\psi(x)$	which leads to $\psi'(x) = A^f\psi(x)$

We should now be familiar that a quantum state in the Bra-Ket formalism is represented by notation $|\psi\rangle$ also known as the Ket. The Ket has a counterpart Bra, which bears its conjugate property. For example, the Bra version of $|p\rangle$ is denoted by $\langle p|$. It is also in the definitions that for a quantum state vector $|p\rangle$, there exists an operator P^v that acts on the vector. The same applies to the position. We may obtain the following:

$$P^v|p\rangle = p|p\rangle$$

$$Q^v|x\rangle = x|x\rangle$$

Note that P^v is the momentum operator and Q^v is the position operator and $|x\rangle$ is the position state vector. By contrast, as shown earlier, we have

$$A^v|\psi\rangle = |\psi'\rangle = |A^v\psi\rangle$$

$$\langle\phi|A^v|\psi\rangle = \langle\phi|A^v\psi\rangle = \langle\phi|\psi'\rangle$$

Problem 1.20 Momentum Operator

Consider P^v an abstract object that must be defined to produce, upon acting on the state $|\psi\rangle$, a real, measurable value. Like the state vectors, the operator too needs to take on different forms, e.g., in the abstract (v), matrix (m), or function (f) spaces. We will focus on the relations in different spaces for the momentum operator, i.e., $P^v \leftrightarrow P^f$.

In the following, considering specifically the momentum operators P^v. Watch the completeness relation being applied as follows:

$$\langle x|P^v|x'\rangle = \int \langle x|P^v|p\rangle\langle p|x'\rangle dp$$

Deduce the following:

(a) $\langle x|P^v|x'\rangle = P^f\delta(x - x')$
(b) $\langle p|P^v|\psi\rangle = P^f\langle p|\psi\rangle$

Note: Superscript v denotes a general vector space, it can take on the matrix P^m or the function P^f forms in their respective spaces.

Solution

(a) Letting P^v act on $|p\rangle$, one has

$$\langle x|P^v|x'\rangle = \int \langle x|P^v|p\rangle\langle p|x'\rangle dp = \int p\langle x|p\rangle\langle p|x'\rangle dp$$

Now, $\langle x|p\rangle$ is written as a function $\phi_p(x)$ as follows:

$$\langle x|p\rangle = \frac{1}{\sqrt{2\pi\hbar}}e^{ikx}$$

Hence, with substitution,

$$\langle x|P^v|x'\rangle = \frac{1}{2\pi\hbar}\int -i\hbar\partial_x\left(e^{ik.(x-x')}\right)dp = -i\hbar\partial_x\int\frac{1}{2\pi}\left(e^{ik\cdot(x-x')}\right)dk$$

With P^f to be constructed as $P^f = -i\hbar\partial_x$,

$$\langle x|P^v|x'\rangle = P^f\int\frac{1}{2\pi}\left(e^{ik.(x-x')}\right)dk$$

Finally,

$$\langle x|P^v|x'\rangle = P^f\delta(x - x')$$

(b) In momentum representation, we will use the momentum eigenvectors $|p\rangle$ as the basis set for the momentum operator in the general space:

$$P^v|p\rangle = p|p\rangle$$

Now, with completeness relation (resolution of identity),

$$\langle p|P^v|\psi\rangle = \int \langle p|P^v|p'\rangle\langle p'|\psi\rangle dp' = p\langle p|\psi\rangle$$

As $\langle p|\psi\rangle = \psi(p)$,

$$\langle p|P^v|\psi\rangle = p\psi(p)$$

Now, in the function space,

$$P^f\psi(p) = p\psi(p)$$

It thus follows that

$$\langle p|P^v|\psi\rangle = P^f\langle p\psi\rangle$$

Problem 1.21 Position Operator

Consider Q^v an abstract object that must be defined to produce, upon acting on the state $|\psi\rangle$, a real, measurable value. Like the state vectors, the operator too needs to take on different forms, e.g., abstract (v), matrix (m), or function (f) spaces. We will focus on the relations in different spaces for the position operator, i.e., $Q^v \leftrightarrow Q^f$.

In the following, show that

$$\langle x|Q^\nu|\psi\rangle = Q^f\langle x|\psi\rangle$$

Solution

Method A

With the expansion theorem, one can write in terms of position vector. Note that the position eigenvalues exist in a continuous spectrum.

$$|\psi\rangle = \int_{-\infty}^{+\infty} A_x|x\rangle dx$$

$$\langle x|Q^\nu|\psi\rangle = \langle x|Q^v|\int A_{x'}|x'\rangle dx'$$

$$= \int \langle x|A_{x'}x'|x'\rangle dx'$$

$$= \int A_{x'}x'\delta(x-x')dx' = A_x x$$

Alternatively, one can also perform the following:

$$\langle x|Q^\nu|\psi\rangle = \langle x|\left(\int |x'\rangle x'\langle x'|dx'\right)\left(\int A_{x''}|x''\rangle dx''\right)$$

where Q^ν has been expressed as $Q^\nu = \int |x\rangle x\langle x|dx$. It thus follows that

$$\langle x|Q^\nu|\psi\rangle = \iint \delta(x-x')x'A_{x''}\delta(x'-x'')dx'dx''$$

$$= \int \delta(x-x')A_{x'}x'dx' = A_x x$$

Note that

$$|\psi\rangle = \int_{-\infty}^{+\infty} A_x|x\rangle dx \rightarrow \langle x'|\psi\rangle = \int_{-\infty}^{+\infty} A_x\delta(x-x')dx = A'_x$$

This leads to

$$A_x = \psi(x) = \langle x|\psi\rangle$$

Therefore, A_x is the probability amplitude of locating the electron in state $|\psi\rangle$ at position x, and the value of A_x changes continuously with x. Finally, one has

$$\langle x|Q^\nu|\psi\rangle = A_x x = \psi(x)x$$

$$\langle x|Q^\nu|\psi\rangle = x\langle x|\psi\rangle = x\psi(x)$$

Now, as

$$Q^f\langle x|\psi\rangle = x\langle x|\psi\rangle$$

$$\langle x|Q^\nu|\psi\rangle = Q^f\langle x|\psi\rangle$$

Method B

With the completeness relation (continuous identity), it can be shown that

$$\langle x|Q^\nu|\psi\rangle = \int \langle x|Q^\nu|x_1\rangle\langle x_1|\psi\rangle dx_1$$

In coordinate representation, the position eigenvectors $\{|x_i\rangle\}$ have been chosen as the basis set. Thus,

$$\langle x|Q^\nu|\psi\rangle = \int \langle x|x_1\rangle x_1\psi(x_1)dx_1 = \int \delta(x - x_1)x_1\psi(x_1)dx_1$$

It thus follows that

$$\langle x|Q^\nu|\psi\rangle = x\psi(x) = x\langle x|\psi\rangle$$

Now, as

$$Q^f\langle x|\psi\rangle = x\langle x|\psi\rangle$$

One has

$$\langle x|Q^\nu|\psi\rangle = Q^f\langle x|\psi\rangle$$

Problem 1.22 Non-commutative Position and Momentum

In the following, show that

$$\langle x|[Q^v, P^v]|x'\rangle = (x - x')\langle x|P^v|x'\rangle$$

where Q^v is the position operator and P^v is the momentum operator, both in general vector space.

Solution

Let's write down the commutative relation as follows:

$$\langle x|[Q^v, P^v]|x'\rangle = \langle x|Q^vP^v|x'\rangle - \langle x|P^vQ^v|x'\rangle$$

We will now proceed with the completeness relation, using the position eigenvectors $|x''\rangle$ as basis set

$$\langle x|[Q^v, P^v]|x'\rangle = \int \langle x|Q^v|x''\rangle\langle x''|P^\nu|x'\rangle - \langle x|P^\nu|x''\rangle\langle x''|Q^v|x'\rangle dx''$$

Note that $\langle x|Q^v|x''\rangle = x''\delta(x - x'')$. It thus follows that

$$\langle x|[Q^v, P^v]|x'\rangle = \int x''\delta(x - x'')\langle x''|P^\nu|x'\rangle - x'\delta(x'' - x')\langle x|P^\nu|x''\rangle dx''$$

Finally,

$$\begin{aligned}\langle x|[Q^v, P^v]|x'\rangle &= x\langle x|P^\nu|x'\rangle - x'\langle x|P^\nu|x'\rangle \\ &= (x - x')\langle x|P^\nu|x'\rangle\end{aligned}$$

Problem 1.23 Explicit: Momentum Operator in Coordinate Representation

We have discussed the relations between forms $A^v \leftrightarrow A^f$, e.g., $\langle\phi|A^v|\psi\rangle = \int_{-\infty}^{+\infty}\phi^*(x)A^f\psi(x)dx$ and $\langle x|A^v|\psi\rangle = A^f\psi(x)$. Discussion will now be shifted to the explicit representation of the operators in the function space. We will focus on the momentum and the position operators

$$P^f \text{ and } Q^f$$

These operators can be expressed in the coordinate or momentum representations. Show that in the coordinate representation, the momentum operator is

$$P^f = -i\hbar\partial_x$$

Note: Notations ∂_x and $\frac{\partial}{\partial x}$ are interchangeable.

Solution

Since $P^f\langle x|\psi\rangle = \langle x|P^v|\psi\rangle$, we will begin with

$$\langle x|P^v|\psi\rangle = \int \langle x|p\rangle\langle p|P^v|p'\rangle\langle p'|\psi\rangle dp dp'$$

where use is made of the completeness relation (resolution of identity) in momentum representation as follows:

$$\sum_i |p_i\rangle\langle p_i| = \int |p\rangle\langle p| dp = 1$$

It thus follows that

$$\begin{aligned}\langle x|P^v|\psi\rangle &= \int \frac{1}{\sqrt{2\pi\hbar}}e^{ik\cdot x}p'\delta(p-p')\psi(p')dpdp' \\ &= -i\hbar\partial_x\int\frac{1}{\sqrt{2\pi\hbar}}e^{ik\cdot x}\psi(p)dp\end{aligned}$$

By inverse Fourier transform,

$$-i\hbar\partial_x \int \frac{1}{\sqrt{2\pi\hbar}} e^{ik\cdot x}\psi(p)dp = -i\hbar\partial_x\psi(x)$$

Note that in the above, use has been made of $\psi(p) = \frac{\psi(k)}{\sqrt{\hbar}}$. Note also that $\langle x|P^v|\psi\rangle = P^f\langle x|\psi\rangle = P^f\psi(x)$. Therefore,

$$P^f\psi(x) = -i\hbar\partial_x\psi(x)$$

Inspecting LHS and RHS,

$$P^f = -i\hbar\partial_x = -i\hbar\frac{\partial}{\partial x}$$

Problem 1.24 Explicit: Position Operator in Momentum Representation

We have discussed the relations between forms $A^v \leftrightarrow A^f$, e.g., $\langle\phi|A^v|\psi\rangle = \int_{-\infty}^{+\infty}\phi^*(x)A^f\psi(x)dx$ and $\langle x|A^v|\psi\rangle = A^f\psi(x)$. Discussion will now be shifted to the explicit expression of the operators in the function space. We will focus on the momentum and the position operators

$$P^f \; and \; Q^f$$

In quantum mechanics, like the momentum, the position of a particle can also be elevated to work like an operator. For instance, an electron is a wave propagating in the crystal structure of materials with translational periodicity. The wavevector is therefore well defined. The position operator helps locate its whereabouts. Show that in the momentum representation, the position operator is

$$Q^f = i\partial_k$$

Note: Notations ∂_k and $\frac{\partial}{\partial k}$ are interchangeable.

Solution

Since $Q^f\langle p|\psi\rangle = \langle p|Q^v|\psi\rangle$, we will begin with

$$\langle p|Q^v|\psi\rangle = \int \langle p|x\rangle\langle x|Q^v|x'\rangle\langle x'|\psi\rangle dxdx'$$

where the resolution of identity is

$$\sum_i |x_i\rangle\langle x_i| = \int |x\rangle\langle x| dx = 1$$

It follows that

$$\begin{aligned}\langle p|Q^v|\psi\rangle &= \int \frac{1}{\sqrt{2\pi\hbar}} e^{-ik\cdot x} x' \delta(x - x')\psi(x') dx dx' \\ &= i\partial_k \int \frac{1}{\sqrt{2\pi\hbar}} e^{-ik\cdot x} \psi(x) dx\end{aligned}$$

By Fourier transform,

$$i\partial_k \int \frac{1}{\sqrt{2\pi\hbar}} e^{-ik\cdot x} \psi(x) dx = i\partial_k \frac{\psi(k)}{\sqrt{\hbar}}$$

Note that in the above, use has been made of $\psi(p) = \frac{\psi(k)}{\sqrt{\hbar}}$. Note also that $\langle p|Q^v|\psi\rangle = Q^f \langle p|\psi\rangle = Q^f \psi(p)$. Therefore,

$$Q^f \langle p|\psi\rangle = i\partial_k \psi(p)$$

Inspecting LHS and RHS,

$$Q^f = i\partial_k = i\frac{\partial}{\partial k}$$

Remarks and Reflections

Note that in the derivation of $P^f = -i\hbar\partial_x$, use is made of the inverse Fourier transform,

$$\psi(x) = \int \frac{1}{\sqrt{2\pi\hbar}} e^{ik\cdot x} \psi(p) dp = \psi(x)$$

On the other hand, in the derivation of $Q^f = i\partial_k$, use is made of the Fourier transform,

$$\psi(k) = \int \frac{1}{\sqrt{2\pi}} e^{-ik\cdot x} \psi(x) dx$$

Problem 1.25 Explicit: Momentum Eigenfunctions in Coordinate Representation

Show that the general form of $\phi^3(x) = \langle \boldsymbol{x}|\hbar\boldsymbol{k}\rangle = ce^{i\boldsymbol{k}\cdot\boldsymbol{x}}$ is derived such that

$$\langle \boldsymbol{x}|\hbar\boldsymbol{k}\rangle = \frac{1}{(2\pi\hbar)^{\frac{3}{2}}}e^{i\boldsymbol{k}\cdot\boldsymbol{x}}$$

Note: $\phi(\boldsymbol{x}) = \phi^3(x) = \phi(x)\phi(y)\phi(z)$ *and* $\delta(\boldsymbol{x}) = \delta^3(x) = \delta(x)\delta(y)\delta(z)$.

Note: $\langle \boldsymbol{x}|\hbar\boldsymbol{k}\rangle = \langle x\hbar k_x\rangle\langle y|\hbar k_y\rangle\langle z|\hbar k_z\rangle$ *and* $\langle \hbar k|x\rangle = \langle x|\hbar k\rangle^*$.

Note: Bold is used to make explicit statements about the vectorial nature of the physical quantities involved.

Solution

Method A

Use is made of two common Identities,

$$\delta(\boldsymbol{w}) = \left(\frac{1}{2\pi}\right)^3 \int e^{i\boldsymbol{w}\cdot\boldsymbol{x}} d^3x \tag{A}$$

$$\langle p|p'\rangle = \langle \hbar k|\hbar k'\rangle = \delta_{pp'} \text{ or } \delta(p - p') \tag{B}$$

With Identity A, one has

$$\begin{aligned}\delta(\boldsymbol{p} - \boldsymbol{p}') &= \left(\frac{1}{2\pi}\right)^3 \int e^{i(\boldsymbol{p}-\boldsymbol{p}')\cdot\boldsymbol{x}} d^3x \\ &= \left(\frac{1}{2\pi\hbar}\right)^3 \int e^{i(\boldsymbol{k}-\boldsymbol{k}')\cdot\hbar\boldsymbol{x}} d^3(\hbar x)\end{aligned}$$

This leads naturally to

$$\delta(\boldsymbol{p} - \boldsymbol{p}') = \frac{1}{\hbar^3}\delta(\boldsymbol{k} - \boldsymbol{k}') \tag{C}$$

consistent with the identity $\delta(ax) = \frac{\delta(x)}{|a|}$ proven earlier. With Identity B, and the general form of $\langle \boldsymbol{x}|\hbar\boldsymbol{k}\rangle = ce^{i\boldsymbol{k}\cdot\boldsymbol{x}}$, now one can also write

$$\begin{aligned}\delta(\boldsymbol{p} - \boldsymbol{p}') &= \int \langle \hbar\boldsymbol{k}'|\boldsymbol{x}\rangle\langle \boldsymbol{x}|\hbar\boldsymbol{k}\rangle d^3x \\ &= \int cc^* e^{i(\boldsymbol{k}-\boldsymbol{k}')\cdot\boldsymbol{x}} d^3x = cc^*(2\pi)^3\delta(\boldsymbol{k} - \boldsymbol{k}')\end{aligned} \tag{D}$$

In the above, use had been made of the closure relation. Comparing C with D, one could thus deduce that $c = \frac{1}{(2\pi\hbar)^{\frac{3}{2}}}$. Therefore,

$$\phi^3(x) = \langle \boldsymbol{x}|\hbar \boldsymbol{k}\rangle = \frac{1}{(2\pi\hbar)^{\frac{3}{2}}} e^{i\boldsymbol{k}\cdot\boldsymbol{x}}$$

Method B

With experience, one alternatively begins with $\langle \boldsymbol{x}|\boldsymbol{p}\rangle = ce^{i\boldsymbol{k}\cdot\boldsymbol{x}}$ and deduces that

$$\delta(\boldsymbol{p} - \boldsymbol{p}') = \int \langle \boldsymbol{p}'|\boldsymbol{x}\rangle\langle \boldsymbol{x}|\boldsymbol{p}\rangle d^3x = \int cc^* e^{i(\boldsymbol{k}-\boldsymbol{k}')\cdot\boldsymbol{x}} d^3x$$

This leads to

$$\delta(\boldsymbol{p} - \boldsymbol{p}') = cc^*(2\pi)^3 \frac{1}{(2\pi)^3} \hbar^3 \int e^{\frac{i}{\hbar}(\boldsymbol{p}-\boldsymbol{p}')\cdot\boldsymbol{x}} d^3 \frac{x}{\hbar^3}$$

$$\delta(\boldsymbol{p} - \boldsymbol{p}') = cc^*(2\pi\hbar)^3 \delta(\boldsymbol{p} - \boldsymbol{p}')$$

Therefore, $c = \frac{1}{(2\pi\hbar)^{\frac{3}{2}}}$.

Chapter 2

Spin Physics

Chapter 2 is dedicated to spinor formalism in quantum mechanics. Spin is presented as a separate chapter because spinor formalism is widely applied in emerging fields like spintronics and graphene, as well as topological systems like topological insulators and Weyl and Dirac systems. It is also a crucial tool for the study of spin-based Berry curvature and the non-abelian quantum transport in the formulation of non-equilibrium Green's function. On top of that, the field of quantum computation and information too is primarily run on spinor physics. These are all new research fields that hover heavily around spinor physics.

Introduction to Spin Physics

Problem 2.01 Quantum Spin States
Problem 2.02 Spin Expectation
Problem 2.03 Spin Operators and Hermiticity
Problem 2.04 Spin Operators and Commutative Property
Problem 2.05 Spin Eigenvalues and Eigenvectors
Problem 2.06 Spin and Magnetism

Spin Transformation

Problem 2.07 Spin Precession: Time Dependence
Problem 2.08 Spin Precession: Heisenberg
Problem 2.09 Spin Precession: Heisenberg and Compact Notation
Problem 2.10 Spinor Transformation
Problem 2.11 Frame Rotation

Introduction to Spin Physics

Problem 2.01 Quantum Spin States

Spin or spinor is an ideal 2D Hilbert space system that allows one to fast check some of the mathematical rules of the vector space and the basis vectors therein, e.g., orthogonality, completeness, and linearity. The quantum spin state is defined on the surface of a Bloch sphere by the variation of angles θ and ϕ as shown in Figure 1.

(a) The qubit spin is given by $|\psi\rangle = \cos\frac{\theta}{2}\,|0\rangle + e^{i\phi}\sin\frac{\theta}{2}\,|1\rangle$ as illustrated in Figure 1. With $e^{i\gamma}$ as an arbitrary phase factor, show that the spin states along axis Y are given by

$$|+y\rangle = \frac{1}{\sqrt{2}}\begin{pmatrix} e^{i\gamma} \\ e^{i\left(\gamma+\frac{\pi}{2}\right)} \end{pmatrix}$$

$$|-y\rangle = \frac{1}{\sqrt{2}}\begin{pmatrix} e^{i\gamma} \\ e^{i\left(\gamma-\frac{\pi}{2}\right)} \end{pmatrix}$$

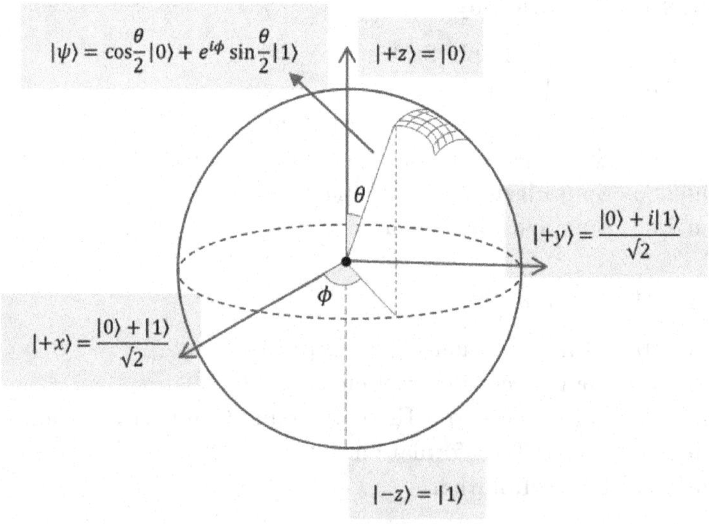

Fig. 1. Bloch sphere, where convention $|+z\rangle = |0\rangle = \binom{1}{0}$, $|-z\rangle = |1\rangle = \binom{0}{1}$ is observed.

(b) Which of the following also represents the spin state $|+y\rangle$? Explain how you arrive at each of your conclusion and show the γ value associated with each of your answers.

$$\frac{1}{\sqrt{2}}\begin{pmatrix}1\\ i\end{pmatrix} \quad \frac{1}{\sqrt{2}}\begin{pmatrix}e^{\frac{i\pi}{4}}\\ e^{-\frac{i\pi}{4}}\end{pmatrix} \quad \frac{1}{\sqrt{2}}\begin{pmatrix}e^{-\frac{i\pi}{4}}\\ e^{\frac{i\pi}{4}}\end{pmatrix} \quad \frac{1}{\sqrt{2}}\begin{pmatrix}e^{-\frac{i\pi}{2}}\\ 1\end{pmatrix} \quad \frac{1}{\sqrt{2}}\begin{pmatrix}1\\ e^{\frac{i\pi}{2}}\end{pmatrix}$$

A	B	C	D	E

(c) Which of the following also represents the spin state $|-y\rangle$? Explain how you arrive at each of your conclusion and show the γ value associated with each of your answers.

$$\frac{1}{\sqrt{2}}\begin{pmatrix}1\\ -i\end{pmatrix} \quad \frac{1}{\sqrt{2}}\begin{pmatrix}e^{\frac{i\pi}{4}}\\ e^{-\frac{i\pi}{4}}\end{pmatrix} \quad \frac{1}{\sqrt{2}}\begin{pmatrix}e^{-\frac{i\pi}{4}}\\ e^{\frac{i5\pi}{4}}\end{pmatrix} \quad \frac{1}{\sqrt{2}}\begin{pmatrix}e^{-\frac{i\pi}{2}}\\ 1\end{pmatrix} \quad \frac{1}{\sqrt{2}}\begin{pmatrix}e^{-\frac{i\pi}{2}}\\ e^{-i\pi}\end{pmatrix}$$

A	B	C	D	E

(d) The spin states of a qubit are given as shown in the following:

$$|+y\rangle = \frac{1}{\sqrt{2}}\begin{pmatrix}e^{i\gamma}\\ e^{i\left(\gamma+\frac{\pi}{2}\right)}\end{pmatrix}, \ |-y\rangle = \frac{1}{\sqrt{2}}\begin{pmatrix}e^{i\gamma}\\ e^{i\left(\gamma-\frac{\pi}{2}\right)}\end{pmatrix}$$

Show that $\langle +y|+y\rangle = 1$, $\langle +y|-y\rangle = \langle -y|+y\rangle = 0, \langle -y|-y\rangle = 1$.

(e) Show with Pauli matrix Y that the eigenvalues are

$$+1 \text{ for } \quad |+y\rangle = \frac{1}{\sqrt{2}}\begin{pmatrix}e^{i\gamma}\\ e^{i\left(\gamma-\frac{\pi}{2}\right)}\end{pmatrix}$$

$$-1 \text{ for } \quad |-y\rangle = \frac{1}{\sqrt{2}}\begin{pmatrix}e^{i\gamma}\\ e^{i\left(\gamma-\frac{\pi}{2}\right)}\end{pmatrix}$$

(f) A qubit is in the spin state of $|\chi\rangle = \begin{pmatrix}\sqrt{0.2}\\ -\sqrt{0.8}\end{pmatrix}$ where $\begin{pmatrix}1\\ 0\end{pmatrix}$ and $\begin{pmatrix}0\\ 1\end{pmatrix}$ represent spins along the $+z$ and $-z$ directions, respectively. Find the expectation value of $\langle \sigma_y \rangle$, i.e., the spin angular momentum to be expected along axis y when spin is in the quantum state of $|\chi\rangle$ as indicated above.

Solution

(a) The arbitrary spin state of a qubit is given by $|\psi\rangle = \cos\frac{\theta}{2}|0\rangle + e^{i\phi}\sin\frac{\theta}{2}|1\rangle$. State $|+y\rangle$ is characterized by $\theta = \frac{\pi}{2}$ and $\phi = \frac{\pi}{2}$, leading to $|+y\rangle = \frac{1}{\sqrt{2}}\begin{pmatrix} 1 \\ e^{\frac{i\pi}{2}} \end{pmatrix}$. Now, impart a phase factor $e^{i\gamma}$ to the eigenstate and

$$|+y\rangle = \frac{1}{\sqrt{2}}\begin{pmatrix} e^{i\gamma} \\ e^{i\left(\gamma+\frac{\pi}{2}\right)} \end{pmatrix}$$

would still be the eigenstate of σ_y. On the other hand, state $|-y\rangle$ is characterized by $\theta = \frac{\pi}{2}$ and $\phi = -\frac{\pi}{2}$, leading to $|-y\rangle = \frac{1}{\sqrt{2}}\begin{pmatrix} 1 \\ e^{-\frac{i\pi}{2}} \end{pmatrix}$. Now, impart a phase factor $e^{i\gamma}$ to the eigenstate and

$$|-y\rangle = \frac{1}{\sqrt{2}}\begin{pmatrix} e^{i\gamma} \\ e^{i\left(\gamma-\frac{\pi}{2}\right)} \end{pmatrix}$$

would still be the eigenstate of σ_y.

(b) With $|+y\rangle = \frac{1}{\sqrt{2}}\begin{pmatrix} e^{i\gamma} \\ e^{i\left(\gamma+\frac{\pi}{2}\right)} \end{pmatrix}$,

Angle of arbitrary phase γ	**Spin state**
Take $\gamma = -\frac{\pi}{2}$	$\|+y\rangle = \frac{1}{\sqrt{2}}\begin{pmatrix} e^{-\frac{i\pi}{2}} \\ 1 \end{pmatrix}$
Take $\gamma = 0$	$\|+y\rangle = \frac{1}{\sqrt{2}}\begin{pmatrix} 1 \\ e^{\frac{i\pi}{2}} \end{pmatrix} = \frac{1}{\sqrt{2}}\begin{pmatrix} 1 \\ i \end{pmatrix}$
Take $\gamma = -\frac{\pi}{4}$	$\|+y\rangle = \frac{1}{\sqrt{2}}\begin{pmatrix} e^{-\frac{i\pi}{4}} \\ e^{\frac{i\pi}{4}} \end{pmatrix}$
Answers are ACDE	

(c) With $|-y\rangle = \frac{1}{\sqrt{2}}\begin{pmatrix} e^{i\gamma} \\ e^{i\left(\gamma-\frac{\pi}{2}\right)} \end{pmatrix}$,

Angle of arbitrary phase γ	**Spin state**
Take $\boldsymbol{\gamma = -\frac{\pi}{2}}$	$\lvert -y\rangle = \frac{1}{\sqrt{2}}\begin{pmatrix} e^{-\frac{i\pi}{2}} \\ e^{-i\pi} \end{pmatrix}$
Take $\boldsymbol{\gamma = 0}$	$\lvert -y\rangle = \frac{1}{\sqrt{2}}\begin{pmatrix} 1 \\ e^{-\frac{i\pi}{2}} \end{pmatrix} = \frac{1}{\sqrt{2}}\begin{pmatrix} 1 \\ -i \end{pmatrix}$
Take $\boldsymbol{\gamma = +\frac{\pi}{4}}$	$\lvert -y\rangle = \frac{1}{\sqrt{2}}\begin{pmatrix} e^{\frac{i\pi}{4}} \\ e^{-\frac{i\pi}{4}} \end{pmatrix}$
Take $\boldsymbol{\gamma = -\frac{\pi}{4}}$	$\lvert -y\rangle = \frac{1}{\sqrt{2}}\begin{pmatrix} e^{-\frac{i\pi}{4}} \\ e^{-\frac{i3\pi}{4}} \end{pmatrix} = \frac{1}{\sqrt{2}}\begin{pmatrix} e^{-\frac{i\pi}{4}} \\ e^{+\frac{i5\pi}{4}} \end{pmatrix}$
Answers are ABCE	

(d) The inner products of the spin states are

$$\langle +y|+y\rangle = \frac{1}{\sqrt{2}}\begin{pmatrix} e^{-i\gamma} & e^{-i\left(\gamma+\frac{\pi}{2}\right)} \end{pmatrix}\frac{1}{\sqrt{2}}\begin{pmatrix} e^{i\gamma} \\ e^{i\left(\gamma+\frac{\pi}{2}\right)} \end{pmatrix} = \frac{1}{2}(1+1) = 1$$

$$\langle +y|-y\rangle = \frac{1}{\sqrt{2}}\begin{pmatrix} e^{-i\gamma} & e^{-i\left(\gamma+\frac{\pi}{2}\right)} \end{pmatrix}\frac{1}{\sqrt{2}}\begin{pmatrix} e^{i\gamma} \\ e^{i\left(\gamma-\frac{\pi}{2}\right)} \end{pmatrix} = \frac{1}{2}\left(1+e^{-i\pi}\right) = 0$$

$$\langle -y|-y\rangle = \frac{1}{\sqrt{2}}\begin{pmatrix} e^{-i\gamma} & e^{-i\left(\gamma-\frac{\pi}{2}\right)} \end{pmatrix}\frac{1}{\sqrt{2}}\begin{pmatrix} e^{i\gamma} \\ e^{i\left(\gamma-\frac{\pi}{2}\right)} \end{pmatrix} = \frac{1}{2}(1+1) = 1$$

$$\langle -y|+y\rangle = \frac{1}{\sqrt{2}}\begin{pmatrix} e^{-i\gamma} & e^{-i\left(\gamma-\frac{\pi}{2}\right)} \end{pmatrix}\frac{1}{\sqrt{2}}\begin{pmatrix} e^{i\gamma} \\ e^{i\left(\gamma+\frac{\pi}{2}\right)} \end{pmatrix} = \frac{1}{2}\left(1+e^{i\pi}\right) = 0$$

(e) The eigenvalues can be found in the following manner:

$$\begin{pmatrix} 0 & -i \\ i & 0 \end{pmatrix}\frac{1}{\sqrt{2}}\begin{pmatrix} e^{i\gamma} \\ e^{i\left(\gamma+\frac{\pi}{2}\right)} \end{pmatrix} = \frac{1}{\sqrt{2}}\begin{pmatrix} -ie^{i\left(\gamma+\frac{\pi}{2}\right)} \\ ie^{i\gamma} \end{pmatrix} = \frac{1}{\sqrt{2}}\begin{pmatrix} e^{-\frac{i\pi}{2}}e^{i\left(\gamma+\frac{\pi}{2}\right)} \\ e^{\frac{i\pi}{2}}e^{i\gamma} \end{pmatrix}$$

$$= \frac{1}{\sqrt{2}}\begin{pmatrix} e^{i\gamma} \\ e^{i\left(\gamma+\frac{\pi}{2}\right)} \end{pmatrix}$$

Therefore,

$$\begin{pmatrix} 0 & -i \\ i & 0 \end{pmatrix} \frac{1}{\sqrt{2}} \begin{pmatrix} e^{i\gamma} \\ e^{i(\gamma+\frac{\pi}{2})} \end{pmatrix} = (+1)\frac{1}{\sqrt{2}} \begin{pmatrix} e^{i\gamma} \\ e^{i(\gamma+\frac{\pi}{2})} \end{pmatrix}$$

$$\begin{pmatrix} 0 & -i \\ i & 0 \end{pmatrix} \frac{1}{\sqrt{2}} \begin{pmatrix} e^{i\gamma} \\ e^{i(\gamma-\frac{\pi}{2})} \end{pmatrix} = \frac{1}{\sqrt{2}} \begin{pmatrix} -ie^{i(\gamma-\frac{\pi}{2})} \\ ie^{i\gamma} \end{pmatrix} = \frac{1}{\sqrt{2}} \begin{pmatrix} e^{-\frac{i\pi}{2}} e^{i(\gamma-\frac{\pi}{2})} \\ e^{\frac{i\pi}{2}} e^{i\gamma} \end{pmatrix}$$

$$= \frac{1}{\sqrt{2}} \begin{pmatrix} e^{i\gamma} e^{-i\pi} \\ e^{i(\gamma+\frac{\pi}{2})} \end{pmatrix} = \frac{1}{\sqrt{2}} \begin{pmatrix} e^{-i\pi} e^{i\gamma} \\ e^{i\pi} e^{i(\gamma-\frac{\pi}{2})} \end{pmatrix}$$

$$= -\frac{1}{\sqrt{2}} \begin{pmatrix} e^{i\gamma} \\ e^{i(\gamma-\frac{\pi}{2})} \end{pmatrix}$$

Therefore,

$$\begin{pmatrix} 0 & -i \\ i & 0 \end{pmatrix} \frac{1}{\sqrt{2}} \begin{pmatrix} e^{i\gamma} \\ e^{i(\gamma-\frac{\pi}{2})} \end{pmatrix} = (-1)\frac{1}{\sqrt{2}} \begin{pmatrix} e^{i\gamma} \\ e^{i(\gamma-\frac{\pi}{2})} \end{pmatrix}$$

(f) The expectation value of the qubit spin along the Y axis is given by $\langle \sigma_y \rangle = \langle \chi \,|\, \sigma_y \,|\, \chi \rangle$. It follows that

$$\langle \sigma_y \rangle = \begin{pmatrix} \sqrt{0.2} \\ -\sqrt{0.8} \end{pmatrix}^{\dagger} \begin{pmatrix} 0 & -i \\ i & 0 \end{pmatrix} \begin{pmatrix} \sqrt{0.2} \\ -\sqrt{0.8} \end{pmatrix}$$

$$\langle \sigma_y \rangle = \begin{pmatrix} \sqrt{0.2} & -\sqrt{0.8} \end{pmatrix} \begin{pmatrix} 0 & -i \\ i & 0 \end{pmatrix} \begin{pmatrix} \sqrt{0.2} \\ -\sqrt{0.8} \end{pmatrix} = 0$$

Problem 2.02 Spin Expectation

From the application point of view, electron spin is gaining foothold in the technologies of magnetism, spintronics, and quantum computation. Electron spin is defined on the surface of a Bloch sphere by the variation of angles θ and ϕ as shown in Figure 1.

(a) Referring to Figure 1, and using $|\psi\rangle = \cos\frac{\theta}{2}\,|0\rangle + e^{i\phi}\sin\frac{\theta}{2}\,|1\rangle$ for an arbitrary spin state, it is evident that $\langle \sigma_y \rangle = 0$ for quantum states of $\theta = 0$, and $\theta = \pi$. When $\theta \neq 0, \pi$, determine the azimuthal angles ϕ for which the expectation value of $\langle \sigma_y \rangle$ is 0.

(b) Based on the results you obtain above, provide a pencil-shade illustration of the spin states on the Bloch sphere.
(c) Using $|\psi\rangle = \cos\frac{\theta}{2}|0\rangle + e^{i\phi}\sin\frac{\theta}{2}|1\rangle$ for an arbitrary spin state, derive the expectation expressions for spin-x, spin-y, and spin-z.
(d) Identify the zone on the Bloch sphere for each of these expectation values to vanish. Explain how those zones are determined and illustrate them on the Bloch sphere.
(e) A spin state oriented at angle (θ, ϕ) is given by $\frac{1}{4}\begin{pmatrix} 2\sqrt{3} \\ i+\sqrt{3} \end{pmatrix}$. Find the expectation values for this spin state along axis X, Y. Find the corresponding angles (θ, ϕ) for the spin state.

Solution

(a) Using $|\psi\rangle = \cos\frac{\theta}{2}|0\rangle + e^{i\phi}\sin\frac{\theta}{2}|1\rangle$, one has

$$\langle\sigma_y\rangle = \begin{pmatrix} \cos\frac{\theta}{2} \\ e^{i\phi}\sin\frac{\theta}{2} \end{pmatrix}^{\dagger} \begin{pmatrix} 0 & -i \\ i & 0 \end{pmatrix} \begin{pmatrix} \cos\frac{\theta}{2} \\ e^{i\phi}\sin\frac{\theta}{2} \end{pmatrix}$$

$$= i\sin\frac{\theta}{2}\cos\frac{\theta}{2}\left(e^{-i\phi} - e^{i\phi}\right) = \sin\theta\sin\phi$$

Therefore, when $\theta \neq 0, \pi$, the azimuthal angles for which $\langle\sigma_y\rangle = 0$ are $\phi = 0$ and $\phi = \pi$.

(b) A pencil-shade illustration of the spin states on the Bloch sphere.

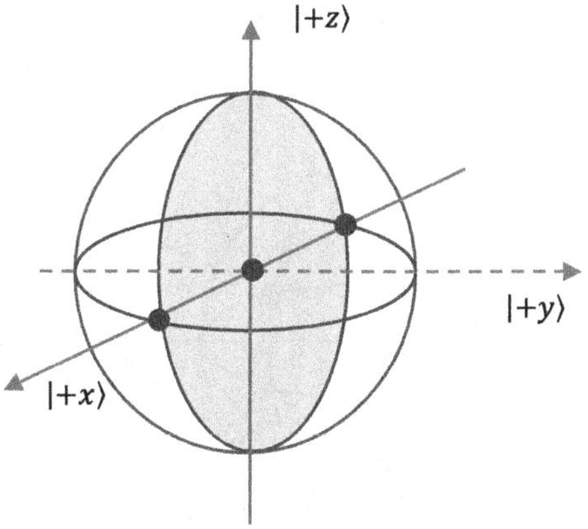

(c) Using $|\psi\rangle = \cos\frac{\theta}{2}|0\rangle + e^{i\phi}\sin\frac{\theta}{2}|1\rangle$ for an arbitrary spin state, the expectation expressions for spin-x, spin-y, and spin-z are

$$\langle\sigma_x\rangle = \begin{pmatrix} \cos\frac{\theta}{2} \\ e^{i\phi}\sin\frac{\theta}{2} \end{pmatrix}^\dagger \begin{pmatrix} 0 & 1 \\ 1 & 0 \end{pmatrix} \begin{pmatrix} \cos\frac{\theta}{2} \\ e^{i\phi}\sin\frac{\theta}{2} \end{pmatrix} = \sin\frac{\theta}{2}\cos\frac{\theta}{2}\left(e^{i\phi} + e^{-i\phi}\right)$$
$$= \sin\theta\cos\phi$$

$$\langle\sigma_y\rangle = \begin{pmatrix} \cos\frac{\theta}{2} \\ e^{i\phi}\sin\frac{\theta}{2} \end{pmatrix}^\dagger \begin{pmatrix} 0 & -i \\ i & 0 \end{pmatrix} \begin{pmatrix} \cos\frac{\theta}{2} \\ e^{i\phi}\sin\frac{\theta}{2} \end{pmatrix} = i\sin\frac{\theta}{2}\cos\frac{\theta}{2}\left(e^{-i\phi} - e^{i\phi}\right)$$
$$= \sin\theta\sin\phi$$

$$\langle\sigma_z\rangle = \begin{pmatrix} \cos\frac{\theta}{2} \\ e^{i\phi}\sin\frac{\theta}{2} \end{pmatrix}^\dagger \begin{pmatrix} 1 & 0 \\ 0 & -1 \end{pmatrix} \begin{pmatrix} \cos\frac{\theta}{2} \\ e^{i\phi}\sin\frac{\theta}{2} \end{pmatrix} = \cos^2\frac{\theta}{2} - \sin^2\frac{\theta}{2} = \cos\theta$$

(d) Zones on the Bloch sphere for each of these expectation values are shown with Bloch sphere illustrations and explanations. Note that $|\psi\rangle = \cos\frac{\theta}{2}|0\rangle + e^{i\phi}\sin\frac{\theta}{2}|1\rangle$.

For $\langle\sigma_x\rangle = \sin\theta\cos\phi = 0$:
When $\theta = 0, \pi$, spin state $|\psi\rangle = |0\rangle, e^{i\phi}|1\rangle$, spin vector is locked to the North and the South poles. But, when $\phi = \frac{\pi}{2}, \frac{3\pi}{2}$, spin state $|\psi\rangle = \cos\frac{\theta}{2}|0\rangle + i\sin\frac{\theta}{2}|1\rangle$ and $\cos\frac{\theta}{2}|0\rangle - i\sin\frac{\theta}{2}|1\rangle$, spin vector is free to rotate in the shaded plane as shown in the following:

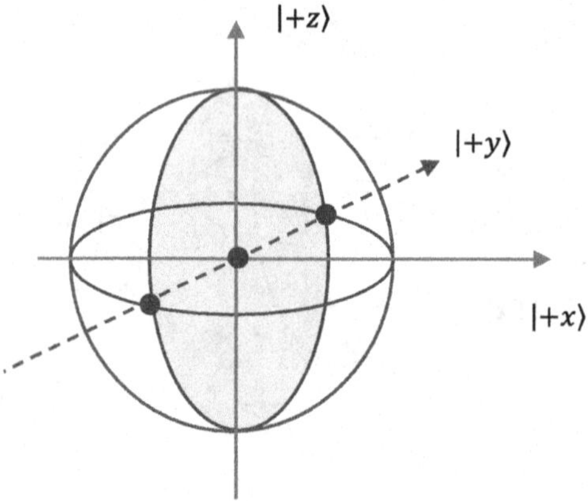

For $\langle \sigma_y \rangle = \sin\theta \sin\phi = 0$:
When $\theta = 0, \pi$, spin state $|\psi\rangle = |0\rangle, e^{i\phi}|1\rangle$, spin vector is locked to the North and the South poles. But, when $\phi = 0, \pi$, spin state $|\psi\rangle = \cos\frac{\theta}{2}|0\rangle + \sin\frac{\theta}{2}|1\rangle$ and $\cos\frac{\theta}{2}|0\rangle - \sin\frac{\theta}{2}|1\rangle$, spin vector is free to rotate in the shaded plane as shown in the following:

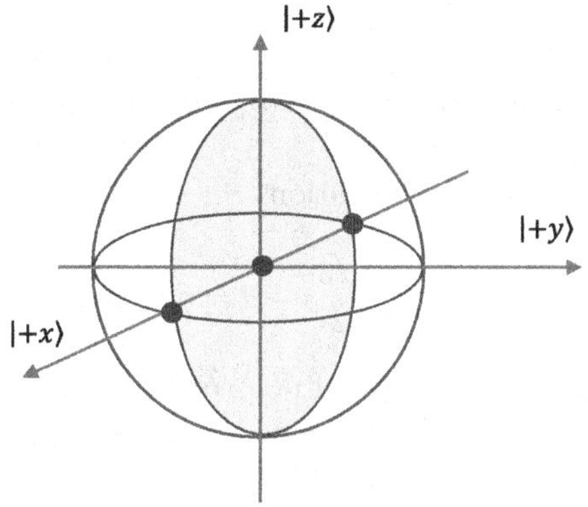

For $\langle \sigma_z \rangle = \cos\theta = 0$:
When $\theta = \frac{\pi}{2}$, spin state $|\psi\rangle = \frac{1}{\sqrt{2}}|0\rangle + e^{i\phi}\frac{1}{\sqrt{2}}|1\rangle$, and spin vector is free to rotate in the shaded plane as shown in the following:

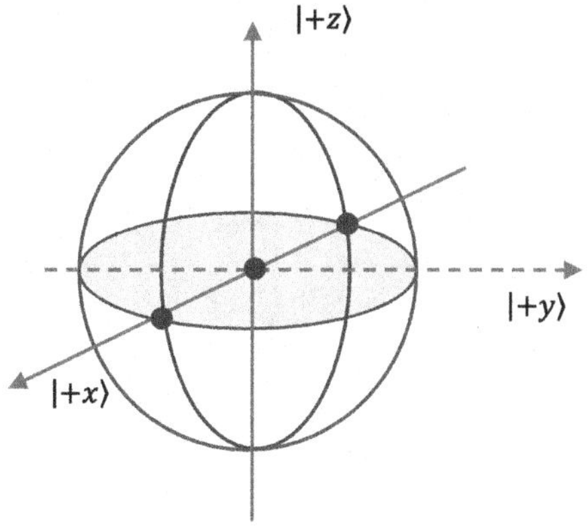

(e) The expectation values for $\frac{1}{4}\begin{pmatrix} 2\sqrt{3} \\ i+\sqrt{3} \end{pmatrix}$ are as follows:

$$\langle \sigma_x \rangle = \frac{1}{4}\begin{pmatrix} 2\sqrt{3} & -i+\sqrt{3} \end{pmatrix}\begin{pmatrix} 0 & 1 \\ 1 & 0 \end{pmatrix}\frac{1}{4}\begin{pmatrix} 2\sqrt{3} \\ i+\sqrt{3} \end{pmatrix}$$

$$= \frac{1}{16}\begin{pmatrix} 2\sqrt{3} & -i+\sqrt{3} \end{pmatrix}\begin{pmatrix} i+\sqrt{3} \\ 2\sqrt{3} \end{pmatrix} = \frac{12}{16} = \frac{3}{4}$$

$$\langle\sigma_y\rangle = \frac{1}{4}\begin{pmatrix}2\sqrt{3} & -i+\sqrt{3}\end{pmatrix}\begin{pmatrix}0 & -i\\ i & 0\end{pmatrix}\frac{1}{4}\begin{pmatrix}2\sqrt{3}\\ i+\sqrt{3}\end{pmatrix}$$
$$= \frac{1}{16}\begin{pmatrix}2\sqrt{3} & -i+\sqrt{3}\end{pmatrix}\begin{pmatrix}1-i\sqrt{3}\\ i2\sqrt{3}\end{pmatrix} = \frac{\sqrt{3}}{4}$$

To find the angles, note that

$$\langle\sigma_x\rangle = \sin\theta\cos\phi = \frac{3}{4} \quad \langle\sigma_y\rangle = \sin\theta\sin\phi = \frac{\sqrt{3}}{4}$$

Take the square of both expectations

$$\sin\theta = \frac{\sqrt{3}}{2} \to \theta = 60, 120$$

For an arbitrary ϕ, the angles of θ correspond to the two spin states as shown in the following figure. Since $\langle\sigma_x\rangle$ and $\langle\sigma_y\rangle$ are both positive, spin state is likely contained in the first quadrant. Therefore,

$$\tan\phi = \frac{1}{\sqrt{3}} \to \phi = 30$$

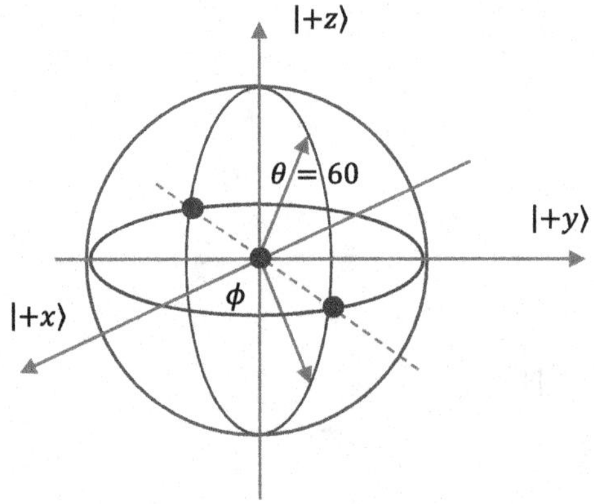

Substituting $\theta = 60$, 120 and $\phi = 30$ into $|\psi\rangle = \cos\frac{\theta}{2}|0\rangle + e^{i\phi}\sin\frac{\theta}{2}|1\rangle$,

$$|\psi\rangle_{60} = \begin{pmatrix}\dfrac{\sqrt{3}}{2}\\ \dfrac{\sqrt{3}}{4}+\dfrac{i}{4}\end{pmatrix} = \frac{1}{4}\begin{pmatrix}2\sqrt{3}\\ \sqrt{3}+i\end{pmatrix}$$

$$|\psi\rangle_{120} = \begin{pmatrix} \frac{1}{2} \\ \frac{3}{4} + \frac{i\sqrt{3}}{4} \end{pmatrix} = \frac{1}{4}\begin{pmatrix} 2 \\ 3 + i\sqrt{3} \end{pmatrix}$$

Therefore, $\theta = 120$ can be ruled out. The answer is $\theta = 60$, $\phi = 30$.

Problem 2.03 Spin Operators and Hermiticity

As in all quantum operators, they are Hermitian. In a single qubit two-level spin system, operators are represented in (2×2) matrices.

(a) Outline the conditions for a (2×2) operator to be Hermitian.
(b) Show that $\begin{pmatrix} 1 & i-1 \\ -i-1 & -1 \end{pmatrix}$, $\begin{pmatrix} 1 & 0 \\ 0 & -1 \end{pmatrix}$, $\begin{pmatrix} 0 & -i \\ i & 0 \end{pmatrix}$ are Hermitian.
(c) Show that $\begin{pmatrix} 0 & e^{i\phi} \\ e^{-i\phi} & e^{-i\phi} \end{pmatrix}$, $\begin{pmatrix} 1+i & i \\ -i & 1-i \end{pmatrix}$, $\begin{pmatrix} 1 & -1 \\ 1 & -1 \end{pmatrix}$ are NOT Hermitian.
(d) Consider the Hermitian matrices of $S = \begin{pmatrix} 1 & i-1 \\ -i-1 & -1 \end{pmatrix}, P = \begin{pmatrix} 1 & 0 \\ 0 & -1 \end{pmatrix}$. Determine if PS and SP are Hermitian.
(e) Check if S and P commute, i.e., $[S, P] = 0$. Show that if SP and PS commute, they will be Hermitian when P and S are individually Hermitian.
(f) Determine that $SP + PS$ is generally always Hermitian. Confirm this with the specific examples of $S = \begin{pmatrix} 1 & i-1 \\ -i-1 & -1 \end{pmatrix}, P = \begin{pmatrix} 1 & 0 \\ 0 & -1 \end{pmatrix}$.

Solution

(a) A Hermitian operator satisfies the following:

$$A = A^\dagger$$

In the case of a (2×2) matrix representation for A,

$$\begin{pmatrix} a & b \\ c & d \end{pmatrix} = \begin{pmatrix} a & b \\ c & d \end{pmatrix}^\dagger = \begin{pmatrix} a^* & c^* \\ b^* & d^* \end{pmatrix}$$

Comparing the matrix components, one has the following:

$$a = a^*,\ d = d^*,\ b = c^*,\ c = b^*$$

Therefore, components a and d must be real. Components b and c must be complex conjugates of one another.

(b) The diagonal components of $\begin{pmatrix} 1 & i-1 \\ -i-1 & -1 \end{pmatrix}$ are real, and $i-1$ is the conjugate of $-i-1$, i.e., the condition $b = c^*$ is satisfied. Matrices $\begin{pmatrix} 1 & 0 \\ 0 & -1 \end{pmatrix}$ and $\begin{pmatrix} 0 & -i \\ i & 0 \end{pmatrix}$ both have diagonal components and satisfy $b = c^*$. Therefore, they are Hermitian.

(c) The diagonal components of $\begin{pmatrix} 0 & e^{i\phi} \\ e^{-i\phi} & e^{-i\phi} \end{pmatrix}$, $\begin{pmatrix} 1+i & i \\ -i & 1-i \end{pmatrix}$ are not real. They are therefore NOT Hermitian. Matrix $\begin{pmatrix} 1 & -1 \\ 1 & -1 \end{pmatrix}$ has real diagonal components. But it doesn't satisfy $b = c^*$. It is, therefore, NOT Hermitian.

(d) Both S and P are individually Hermitian, now

$$SP = \begin{pmatrix} 1 & i-1 \\ -i-1 & -1 \end{pmatrix} \begin{pmatrix} 1 & 0 \\ 0 & -1 \end{pmatrix} = \begin{pmatrix} 1 & 1-i \\ -1-i & -1 \end{pmatrix}$$

$$PS = \begin{pmatrix} 1 & 0 \\ 0 & -1 \end{pmatrix} \begin{pmatrix} 1 & i-1 \\ -i-1 & -1 \end{pmatrix} = \begin{pmatrix} 1 & i-1 \\ 1+i & 1 \end{pmatrix}$$

Both PS and SP do not satisfy $b = c^*$, $c = b^*$. They are, therefore, NOT Hermitian.

(e) In the above, $SP \neq PS$. Therefore, $[S, P] \neq 0$. Let's now check what happens if $[S, P] = 0$. First off, let's check the following:

$$(SP)^\dagger = P^\dagger S^\dagger = PS$$
$$(PS)^\dagger = S^\dagger P^\dagger = SP$$

From the above, we can conclude that $[S, P] = 0 \rightarrow (SP)^\dagger = SP$, $(PS)^\dagger = PS$.

Therefore, when $[S, P] \neq 0$, PS and SP will NOT be Hermitian even though S and P are both Hermitian. By contrast, when $[S, P] = 0$, PS and SP will be Hermitian when S and P are Hermitian.

(f) We need to check that $(PS + SP)^\dagger = PS + SP$

$$(PS + SP)^\dagger = S^\dagger P^\dagger + P^\dagger S^\dagger$$

When S and P are individually Hermitian, the above indeed leads to

$$(PS + SP)^\dagger = PS + SP$$

This is indeed the case with this example.

$$PS + SP = \begin{pmatrix} 2 & 0 \\ 0 & 2 \end{pmatrix}$$

Remarks and Reflections

Repeat the Hermiticity check for SP, PS, $SP + PS$ with $S = \begin{pmatrix} 1 & i-1 \\ -i-1 & -1 \end{pmatrix}$, $P = \begin{pmatrix} 0 & -i \\ i & 0 \end{pmatrix}$

Problem 2.04 Spin Operators and Commutative Property

Evaluate the following commutative property involving the spin Pauli matrices:

(a) If $A = \sigma_x \sigma_y$ and $B = \sigma_y \sigma_z$, find the commutation of $[A, B]$.
(b) Likewise, find the commutation of $[\sigma_z, \sigma_y \sigma_z]$.
(c) Show that

$$\boldsymbol{\sigma}^* = \boldsymbol{\sigma}^T$$

Note: The Pauli matrices are given as

$$\sigma_x = \begin{pmatrix} 0 & 1 \\ 1 & 0 \end{pmatrix}, \sigma_y = \begin{pmatrix} 0 & -i \\ i & 0 \end{pmatrix}, \sigma_z = \begin{pmatrix} 1 & 0 \\ 0 & -1 \end{pmatrix}$$

Solution

(a) Note that the Pauli matrices are related to one another in the following manner:

$$\sigma_x \sigma_y = i\sigma_z, \ \sigma_y \sigma_z = i\sigma_x$$

The above follows from the more general relation of $\sigma_a \sigma_b = i\sigma_c \epsilon_{abc} + I\delta_{ab}$. Now,

$$[A, B] = i\sigma_z (i\sigma_x) - i\sigma_x (i\sigma_z) = -(i)\,\sigma_y + (-i)\,\sigma_y = -2i\sigma_y$$

Note that ϵ_{abc} is known as the anti-symmetric Levi-Civita symbol with

$$\epsilon_{123} = \epsilon_{231} = \epsilon_{312} = 1, \quad \epsilon_{213} = \epsilon_{321} = \epsilon_{132} = -1$$

All other permutations of subscripts 1, 2, 3 are zero. On the other hand, δ_{ab} is the Kronecker delta with

$$\delta_{ab} = 1\, for\, a = b, \quad \delta_{ab} = 0\, for\, a \neq b$$

(b) Taking note that $\sigma_y\sigma_z = i\sigma_x$, one has

$$[\sigma_z, \sigma_y\sigma_z] = [\sigma_z, i\sigma_x] = i\,(\sigma_z\sigma_x - \sigma_x\sigma_z)$$

Making use of $\sigma_a\sigma_b = i\sigma_c\epsilon_{abc} + I\delta_{ab}$, one now has

$$[\sigma_z, \sigma_y\sigma_z] = i\,(i\sigma_y + i\sigma_y) = -2\sigma_y$$

(c) To show that $\boldsymbol{\sigma}^* = \boldsymbol{\sigma}^T$, examine the Pauli matrices individually

$$\sigma_x = \begin{pmatrix} 0 & 1 \\ 1 & 0 \end{pmatrix}, \sigma_y = \begin{pmatrix} 0 & -i \\ i & 0 \end{pmatrix}, \sigma_z = \begin{pmatrix} 1 & 0 \\ 0 & -1 \end{pmatrix}$$

$$(\sigma_x)^* = \sigma_x, (\sigma_x)^T = \sigma_x$$

$$(\sigma_y)^* = -\sigma_y, (\sigma_y)^T = -\sigma_y$$

$$(\sigma_z)^* = \sigma_z, (\sigma_y)^T = -\sigma_z$$

It thus follows that $\boldsymbol{\sigma}^* = \boldsymbol{\sigma}^T$.

Problem 2.05 Spin Eigenvalues and Eigenstates

Referring to Figure 2, a magnetic field $\boldsymbol{B}$ lying in the $X - Y$ plane is pointing in the direction of $\boldsymbol{e_B}$ with ϕ being its azimuthal angle. Electron spin would be aligned anti-parallel to the field to produce the lowest energy. Therefore, the Hamiltonian of the system is $H = \boldsymbol{S}.\boldsymbol{B}$.

Note:: The Pauli matrices are $\sigma_x = \begin{pmatrix} 0 & 1 \\ 1 & 0 \end{pmatrix}, \sigma_y = \begin{pmatrix} 0 & -i \\ i & 0 \end{pmatrix}, \sigma_z = \begin{pmatrix} 1 & 0 \\ 0 & -1 \end{pmatrix}$, *and* $\boldsymbol{S} = \frac{\hbar}{2}\boldsymbol{\sigma}$.

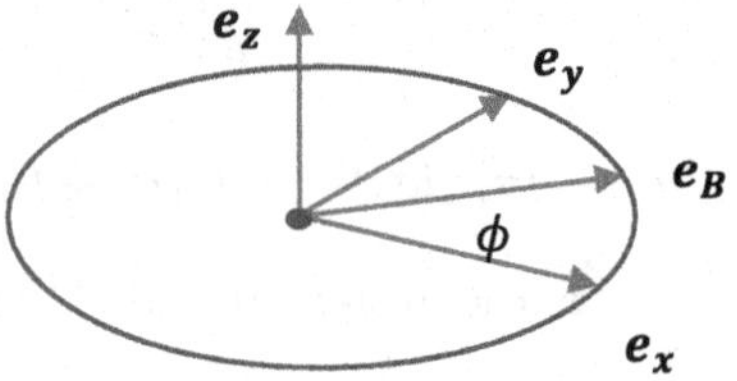

Fig. 2. A 2D system in which the magnetic field lies in the X–Y plane.

(a) Show that the spin operator is given by $S = S_x \cos\phi + S_y \sin\phi$.
(b) Using the determinant or other methods, show that the eigenvalues for the spin are $\alpha_e = \pm\frac{\hbar}{2}$.
(c) Show that the spin states are given by eigenvectors: $\eta_+ = \frac{1}{\sqrt{2}}\begin{pmatrix} e^{-\frac{i\phi}{2}} \\ e^{\frac{i\phi}{2}} \end{pmatrix}$, $\eta_- = \frac{1}{\sqrt{2}}\begin{pmatrix} e^{-\frac{i\phi}{2}} \\ -e^{\frac{i\phi}{2}} \end{pmatrix}$.

Solution

(a) The Hamiltonian of the system is

$$\begin{aligned} H &= \boldsymbol{S}.\boldsymbol{B} \\ &= \boldsymbol{S}.\left(B\cos\phi\,\boldsymbol{e_x} + B\sin\phi\,\boldsymbol{e_y}\right) = B\,\left(s_x\cos\phi + s_y\sin\phi\right) \end{aligned}$$

(b) The eigenvalues for the spin operators are found in this equation

$$(S_x\cos\phi + S_y\sin\phi)\begin{pmatrix} u \\ v \end{pmatrix} = \frac{\hbar}{2}\lambda\begin{pmatrix} u \\ v \end{pmatrix}$$

Using the determinant approach,

$$Det\,|\sigma_x\cos\phi + \sigma_y\sin\phi - \lambda I\,| = 0$$

Explicitly,

$$\begin{aligned} &Det\left|\begin{pmatrix} 0 & \cos\phi - i\sin\phi \\ \cos\phi + i\sin\phi & 0 \end{pmatrix} - \lambda\begin{pmatrix} 1 & 0 \\ 0 & 1 \end{pmatrix}\right| = 0 \to \\ &Det\left|\begin{pmatrix} -\lambda & e^{-i\phi} \\ e^{i\phi} & -\lambda \end{pmatrix}\right| = 0 \\ &\lambda^2 - 1 = 0 \to \lambda = \pm 1 \\ &\alpha_e = \frac{\hbar}{2}\lambda = \pm\frac{\hbar}{2} \end{aligned}$$

(c) The eigenstates are derived as follows:

$$(S_x\cos\phi + S_y\sin\phi)\begin{pmatrix} u \\ v \end{pmatrix} = \pm\frac{\hbar}{2}\begin{pmatrix} u \\ v \end{pmatrix}$$

In matrix form,

$$\begin{pmatrix} 0 & \cos\phi - i\sin\phi \\ \cos\phi + i\sin\phi & 0 \end{pmatrix}\begin{pmatrix} u \\ v \end{pmatrix} = \pm\begin{pmatrix} u \\ v \end{pmatrix}$$

$$\begin{pmatrix} 0 & e^{-i\phi} \\ e^{i\phi} & 0 \end{pmatrix}\begin{pmatrix} u \\ v \end{pmatrix} = \pm\begin{pmatrix} u \\ v \end{pmatrix}$$

Recalling the eigenvalues: For $\lambda = +1$, $v = e^{i\phi}u$. Assign $u = 1$,

$$\eta_+ = N\begin{pmatrix} 1 \\ e^{i\phi} \end{pmatrix} = N\begin{pmatrix} e^{-\frac{i\phi}{2}} \\ e^{\frac{i\phi}{2}} \end{pmatrix}$$

For $\lambda = -1$, $v = -e^{i\phi}u$. Assign $u = 1$,

$$\eta_- = N\begin{pmatrix} 1 \\ -e^{i\phi} \end{pmatrix} = N\begin{pmatrix} e^{-\frac{i\phi}{2}} \\ -e^{\frac{i\phi}{2}} \end{pmatrix}$$

$$\langle\eta_+|\eta_+\rangle = 1 \rightarrow N^2\begin{pmatrix} e^{\frac{i\phi}{2}} & e^{-\frac{i\phi}{2}} \end{pmatrix}\begin{pmatrix} e^{-\frac{i\phi}{2}} \\ e^{\frac{i\phi}{2}} \end{pmatrix} = 1 \rightarrow N = \frac{1}{\sqrt{2}}$$

Therefore,

$$\eta_+ = \frac{1}{\sqrt{2}}\begin{pmatrix} e^{-\frac{i\phi}{2}} \\ e^{\frac{i\phi}{2}} \end{pmatrix}, \eta_- = \frac{1}{\sqrt{2}}\begin{pmatrix} e^{-\frac{i\phi}{2}} \\ -e^{\frac{i\phi}{2}} \end{pmatrix}$$

Problem 2.06 Spin and Magnetism

Show that the spin energy in both the classical and the quantum pictures are equivalent:

$$\boldsymbol{E} = -\boldsymbol{m_C}.\boldsymbol{b} \leftrightarrow -\langle\boldsymbol{m_S}.\boldsymbol{b}\rangle = -\langle\boldsymbol{m_S}\rangle.\boldsymbol{b}$$

where $m_C = \sqrt{m_x^2 + m_y^2}$ is the classical magnetic moment, and θ is the angle between m and b the magnetic field. In the quantum picture, $\boldsymbol{m_S} = -\frac{eg}{2m}\boldsymbol{S}$, while $\boldsymbol{S} = \frac{\hbar}{2}\boldsymbol{\sigma}$ is the spin operator and $\boldsymbol{\sigma}$ is the Pauli matrix.

Solution

Consider a 2D plane of $X-Y$, where the magnetic moment m and magnetic field b are contained as shown on the left of Figure 3. In the classical picture,

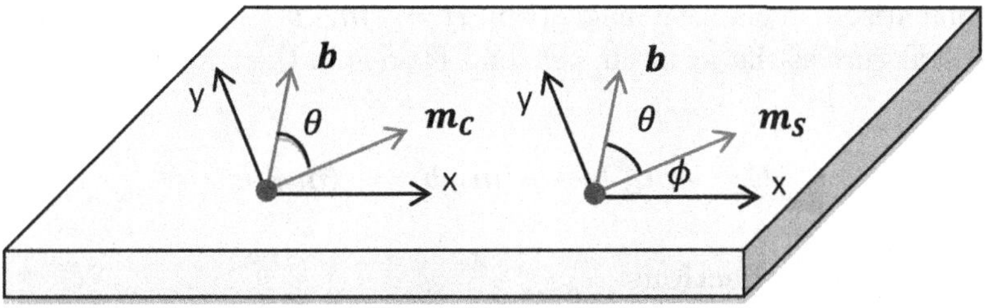

Fig. 3. A 2D plane containing the magnetic fields and magnetic moments.

energy H is

$$E = -\boldsymbol{m_C}.\boldsymbol{b} = -(m_x b_x + m_y b_y) = -m_C b \cos\theta$$

Now, consider a single quantum spin to align along m and to replace m as shown on the right. In the quantum picture, energy H is

$$H = \boldsymbol{S}.\boldsymbol{b}$$

Take the spin operator along azimuthal angle ϕ from axis x, and the operator is given by

$$S = S_x \cos\phi + S_y \sin\phi$$

The eigenvalues are $\lambda = \pm 1$ and the quantum vector of the spin states are given by eigenvectors:

$$\eta_+ = \frac{1}{\sqrt{2}} \begin{pmatrix} e^{-\frac{i\phi}{2}} \\ e^{\frac{i\phi}{2}} \end{pmatrix}, \ \eta_- = \frac{1}{\sqrt{2}} \begin{pmatrix} e^{-\frac{i\phi}{2}} \\ -e^{\frac{i\phi}{2}} \end{pmatrix}$$

Now, find the expectation of $\boldsymbol{S} = S_x \boldsymbol{e_x} + S_y \boldsymbol{e_y}$

$$\langle \eta_+ \mid \boldsymbol{s} \mid \eta_+ \rangle = \frac{\hbar}{2} (\cos\phi \boldsymbol{e_x} + \sin\phi \boldsymbol{e_y}) = \frac{\hbar}{2} \boldsymbol{e_s}$$

where $\boldsymbol{e_s}$ is the classical vector of the spin state. Therefore, going back to $H = \boldsymbol{S}.\boldsymbol{b} = -\langle \boldsymbol{m_S} \rangle.\boldsymbol{b}$, one has

$$H = \left\langle \frac{eg\hbar}{4m} \right\rangle \boldsymbol{e_s}.\boldsymbol{b} = \frac{eg\hbar}{4m} b \cos\theta$$

By comparison, in classical magnetism, $H = -\boldsymbol{m_C}.\boldsymbol{b} = -m_C b\, cos\theta$ has the same structure as the quantum version. Therefore, the two are equivalent, i.e.,

$$\boldsymbol{E} = -\boldsymbol{m_C}.\boldsymbol{b} \leftrightarrow -\langle \boldsymbol{m_S}.\boldsymbol{b} \rangle = -\langle \boldsymbol{m_S} \rangle.\boldsymbol{b}$$

Remarks and Reflections

Consider $\boldsymbol{b} = b\,\boldsymbol{e_n} = b_x\boldsymbol{e_x} + b_y\boldsymbol{e_y}$, and $\boldsymbol{e_b} = cos\,\phi\boldsymbol{e_x} + sin\,\phi\boldsymbol{e_y}$ $\boldsymbol{e_b}.\boldsymbol{e_b} = 1$. In three dimensions,

$$\boldsymbol{e_b}.\boldsymbol{e_b} = \left(\frac{b_x}{b}\boldsymbol{e_x} + \frac{b_y}{b}\boldsymbol{e_y} + \frac{b_z}{b}\boldsymbol{e_z}\right) \cdot \left(\frac{b_x}{b}\boldsymbol{e_x} + \frac{b_y}{b}\boldsymbol{e_y} + \frac{b_z}{b}\boldsymbol{e_z}\right)$$

$$= \frac{b_x^2 + b_y^2 + b_z^2}{b^2} = 1$$

$$b_x = b\sin\theta\cos\phi, \quad b_y = b\sin\theta\sin\phi, \quad b_z = b\cos\theta$$

Spin Transformation

Problem 2.07 Spin Precession: Time Dependence

Consider a spin pointing at a specific spinor direction $\psi(0) = \binom{a}{b}$, at an angular deviation from the magnetic field. Physically, the spin would start to precess about the magnetic field.

(a) Show that as the spinor function starts to precess, its time-dependent expression is given by

$$\psi(t) = \begin{pmatrix} ae^{-iw_0t} \\ be^{iw_0t} \end{pmatrix}$$

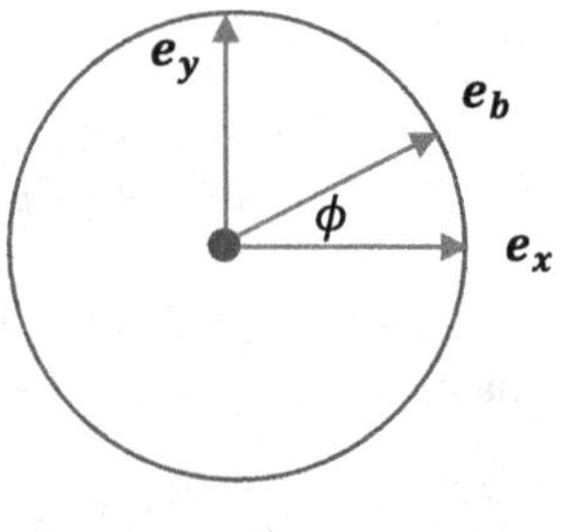

Fig. 4. Top view of a 2D plane in which the magnetic field lies in the X–Y plane.

(b) Show that the precession operator is given by

$$U = \exp\left(-i\boldsymbol{\sigma}.\boldsymbol{e_b}\frac{\omega t}{2}\right)$$

Solution

(a) The Schrodinger equation for the time-dependent spin state is

$$i\hbar\frac{d\,\psi(t)}{dt} = \frac{eg\hbar}{4m}\boldsymbol{\sigma}.\boldsymbol{B}\,\psi(t)$$

The spin is not aligned with the B field. It is, therefore, not a good quantum number and the spin would start to precess about the B field. Let the B field axis be z. We can write $\psi(t)$ in a linear superposition in the z basis, i.e., the basis along B, as follows:

$$\psi(t) = \begin{pmatrix} a(t) \\ b(t) \end{pmatrix} = a\begin{pmatrix} 1 \\ 0 \end{pmatrix} e^{-i\omega_a t} + b\begin{pmatrix} 0 \\ 1 \end{pmatrix} e^{-i\omega_b t}$$

At time 0, before precessions start,

$$\psi(0) = a\begin{pmatrix} 1 \\ 0 \end{pmatrix} + b\begin{pmatrix} 0 \\ 1 \end{pmatrix}$$

Since the bases are the eigenstates of σ^z, we shall examine each eigenstate separately for

$$i\hbar\frac{d}{dt}\psi(t) = \frac{eg\hbar}{4m}\sigma^z B\,\psi(t)$$

The results are

$$\psi(t) = e^{-i\omega_a t}\begin{pmatrix} 1 \\ 0 \end{pmatrix} \rightarrow \hbar\omega_a = \frac{eg\hbar}{4m}B = \hbar\omega_0$$

$$\psi(t) = e^{-i\omega_b t}\begin{pmatrix} 0 \\ 1 \end{pmatrix} \rightarrow \hbar\omega_b = -\frac{eg\hbar}{4m}B = -\hbar\omega_0$$

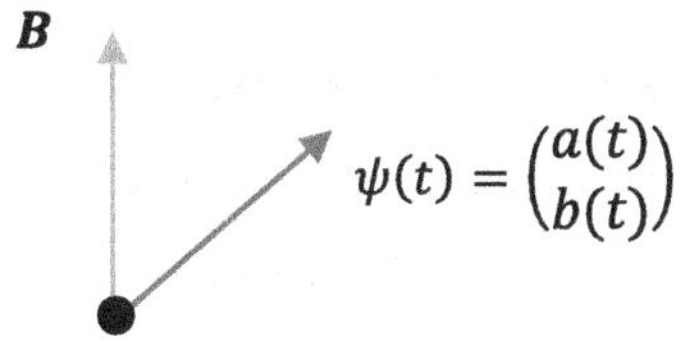

Fig. 5. A spinor function precessing about an external magnetic field.

Note that $\omega_0 = \frac{eg}{4m}B$. Finally,

$$\psi(t) = a\begin{pmatrix}1\\0\end{pmatrix}e^{-i\omega_0 t} + b\begin{pmatrix}0\\1\end{pmatrix}e^{i\omega_0 t} = \begin{pmatrix}ae^{-i\omega_0 t}\\ be^{i\omega_0 t}\end{pmatrix}$$

(b) Let $U = \exp\left(-i\boldsymbol{\sigma}.\boldsymbol{e_b}\frac{\omega_C t}{2}\right)$ act on the initial state of the wavefunction

$$\psi(t) = U\psi(0)$$

where $\omega_C = \frac{eg}{2m}B$. Call the B axis z and write the initial state in a linear superposition of the z basis states. One is led to the following:

$$U = \exp\left(-i\sigma^z\frac{\omega_C t}{2}\right)$$

$$\psi(t) = U\left[a\begin{pmatrix}1\\0\end{pmatrix} + b\begin{pmatrix}0\\1\end{pmatrix}\right]$$

Expand U

$$\exp\left(-i\sigma^z\frac{\omega_C t}{2}\right) = 1 + \left(-i\sigma^z\frac{\omega_C t}{2}\right) + \frac{1}{2}\left(-i\sigma^z\frac{\omega_C t}{2}\right) \times \left(-i\sigma^z\frac{\omega_C t}{2}\right) + \cdots$$

And the following will be apparent:

$$U\,a\begin{pmatrix}1\\0\end{pmatrix} = \exp\left(-i\frac{\omega_C t}{2}\right)\begin{pmatrix}a\\0\end{pmatrix}$$

$$U\,b\begin{pmatrix}0\\1\end{pmatrix} = \exp\left(+i\frac{\omega_C t}{2}\right)\begin{pmatrix}0\\b\end{pmatrix}$$

Noting that $\omega_C = \frac{eg}{2m}B = 2\omega_0$

$$\psi(t) = U\psi(0) = \begin{pmatrix}ae^{-i\omega_0 t}\\ be^{i\omega_0 t}\end{pmatrix}$$

Problem 2.08 Spin Precession: Heisenberg

The Hamiltonian of an electron in a magnetic field is given by

$$H = \mu\,\boldsymbol{\sigma}.\boldsymbol{B}$$

where μ is a constant denoting the magnetic moment of the electron and $\boldsymbol{\sigma} = (\sigma_x, \sigma_y, \sigma_z)$ are the Pauli spin matrices along the three Cartesian axes.

The magnetic field is applied along the Z axis, i.e., $\boldsymbol{B} = B\boldsymbol{e_z}$. Considering Heisenberg's equation of motion,

(a) evaluate $\frac{d\sigma_z}{dt}$,
(b) show that $\frac{d\sigma_x}{dt} = -\gamma\left(\boldsymbol{\sigma}\times\boldsymbol{B}\right).\boldsymbol{e_x}$.

Note: $\mu = \frac{eg\hbar}{4m} = \gamma\frac{\hbar}{2}$, $\gamma = \frac{eg}{2m}$.

Solution

(a)

$$H = \mu\,\sigma_z B$$

$$\frac{d\sigma_z}{dt} = \frac{i}{\hbar}\left[\mu\,\sigma_z B,\, \sigma_z\right] = 0$$

(b) LHS:

$$\frac{d\sigma_x}{dt} = \frac{i}{\hbar}\left[\mu\sigma_z B,\, \sigma_x\right] = \frac{i}{\hbar}\mu B\left[\sigma_z,\, \sigma_x\right] = -\frac{2\mu B}{\hbar}\sigma_y$$

RHS:

$$-\gamma\left(\boldsymbol{\sigma}\times\boldsymbol{B}\right).\boldsymbol{e_x} = -\gamma\left(\boldsymbol{\sigma}\times B\boldsymbol{e_z}\right).\boldsymbol{e_x} = -\gamma\boldsymbol{\sigma}.\left(B\boldsymbol{e_z}\times\boldsymbol{e_x}\right)$$

$$= -\gamma B\boldsymbol{\sigma}.\boldsymbol{e_y} = -\gamma B\sigma_y$$

Now,

$$\mu = \frac{eg\hbar}{4m} = \gamma\frac{\hbar}{2} \rightarrow \gamma = \frac{2\mu}{\hbar}$$

Finally,

$$-\gamma\left(\boldsymbol{\sigma}\times\boldsymbol{B}\right).\boldsymbol{e_x} = -\frac{2\mu B}{\hbar}\sigma_y \rightarrow \frac{d\sigma_x}{dt} = -\gamma\left(\boldsymbol{\sigma}\times\boldsymbol{B}\right).\boldsymbol{e_x}$$

Problem 2.09 Spin Precession: Heisenberg and Compact Notation

In the following, vector algebra in quantum physics and tensor identities will be used in the problem of spin. It is given that

$$H = \gamma\,\boldsymbol{S}.\boldsymbol{B}$$

And $\boldsymbol{S}$ are the spin vectors of Pauli matrices. Show that

(a) $\frac{d\boldsymbol{S}}{dt} = \frac{i}{\hbar}[\gamma\boldsymbol{S}.\boldsymbol{B},\, \boldsymbol{S}] = -\boldsymbol{S}\times\boldsymbol{B}$
(b) $\frac{dS_z}{dt} = \frac{i}{\hbar}[\gamma\,\boldsymbol{S}.\boldsymbol{B},\, S_z] = -(\boldsymbol{S}\times\boldsymbol{B}).\boldsymbol{e_z}$

Solution

(a) One can rewrite vector quantities and their algebra in the following manner:

$$\boldsymbol{S} = S_a \boldsymbol{e_a}$$

$$\boldsymbol{S}.\boldsymbol{A} = S_a A_a$$

where S_a is the component vector and $\boldsymbol{e_a}$ is the basis vector. The above is written in the Einstein convention, in which pairs are summed over identical indices. The commutation relation is

$$[A, B] = AB - BA$$

Therefore,

$$\frac{i}{\hbar}[\gamma \boldsymbol{S}.\boldsymbol{B}, \boldsymbol{S}] = \frac{i}{\hbar}\gamma\left(\left(S_a B_a\right) S_b \boldsymbol{e_b} - \left(S_b \boldsymbol{e_b}\right) S_a B_a\right)$$

$$= \frac{i}{\hbar}\gamma\left(S_a S_b - S_b S_a\right) \boldsymbol{e_b} B_a$$

Observe that $\frac{i}{\hbar}[\gamma \boldsymbol{S}.\boldsymbol{B}, \boldsymbol{S}]$ is a vector quantity. Now, in spin physics, S_a is a Pauli matrix. Recalling the Pauli matrix identity that

$$[\sigma_a, \sigma_b] = i\sigma_c \varepsilon_{abc} + \delta_{ab} \rightarrow [\sigma_a, \sigma_b] = 2i\sigma_c \varepsilon_{abc}$$

It follows that

$$[S_a, S_b] = i\hbar S_c \varepsilon_{abc}$$

Thus,

$$\frac{i}{\hbar}[\gamma \boldsymbol{S}.\boldsymbol{B}, \boldsymbol{S}] = \frac{i}{\hbar}\gamma\left(i\hbar S_c \varepsilon_{abc}\right) \boldsymbol{e_b} B_a$$

$$= -\gamma\, S_c B_a\, \varepsilon_{abc}\, \boldsymbol{e_b} = -\gamma(\boldsymbol{S} \times \boldsymbol{B})$$

Finally,

$$\frac{d\boldsymbol{S}}{dt} = -\gamma\left(\boldsymbol{S} \times \boldsymbol{B}\right)$$

(b) In Einstein notation, one can rewrite vector quantities and their algebra in the following manner:

$$\boldsymbol{S} = S_a \boldsymbol{e_a}$$

$$\boldsymbol{S}.\boldsymbol{A} = S_a A_a$$

Note, however, that unlike the previous problem, the following algebraic operation leads to a scalar quantity.

$$\begin{aligned}\frac{i}{\hbar}\left[\gamma \boldsymbol{S}.\boldsymbol{B}, S_z\right] &= \frac{i}{\hbar}\gamma((S_a B_a)S_z - (S_z)S_a B_a) \\ &= \frac{i}{\hbar}\gamma(S_a S_z - S_z S_a)B_a\end{aligned}$$

Like in the previous case, S_a is a Pauli matrix. Recalling the Pauli matrix identity that

$$[S_a, S_b] = i\hbar S_c \varepsilon_{abc}$$

It follows that

$$\begin{aligned}\frac{i}{\hbar}\left[\gamma \boldsymbol{S}.\boldsymbol{B}, S_z\right] &= \frac{i}{\hbar}\gamma i\hbar S_c \varepsilon_{azc} B_a = -\gamma S_c B_a \varepsilon_{abc}\delta_{bz} \\ &= -\gamma S_c B_a \varepsilon_{abc} \boldsymbol{e_b}.\boldsymbol{e_z} = -\gamma(\boldsymbol{S} \times \boldsymbol{B}).\boldsymbol{e_z}\end{aligned}$$

Finally,

$$\frac{dS_z}{dt} = -\gamma(\boldsymbol{S} \times \boldsymbol{B}).\boldsymbol{e_z}$$

Remarks and Reflections

In spin and magnetic physics, $\boldsymbol{S}$ represents the electron spin and $\boldsymbol{B}$ the magnetic field. Thus, $\boldsymbol{S}.\boldsymbol{B}$ is the magnetic Zeeman energy. It is also commonly written as $\boldsymbol{\sigma}.\boldsymbol{B}$. Term $(\boldsymbol{S} \times \boldsymbol{B})$ would take on the physical meaning of the electron spin precessing about the magnetic field. For reference, the well-known Landau–Lifshitz–Gilbert equation has the form as follows:

$$\frac{d\boldsymbol{M}}{dt} = -\gamma(\boldsymbol{M} \times \boldsymbol{B})$$

where $\boldsymbol{M}$ precesses about magnetic field $\boldsymbol{B}$. The above shows a remarkable consistency between quantum and classical physics.

Recalling previous descriptions of a magnetic moment precessing about the magnetic field, where $\frac{d\boldsymbol{M}}{dt} = -\gamma(\boldsymbol{M} \times \boldsymbol{B})$, the classical analogy for $\frac{dS_z}{dt}$ would be the z component of the equation of motion, i.e.,

$$\frac{d\boldsymbol{M}}{dt}.\boldsymbol{e_z} = -\gamma(\boldsymbol{M} \times \boldsymbol{B}).\boldsymbol{e_z}$$

Problem 2.10 Spinor Transformation

Spin rotation is the most important form of transformation in applied physics. It is the foundational physics of the quantum gates. It is the physics of the fame rotation commonly applied to study the gauge and topology theory of modern systems like graphene, Weyl, Rashba, and topological surfaces.

(a) Prove that

$$(\boldsymbol{\sigma}.\boldsymbol{A})\,(\boldsymbol{\sigma}.\boldsymbol{B}) = (\boldsymbol{A}.\boldsymbol{B})\,I + i\boldsymbol{\sigma}.(\boldsymbol{A} \times \boldsymbol{B})$$

(b) Derive the RHS of the spin rotation operator

$$\exp\left(-\frac{\theta}{2}\boldsymbol{\sigma}.\boldsymbol{e_m}\right) = \cos\frac{\theta}{2}I - i\boldsymbol{\sigma}.\boldsymbol{e_m}\sin\frac{\theta}{2}$$

(c) Show explicitly the matrix representation for the operator.

Note that $\boldsymbol{e_m}$ is the unit vector for the axis of rotation. $\boldsymbol{\sigma}$ is the vector set of the Pauli matrices.

Solution

(a) In tensor notation,

$$(\sigma^a A_a)\,(\sigma^b B_b) = \sigma^a\sigma^b A_a B_b$$

Making use of $\sigma^a\sigma^b = I\delta_{ab} + i\sigma^c\varepsilon_{abc}$,

$$\sigma^a\sigma^b A_a B_b = A_a B_a + i\sigma^c\varepsilon_{abc}A_a B_b$$

In vector notation,

$$A_a B_a \to \boldsymbol{A}.\boldsymbol{B}$$

$$\sigma^c\varepsilon_{abc}A_a B_b \to \boldsymbol{\sigma}.(\boldsymbol{A} \times \boldsymbol{B})$$

Therefore,

$$(\boldsymbol{\sigma}.\boldsymbol{A})\,(\boldsymbol{\sigma}.\boldsymbol{B}) = (\boldsymbol{A}.\boldsymbol{B})\,I + i\boldsymbol{\sigma}.(\boldsymbol{A}\times\boldsymbol{B})$$

(b) Note that

$$\exp\left(-i\frac{\theta}{2}\boldsymbol{\sigma}.\boldsymbol{e_m}\right) = \sum_{n=0}^{\infty}\frac{\left(-i\frac{\theta}{2}\boldsymbol{\sigma}.\boldsymbol{e_m}\right)^n}{n!}$$

To examine $(\boldsymbol{\sigma}.\boldsymbol{e_m})^2$, use is made of $(\boldsymbol{\sigma}.\boldsymbol{A})\,(\boldsymbol{\sigma}.\boldsymbol{B}) = (\boldsymbol{A}.\boldsymbol{B})\,I + i\boldsymbol{\sigma}.(\boldsymbol{A}\times \boldsymbol{B})$, which leads to

$$(\boldsymbol{\sigma}.\boldsymbol{e_m})^2 = \boldsymbol{e_m}.\boldsymbol{e_m}I + i\boldsymbol{\sigma}.(\boldsymbol{e_m}\times\boldsymbol{e_m}) = I$$

As $\boldsymbol{e_m} = \sin\theta_m\cos\phi_m\,\boldsymbol{e_x} + \sin\theta_m\sin\phi_m\,\boldsymbol{e_y} + \cos\theta_m\boldsymbol{e_z}$, where (θ_m, ϕ_m) characterize the axis of rotation, one deduces that $(\boldsymbol{\sigma}.\boldsymbol{e_m})^n$ for even n is I. Now, for odd value of n, $(\boldsymbol{\sigma}.\boldsymbol{e_m})^n$ can always be expressed as

$$(\boldsymbol{\sigma}.\boldsymbol{e_m})^{2n}\,(\boldsymbol{\sigma}.\boldsymbol{e_m}) = (\boldsymbol{\sigma}.\boldsymbol{e_m})$$

Looking in $(-i)^n$ for even terms,

$$(-i)^{2n} = (-1)^{2n}\,i^{2n} = (-1)^n$$

For odd terms,

$$(-i)^{2n+1} = (-i)^{2n}\,(-i) = -i\,(-1)^n$$

Finally, expression $\sum_{n=0}^{\infty}\frac{(-i\frac{\theta}{2}\boldsymbol{\sigma}.\boldsymbol{e_m})^n}{n!}$ can be broken up into the even and the odd series as follows:

$$\sum_{n=0}^{\infty}\frac{\left(-i\frac{\theta}{2}\boldsymbol{\sigma}.\boldsymbol{e_m}\right)^n}{n!} = \sum_{n=0}^{\infty}\frac{(-1)^n}{(2n)!}\left(\frac{\theta}{2}\right)^{2n} I + \sum_{n=0}^{\infty}\frac{-i(-1)^n}{(2n+1)!}\left(\frac{\theta}{2}\right)^{2n+1}\boldsymbol{\sigma}.\boldsymbol{e_m}$$

For ease of reference, all the odd and even terms above are summarized in Table 1.

Let's now drop in on the series expansion for $\cos\frac{\theta}{2}$ and $\sin\frac{\theta}{2}$

$$\cos\frac{\theta}{2} = \sum_{n=0}^{\infty}\frac{(-1)^n}{(2n)!}\left(\frac{\theta}{2}\right)^{2n}, \qquad \sin\frac{\theta}{2} = \sum_{n=0}^{\infty}\frac{(-1)^n}{(2n+1)\,!}\left(\frac{\theta}{2}\right)^{2n+1}$$

Therefore,

$$\exp\left(-i\frac{\theta}{2}\boldsymbol{\sigma}.\boldsymbol{e_m}\right) = \cos\frac{\theta}{2}\,I - i\boldsymbol{\sigma}.\boldsymbol{e_m}\sin\frac{\theta}{2}$$

Table 1. Summary of the even and odd terms in the derivation for $\exp\left(-i\frac{\theta}{2}\boldsymbol{\sigma}.\boldsymbol{e_n}\right)$.

	even terms	odd terms
$(-i)^n$	$(-1)^n$	$-i\,(-1)^n$
$(\boldsymbol{\sigma}.\boldsymbol{e_m})^n$	I	$(\boldsymbol{\sigma}.\boldsymbol{e_m})$
Combining the above		
$\exp\left(-i\frac{\theta}{2}\boldsymbol{\sigma}.\boldsymbol{e_n}\right)$	$\sum_{n=0}^{\infty}\frac{(-1)^n}{(2n)!}\left(\frac{\theta}{2}\right)^{2n} I$	$\sum_{n=0}^{\infty}\frac{-i\,(-1)^n}{(2n+1)!}\left(\frac{\theta}{2}\right)^{2n+1}\boldsymbol{\sigma}.\boldsymbol{e_m}$

(c) Matrix representation:
Note that $\boldsymbol{e_m} = \sin\theta_m\cos\phi_m\,\boldsymbol{e_x} + \sin\theta_m\sin\phi_m\,\boldsymbol{e_y} + \cos\theta_m\boldsymbol{e_z}$, and (θ_m, ϕ_m) are the angles of the axis of rotation. The Pauli matrices are

$$\sigma^x = \begin{pmatrix} 0 & 1 \\ 1 & 0 \end{pmatrix},\ \sigma^y = \begin{pmatrix} 0 & -i \\ i & 0 \end{pmatrix},\ \sigma^z = \begin{pmatrix} 1 & 0 \\ 0 & -1 \end{pmatrix}$$

The matrix representation for $\exp\left(-i\frac{\theta}{2}\boldsymbol{\sigma}.\boldsymbol{e_m}\right) = \cos\frac{\theta}{2}\,I - i\boldsymbol{\sigma}.\boldsymbol{e_m}\sin\frac{\theta}{2}$ is

$$\exp\left(-i\frac{\theta}{2}\boldsymbol{\sigma}.\boldsymbol{e_m}\right) = \cos\frac{\theta}{2}\begin{pmatrix} 1 & 0 \\ 0 & 1 \end{pmatrix} - i\sin\frac{\theta}{2}\begin{pmatrix} \cos\theta_m & \sin\theta_m e^{-i\phi_m} \\ \sin\theta_m e^{i\phi_m} & -cos\theta_m \end{pmatrix}$$

$$= \begin{pmatrix} \cos\frac{\theta}{2} - i\sin\frac{\theta}{2}\cos\theta_m & -i\sin\frac{\theta}{2}\sin\theta_m e^{-i\phi_m} \\ -i\sin\frac{\theta}{2}\sin\theta_m e^{i\phi_m} & \cos\frac{\theta}{2} + i\sin\frac{\theta}{2}\cos\theta_m \end{pmatrix}$$

Remarks and Reflections

One can work out from the above that the operators that rotate a spin state by an angle θ about axis X, Y, and Z are given by

$$\left(\theta_m = \frac{\pi}{2}, \phi_m = 0\right) \to R_x = \begin{pmatrix} \cos\frac{\theta}{2} & -i\sin\frac{\theta}{2} \\ -i\sin\frac{\theta}{2} & \cos\frac{\theta}{2} \end{pmatrix}$$

$$\left(\theta_m = \frac{\pi}{2}, \phi_m = \frac{\pi}{2}\right) \to R_y = \begin{pmatrix} \cos\frac{\theta}{2} & -\sin\frac{\theta}{2} \\ \sin\frac{\theta}{2} & \cos\frac{\theta}{2} \end{pmatrix}$$

$$(\theta_m = 0) \to R_z = \begin{pmatrix} e^{-\frac{i\theta}{2}} & 0 \\ 0 & e^{\frac{i\theta}{2}} \end{pmatrix}$$

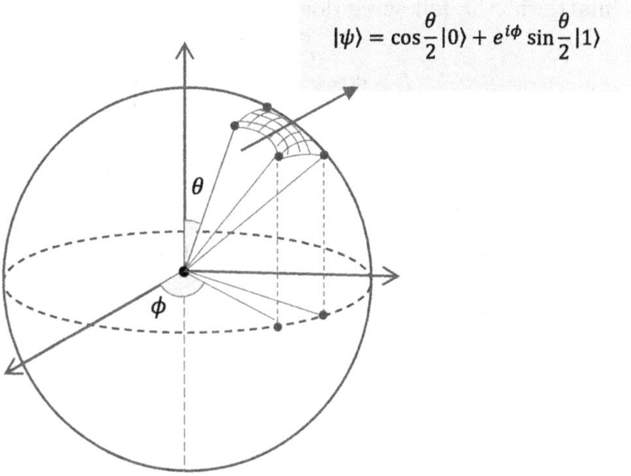

Fig. 6. A 3D visualization of the spin traversing the surface of the Bloch sphere.

Now, let R_z act on spin-X state $\frac{1}{\sqrt{2}}\begin{pmatrix}1\\1\end{pmatrix}$ by $\frac{\pi}{2}$

$$\begin{pmatrix} e^{-\frac{i\pi}{4}} & 0 \\ 0 & e^{\frac{i\pi}{4}} \end{pmatrix} \frac{1}{\sqrt{2}}\begin{pmatrix}1\\1\end{pmatrix} = \frac{1}{\sqrt{2}}e^{-\frac{i\pi}{4}}\begin{pmatrix}1\\e^{\frac{i\pi}{2}}\end{pmatrix} = \frac{1}{\sqrt{2}}e^{-\frac{i\pi}{4}}\begin{pmatrix}1\\i\end{pmatrix}$$

Spin state X has been rotated by 90 degrees anti-clockwise to take on spin state Y.

Problem 2.11 Frame Rotation

In modern spinor systems like graphene, spintronics, and topological surfaces, it is common to apply a rotational transformation to study the Berry–Pancharatnam phase of the systems. One simple example is to rotate the frame about an axis in the X–Y plane (see Bloch sphere in the following). Such a frame rotation would be mathematically described by

$$H' = U^{\dagger}HU$$

In this problem, the physical significance of the U matrix is investigated.

(a) Shown in the following is a unitary matrix commonly used to perform a local gauge transformation for modern spinor systems. Show that the unitary matrix in the following does rotate spin (θ, ϕ) to the z axis.

$$U = \begin{pmatrix} \cos\frac{\theta}{2} & \sin\frac{\theta}{2}e^{-i\phi} \\ -\sin\frac{\theta}{2}e^{i\phi} & \cos\frac{\theta}{2} \end{pmatrix}$$

(b) The unitary matrix above rotates spin (θ, ϕ) about an axis $\boldsymbol{e_m}$ lying in the X–Y plane. Find the angles (θ_m, ϕ_m) and determine $\boldsymbol{e_m}$.

Note: $\boldsymbol{e_m} = \sin\theta_m \cos\phi_m\, \boldsymbol{e_x} + \sin\theta_m \sin\phi_m\, \boldsymbol{e_y} + \cos\theta_m \boldsymbol{e_z}$, *where* (θ_m, ϕ_m) *are the angles of the axis of rotation.*

Solution

(a) Let the matrix act on a spin oriented at angles (θ, ϕ) and obtain

$$\begin{pmatrix} \cos\frac{\theta}{2} & \sin\frac{\theta}{2}e^{-i\phi} \\ -\sin\frac{\theta}{2}e^{i\phi} & \cos\frac{\theta}{2} \end{pmatrix} \begin{pmatrix} \cos\frac{\theta}{2} \\ e^{i\phi}\sin\frac{\theta}{2} \end{pmatrix} = \begin{pmatrix} 1 \\ 0 \end{pmatrix}$$

(b) The unitary operator that rotates the spin about axis $\boldsymbol{e_m}$ is given by

$$\exp\left(-i\frac{\theta}{2}\boldsymbol{\sigma}.\boldsymbol{e_m}\right) = \cos\frac{\theta}{2} I - i\boldsymbol{\sigma}.\boldsymbol{e_m} \sin\frac{\theta}{2}$$

$$= \begin{pmatrix} \cos\frac{\theta}{2} - i\sin\frac{\theta}{2}\cos\theta_m & -i\sin\frac{\theta}{2}\sin\theta_m e^{-i\phi_m} \\ -i\sin\frac{\theta}{2}\sin\theta_m e^{i\phi_m} & \cos\frac{\theta}{2} + i\sin\frac{\theta}{2}\cos\theta_m \end{pmatrix}$$

As the axis of rotation lies in the X–Y plane, $\theta_m = 90$, thus

$$\begin{pmatrix} \cos\frac{\theta}{2} & -i\sin\frac{\theta}{2}e^{-i\phi_m} \\ -i\sin\frac{\theta}{2}e^{i\phi_m} & \cos\frac{\theta}{2} \end{pmatrix} \begin{pmatrix} \cos\frac{\theta}{2} \\ e^{i\phi}\sin\frac{\theta}{2} \end{pmatrix} = \begin{pmatrix} 1 \\ 0 \end{pmatrix}$$

Solving the above leads to two equations

$$\left(\cos\frac{\theta}{2}\right)^2 - i\left(\sin\frac{\theta}{2}\right)^2 e^{i(\phi-\phi_m)} = 1,$$

$$i\sin\frac{\theta}{2}\cos\frac{\theta}{2}e^{i\phi_m} = \sin\frac{\theta}{2}\cos\frac{\theta}{2}e^{i\phi}$$

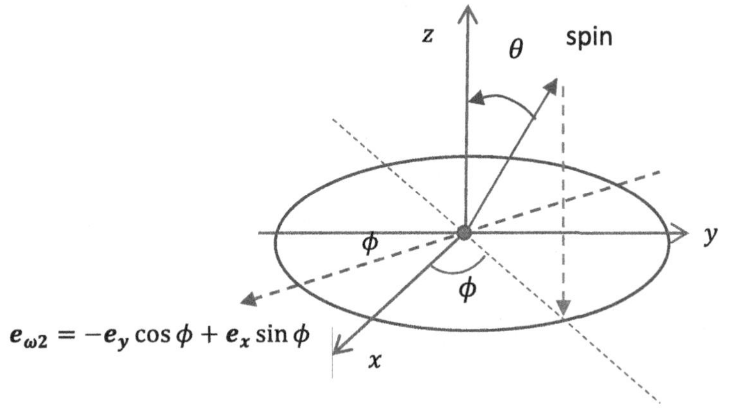

Fig. 7. Spin rotates about the axis of rotation by θ and reaches the Z axis.

Both equations lead to

$$e^{i(\phi-\phi_m)} = i \rightarrow \phi_m = \phi + 3\pi/2$$

Therefore, referring to Figure 7, the angles for $\boldsymbol{e_m}$ are $(\theta_m = \pi/2, \phi_m = \phi + 3\pi/2)$. Substituting these angles into $\boldsymbol{e_m}$ leads to

$$\boldsymbol{e_m} = \cos(\phi + 3\pi/2)\,\boldsymbol{e_x} + \sin(\phi + 3\pi/2)\,\boldsymbol{e_y} = \sin\phi\,\boldsymbol{e_x} - \cos\phi\,\boldsymbol{e_y}$$

Remarks and Reflections

It has been shown in (b) that the unitary matrix $U_2 = \exp\left(\frac{-i}{2}\boldsymbol{\sigma}.\boldsymbol{\omega_{m2}}t\right) = \left(\cos\frac{\theta}{2} - i\boldsymbol{\sigma}.\boldsymbol{e_{m2}}\sin\frac{\theta}{2}\right)$ rotates the spin about axis $\boldsymbol{e_{m2}}$. Rotation is in an anti-clockwise manner. Likewise, one can work out that the unitary matrix $U_1 = \exp\left(\frac{i}{2}\boldsymbol{\sigma}.\boldsymbol{\omega_{m1}}t\right) = \left(\cos\frac{\theta}{2} + i\boldsymbol{\sigma}.\boldsymbol{e_{m1}}\sin\frac{\theta}{2}\right)$ rotates the spin about axis $\boldsymbol{e_{\omega 1}}$ in a clockwise fashion. Note that $\boldsymbol{\omega_{m2}}t = \theta t\,\boldsymbol{e_{m2}}$.

Referring to Figure 8 and by choosing $\boldsymbol{e_{m1}} = \boldsymbol{e_y}\cos\phi - \boldsymbol{e_x}\sin\phi$ (X–Y plane), one could arrive at

$$U_1 = \exp\left(\frac{i}{2}\boldsymbol{\sigma}.\boldsymbol{\omega_{m1}}t\right) = \begin{pmatrix} \cos\dfrac{\theta}{2} & \sin\dfrac{\theta}{2}e^{-i\phi} \\ -sin\dfrac{\theta}{2}e^{i\phi} & \cos\dfrac{\theta}{2} \end{pmatrix}$$

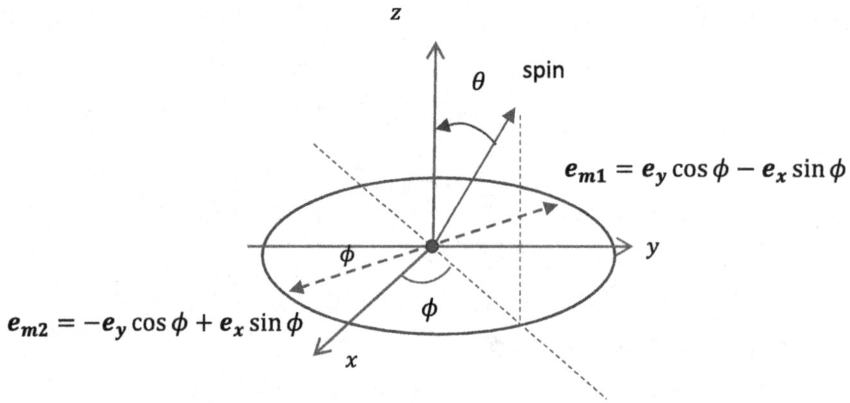

Fig. 8. Spin rotates about the axis of rotation by θ and reaches the Z axis.

Referring to Figure 8 and by choosing $\boldsymbol{e_{m2}} = -\boldsymbol{e_y} \cos\phi + \boldsymbol{e_x} \sin\phi$ (X–Y plane), one could arrive at

$$U_2 = \exp\left(\frac{-i}{2}\boldsymbol{\sigma}.\boldsymbol{\omega}_{m2}t\right) = \begin{pmatrix} \cos\frac{\theta}{2} & \sin\frac{\theta}{2}e^{-i\phi} \\ -sin\frac{\theta}{2}e^{i\phi} & \cos\frac{\theta}{2} \end{pmatrix}$$

Therefore, $U_1 = U_2$ is the matrix that rotates the spin to axis Z about an axis in the X–Y plane. Rotation continues at a constant ϕ.

Chapter 3

Second Quantization and Applications

Second quantization is the foundation of quantum field theory. Recently, it has also been used very extensively in research fields like quantum transport, quantum optics, quantum nanoscience, and so forth. This chapter deals with second quantization and its applications in mainly the context of nanoscience and condensed matter physics. Particular attention is also paid to its application in the non-equilibrium Green's function — a modern topic for quantum transport in electronic and spintronic devices. It is suitable for postgraduate or senior undergraduate students taking courses in advanced quantum mechanics.

Second Quantization

Problem 3.01 Second Quantization 1
Problem 3.02 Second Quantization 11
Problem 3.03 Field Operators
Problem 3.04 Second-Quantized Kinetic Energy
Problem 3.05 Quantized Kinetic Energy for Discrete System
Problem 3.06 Second Quantized Electron Spin
Problem 3.07 Second Quantized Spin–Orbit Coupling 1
Problem 3.08 Second Quantized Spin–Orbit Coupling 11

Dirac Delta Calculus

Problem 3.09 Dirac Delta and Fourier Transform
Problem 3.10 Dirac Delta Calculus
Problem 3.11 Dirac Delta Representations
Problem 3.12 Vector Calculus and Dirac Delta

Condensed Matter Applications

Problem 3.13 Electron Background Energy
Problem 3.14 Electron Interaction Formulation
Problem 3.15 Electron–Electron Energy: Direct
Problem 3.16 Electron–Electron Energy: Exchange
Problem 3.17 Scattering Strength of Electron Interaction
Problem 3.18 Quantifying the Exchange Energy
Problem 3.19 Kinetic Energy per Particle

Appendix 3A. Dirac Delta Identities
Appendix 3B. Vector Calculus Identities
Appendix 3C. Fourier Transform Identities

Second Quantization

Problem 3.01 Second Quantization 1

The fermionic physics (statistics) is best illuminated by the relationships of the second-quantized operators as shown on the left column of Table 1. Making use of these basic relationships, prove the relationship between second-quantized operators listed on the right column of Table 1.

Table 1.

Fermionic Physics	Prove the Following
1. $\{c_i, c_j^\dagger\} = \delta_{ij}$	(a) $[n_j, c_j] = -c_j$
2. $\{c_i, c_j\} = 0$	(b) $[n_j, c_k] = 0$
3. $\{c_i^\dagger, c_j^\dagger\} = 0$	(c) $[n_j, c_j^\dagger] = c_j^\dagger$
4. $(c_j)^2 = 0$	(d) $[n_j, n_k] = 0$
5. $(c_j^\dagger)^2 = 0$	(e) $[n_j n_k, c_j] = -n_k c_j$
6. $n_j = c_j^\dagger c_j$	

Solution

(a) To show $[n_j, c_j] = -c_j$,

$$\begin{aligned}[n_j, c_j] &= c_j^\dagger c_j c_j - c_j c_j^\dagger c_j \\ &= -c_j c_j^\dagger c_j \\ &= -(1 - c_j^\dagger c_j) c_j \\ &= -c_j + c_j^\dagger c_j c_j = -c_j\end{aligned}$$	where use has been made of $\{c_i, c_j^\dagger\} = \delta_{ij}$ and $(c_j)^2 = 0$.

(b) To show $[n_j, c_k] = 0$,

$[n_j, c_k] = n_j c_k - c_k n_j$ $= c_j^\dagger c_j c_k - c_k n_j$ $= c_k c_j^\dagger c_j - c_k n_j = 0$	where use has been made of $n_j = c_j^\dagger c_j$, $\{c_i, c_j\} = 0$ and $\{c_i, c_j^\dagger\} = \delta_{ij}$.

(c) To show $[n_j, c_j^\dagger] = c_j^\dagger$,

$[n_j, c_j^\dagger] = c_j^\dagger c_j c_j^\dagger - c_j^\dagger c_j^\dagger c_j$ $= c_j^\dagger c_j c_j^\dagger$ $= c_j^\dagger(1 - c_j^\dagger c_j)$ $= c_j^\dagger - c_j^\dagger c_j^\dagger c_j = c_j^\dagger$	where use has been made of $\{c_i, c_j^\dagger\} = \delta_{ij}$ and $(c_j^\dagger)^2 = 0$.

(d) To show $[n_j, n_k] = 0$,

$[n_j, n_k] = c_j^\dagger c_j c_k^\dagger c_k - c_k^\dagger c_k c_j^\dagger c_j$ $= -c_j^\dagger c_k^\dagger c_j c_k - c_k^\dagger c_k c_j^\dagger c_j$ $= c_k^\dagger c_j^\dagger c_j c_k - c_k^\dagger c_k c_j^\dagger c_j$ $= -c_k^\dagger c_j^\dagger c_k c_j - c_k^\dagger c_k c_j^\dagger c_j$ $= c_k^\dagger c_k c_j^\dagger c_j - c_k^\dagger c_k c_j^\dagger c_j$ $= 0$	where use has been made of $n_j = c_j^\dagger c_j$, $\{c_i, c_j\} = 0$, $\{c_i^\dagger, c_j^\dagger\} = 0$ and $\{c_i, c_j^\dagger\} = \delta_{ij}$.

(e) To show $[n_j n_k, c_j] = -n_k c_j$,

$[n_j n_k, c_j] = n_j n_k c_j - c_j n_j n_k$ Since indices $j \neq k$, $[n_j n_k, c_j] = n_j c_j n_k - c_j n_j n_k$ $= [n_j, c_j] n_k$ Referring to (a) where $[n_j, c_j] = -c_j$, $[n_j n_k, c_j] = -c_j n_k$ Once again, since indices $j \neq k$, $[n_j n_k, c_j] = -n_k c_j$	

Remarks and Reflections

From (a) and (c), one also has $[c_j, n_j] = c_j$ and $[c_j^\dagger, n_j] = -c_j^\dagger$. It can also be shown that

$$[c_v^\dagger c_\mu, c_{v'}^\dagger c_{\mu'}] = c_v^\dagger c_{\mu'} \delta_{\mu v'} - c_{v'}^\dagger c_\mu \delta_{v\mu'}$$

Problem 3.02 Second Quantization 11

The second quantized operators are defined from the first quantized operators of A^v and B^v as follows:

$$A = \sum_{ij} \langle i|A^v|j\rangle a_i^\dagger a_j, \quad B = \sum_{ij} \langle i|B^v|j\rangle a_i^\dagger a_j$$

Show that if A^v and B^v commute, i.e., $[A^v, B^v] = 0$, so does A and B.

Note: Superscript v denotes a general vector space. The operator takes on different forms in their respective spaces.

Solution

$$AB = \sum_{ijkl} a_i^\dagger \langle i|A^v|j\rangle a_j a_k^\dagger \langle k|B^v|l\rangle a_l$$

Now, use is made of $a_j a_k^\dagger - \zeta a_k^\dagger a_j = \delta_{jk}$, where $\zeta = \pm$ in the case of bosons/fermions. Hence,

$$\begin{aligned} AB &= \sum_{ijl} a_l^\dagger \langle i|A^v|j\rangle (\zeta a_k^\dagger a_j + \delta_{jk}) \langle k|B^v|l\rangle a_l \\ &= \sum_{ijkl} a_l^\dagger \langle i|A^v|j\rangle \zeta a_k^\dagger a_j \langle k|B|l\rangle a_l + a_l^\dagger \langle i|A^v|k\rangle \langle k|B|l\rangle a_l \end{aligned}$$

With $\{a_i, a_j\} = 0$, $\{a_i^\dagger, a_j^\dagger\} = 0$,

$$\begin{aligned} AB &= \sum_{ijkl} a_k^\dagger \langle k|B^v|l\rangle \zeta a_i^\dagger a_l \langle i|A^v|j\rangle a_j + a_l^\dagger \langle i|A^v|k\rangle \langle k|B^v|l\rangle a_l \\ &= \sum_{ijkl} a_k^\dagger \langle k|B^v|l\rangle (a_l a_i^\dagger - \delta_{il}) \langle i|A^v|j\rangle a_j + a_l^\dagger \langle i|A^v|k\rangle \langle k|B^v|l\rangle a_l \end{aligned}$$

Shuffling,

$$AB = \sum_{ijkl} a_k^\dagger \langle k|B^v|l\rangle a_l a_i^\dagger \langle i|A^v|j\rangle a_j + a_l^\dagger \langle i|A^v|k\rangle\langle k|B^v|l\rangle a_l$$
$$- a_k^\dagger \langle k|B^v|l\rangle\langle l|A^v|j\rangle a_j = BA + \sum_{nm} \langle n|A^v B^v|m\rangle$$
$$- \sum_{nm} \langle n|B^v A^v|m\rangle$$

It thus follows that

$$[A^v, B^v] = 0 \;\; \rightarrow \;\; [A, B] = 0$$

Problem 3.03 Field Operators

The quantum fields that describe particles in a system are written as follows:

$$\psi(\boldsymbol{r}) = \frac{1}{\sqrt{V}} \sum_{\boldsymbol{k}} a_{\boldsymbol{k}} e^{i\boldsymbol{k}.\boldsymbol{r}} = \sum_{\boldsymbol{k}} a_{\boldsymbol{k}} \phi_{\boldsymbol{k}}(\boldsymbol{r})$$
$$\psi^\dagger(\boldsymbol{r}) = \frac{1}{\sqrt{V}} \sum_{\boldsymbol{k}} a_{\boldsymbol{k}}^\dagger e^{-i\boldsymbol{k}.\boldsymbol{r}} = \sum_{\boldsymbol{k}} a_{\boldsymbol{k}}^\dagger \phi_{\boldsymbol{k}}^\dagger(\boldsymbol{r})$$

(a) Show that the total number of particles is given by

$$N = \int \psi^\dagger(\boldsymbol{r})\psi(\boldsymbol{r}) d^3r = \sum_k a_{\boldsymbol{k}}^\dagger a_{\boldsymbol{k}}$$

(b) Show by Fourier transform of $\psi(r)$ and $\psi^\dagger(r)$ that

$$a_{\boldsymbol{k}}^\dagger = \frac{1}{\sqrt{V}} \int \psi^\dagger(\boldsymbol{r}) e^{i\boldsymbol{k}.\boldsymbol{r}} d^3r$$
$$a_{\boldsymbol{k}} = \frac{1}{\sqrt{V}} \int \psi(\boldsymbol{r}) e^{-i\boldsymbol{k}.\boldsymbol{r}} d^3r$$

Note: r, k stand for position and momentum coordinates, respectively. Integration is carried out for range $(-\infty, +\infty)$ and V is the volume containing the particles.

Note: $(\boldsymbol{r}, \boldsymbol{k})$ are vector quantities. The integrals of $\int d^3r = \int dx\, dy\, dz = \int d\boldsymbol{r}$ have the same meaning. Sometimes, they are written like $\int d^3x = \int d\boldsymbol{x}$, $\int d^3y = \int d\boldsymbol{y}$. Likewise, the integrals of $\int d^3k = \int dk_x\, dk_y\, dk_z = \int d\boldsymbol{k}$ have the same meaning.

Solution

(a) The total number of electron is given by

$$N = \int \psi^\dagger(\boldsymbol{r})\psi(\boldsymbol{r})d^3r = \int \left(\frac{1}{\sqrt{V}}\sum_{\boldsymbol{k}'} a_{\boldsymbol{k}'}^\dagger e^{-i\boldsymbol{k}'.\boldsymbol{r}}\right)\left(\frac{1}{\sqrt{V}}\sum_{\boldsymbol{k}} a_{\boldsymbol{k}} e^{i\boldsymbol{k}.\boldsymbol{r}}\right) d^3r$$

$$= \frac{1}{V}\int \sum_{\boldsymbol{k}\boldsymbol{k}'} a_{\boldsymbol{k}'}^\dagger a_{\boldsymbol{k}} e^{i(\boldsymbol{k}-\boldsymbol{k}').\boldsymbol{r}} d^3r$$

Observe that momentum indices are denoted by k and k' to allow for all possible pairings. In the end, only identical pairings are retained as shown in the following:

$$N = \frac{1}{V}\sum_{\boldsymbol{k}\boldsymbol{k}'} a_{\boldsymbol{k}'}^\dagger a_{\boldsymbol{k}}(2\pi)^3\delta^3(k-k')$$

$$= \frac{1}{V}\frac{V}{(2\pi)^3}\int \sum_{\boldsymbol{k}'} a_{\boldsymbol{k}'}^\dagger a_{\boldsymbol{k}}(2\pi)^3\delta^3(k-k')d^3k' = \sum_{\boldsymbol{k}} a_{\boldsymbol{k}}^\dagger a_{\boldsymbol{k}}$$

In the above, use has been made of one of the following:

Volume summation (Continuous)	*Dirac delta function (Continuous)*
$\sum_{\boldsymbol{k}} \to \int d^3k \frac{V}{(2\pi)^3}$	$\frac{1}{(2\pi)^3}\int e^{i(\boldsymbol{k}-\boldsymbol{k}').\boldsymbol{r}} d^3r = \delta^3(k-k')$
Volume summation (Discrete)	*Dirac delta function (Discrete)*
$\sum_{\boldsymbol{k}}$	$\int e^{i(\boldsymbol{k}-\boldsymbol{k}').\boldsymbol{r}} d^3r = V\delta_{\boldsymbol{k}\boldsymbol{k}'}^3$

(b) By Fourier transform of $\psi(\boldsymbol{r})$,

$$FT\{\psi(\boldsymbol{r})\} = \left(\frac{1}{\sqrt{2\pi}}\right)^3 \int \psi(\boldsymbol{r})e^{-i\boldsymbol{k}.\boldsymbol{r}} d^3r$$

With $\psi(\boldsymbol{r}) = \frac{1}{\sqrt{V}}\sum_{\boldsymbol{k}} a_{\boldsymbol{k}} e^{i\boldsymbol{k}.\boldsymbol{r}}$,

$$FT\{\psi(\boldsymbol{r})\} = \left(\frac{1}{\sqrt{2\pi}}\right)^3 \frac{1}{\sqrt{V}}\sum_{\boldsymbol{k}'} a_{\boldsymbol{k}'}(2\pi)^3\delta^3(k'-k)$$

With $\sum_{\boldsymbol{k}} \to \int d^3k \frac{V}{(2\pi)^3}$,

$$FT\{\psi(\boldsymbol{r})\} = \left(\frac{1}{\sqrt{2\pi}}\right)^3 \sqrt{V} a_{\boldsymbol{k}}$$

Therefore,

$$\left(\frac{1}{\sqrt{2\pi}}\right)^3 \sqrt{V} a_{\boldsymbol{k}} = \left(\frac{1}{\sqrt{2\pi}}\right)^3 \int \psi(\boldsymbol{r}) e^{-i\boldsymbol{k}.\boldsymbol{r}} d^3r$$

$$\to a_{\boldsymbol{k}} = \frac{1}{\sqrt{V}} \int \psi(\boldsymbol{r}) e^{-i\boldsymbol{k}.\boldsymbol{r}} d^3r$$

Following the same procedures,

$$a_{\boldsymbol{k}}^\dagger = \frac{1}{\sqrt{V}} \int \psi^\dagger(\boldsymbol{r}) e^{i\boldsymbol{k}.\boldsymbol{r}} d^3r$$

Remarks and Reflections

Take note that the total particle is

$$N = \int \psi^\dagger(\boldsymbol{r})\psi(\boldsymbol{r}) d^3r = \sum_{\boldsymbol{k}} a_{\boldsymbol{k}}^\dagger a_{\boldsymbol{k}}$$

And the particle density is

$$\rho(\boldsymbol{r}) = \psi^\dagger(\boldsymbol{r})\psi(\boldsymbol{r}) = \frac{1}{V} \sum_{\boldsymbol{k}\boldsymbol{k}'} a_{\boldsymbol{k}'}^\dagger a_{\boldsymbol{k}} e^{i(\boldsymbol{k}-\boldsymbol{k}').\boldsymbol{r}}$$

With the Fourier components $a_{\boldsymbol{k}}$ and $\boldsymbol{a}_{\boldsymbol{k}}^\dagger$, one could also show that $\sum_{\boldsymbol{k}} a_{\boldsymbol{k}}^\dagger a_{\boldsymbol{k}}$ leads to $\int \psi^\dagger(\boldsymbol{r})\psi(\boldsymbol{r}) d^3r$

$$\sum_{\boldsymbol{k}} a_{\boldsymbol{k}}^\dagger a_{\boldsymbol{k}} = \sum_{\boldsymbol{k}} \int \left(\frac{1}{\sqrt{V}} \psi^\dagger(r) e^{i\boldsymbol{k}.\boldsymbol{r}}\right) \left(\frac{1}{\sqrt{V}} \psi(\boldsymbol{r}') e^{-i\boldsymbol{k}.\boldsymbol{r}'}\right) d^3r d^3r'$$

By analogy to the momentum indices, the spatial indices are denoted by r and r' to allow for all possible pairings. Once again, only identical pairings are retained due to the Dirac delta functions. One would now have

$$\sum_{\boldsymbol{k}} a_{\boldsymbol{k}}^\dagger a_{\boldsymbol{k}} = \frac{V}{(2\pi)^3}\left(\frac{1}{V}\right)\int \psi^\dagger(\boldsymbol{r})e^{ik(r-r')}d^3k\psi(\boldsymbol{r}')d^3rd^3r'$$

$$= \int \psi^\dagger(\boldsymbol{r})\psi(\boldsymbol{r}')\delta^3(r-r')d^3rd^3r' = \int \psi^\dagger(\boldsymbol{r})\psi(\boldsymbol{r})d^3r$$

Note that if one writes $N = \sum_{\boldsymbol{k}} a_{\boldsymbol{k}}^\dagger a_{\boldsymbol{k}} = \sum_{\boldsymbol{k}} n_{\boldsymbol{k}}$, one can also write in discrete form

$$N = \int \psi^\dagger(\boldsymbol{r})\psi(\boldsymbol{r})d^3r = \sum_i a_i^\dagger a_i$$

One should, therefore, note that $a_{\boldsymbol{k}}^\dagger a_{\boldsymbol{k}} = n_{\boldsymbol{k}}$ and $a_i^\dagger a_i = n_i$. It thus follows that

$$\sum_{\boldsymbol{k}} a_{\boldsymbol{k}}^\dagger a_{\boldsymbol{k}} = \sum_i a_i^\dagger a_i$$

The number of particles conserves irrespective of the basis used in calculations. But take note that $n_{\boldsymbol{k}} \neq n_i$.

Problem 3.04 Second-Quantized Kinetic Energy

In the quantum mechanics section, momentum, position, and kinetic energy have been studied. In this section, we will look into physical quantities in their second-quantized form. Given the kinetic energy in second-quantized form

$$H_{KE} = \sum_{\boldsymbol{k}\boldsymbol{k}'\sigma\sigma'} \langle \boldsymbol{k}\sigma|H_{KE}^v|\boldsymbol{k}'\sigma'\rangle a_{\boldsymbol{k}\sigma}^\dagger a_{\boldsymbol{k}'\sigma'}$$

where k is the momentum and σ is the spin quantum number, show the expression in

(a) momentum space,
(b) real space

Note: Superscripts v and f represent operators in general vector and function forms, respectively.

Solution

(a) In momentum space,

$$H_{KE} = \sum_{kk'\sigma\sigma'} \langle \boldsymbol{k}\sigma | H_{KE}^{v} | \boldsymbol{k}'\sigma' \rangle a_{\boldsymbol{k}\sigma}^{\dagger} a_{\boldsymbol{k}'\sigma'}$$

$$= \sum_{kk'\sigma\sigma'} \langle \sigma | \sigma' \rangle \langle \boldsymbol{k} | \boldsymbol{k}' \rangle E_{\boldsymbol{k}'\sigma'} a_{\boldsymbol{k}\sigma}^{\dagger} a_{\boldsymbol{k}'\sigma'}$$

With $\langle \sigma | \sigma' \rangle = \delta_{\sigma\sigma'}$ and $\langle \boldsymbol{k} | \boldsymbol{k}' \rangle = \delta_{\boldsymbol{k}\boldsymbol{k}'}$, it follows that

$$H_{KE} = \sum_{kk'\sigma\sigma'} \delta_{\boldsymbol{k}\boldsymbol{k}'} \delta_{\sigma\sigma'} E_{\boldsymbol{k}'\sigma'} a_{\boldsymbol{k}\sigma}^{\dagger} a_{\boldsymbol{k}'\sigma'}$$

Note that the operator H_{KE}^{f} has no effect on the spin. Therefore, the kinetic energy in momentum space is

$$H_{KE} = \sum_{\boldsymbol{k}\sigma} E_{\boldsymbol{k}\sigma} a_{\boldsymbol{k}\sigma}^{\dagger} a_{\boldsymbol{k}\sigma}$$

(b) In real space, with the completeness relation,

$$H_{KE} = \sum_{kk'\sigma\sigma'} \langle \boldsymbol{k}\sigma | H_{KE}^{v} | \boldsymbol{k}'\sigma' \rangle a_{\boldsymbol{k}\sigma}^{\dagger} a_{\boldsymbol{k}'\sigma'}$$

$$= \int \sum_{kk'\sigma\sigma'} \langle \boldsymbol{k}\sigma | \boldsymbol{r} \rangle \langle \boldsymbol{r} | H_{KE}^{v} | \boldsymbol{r}' \rangle \langle \boldsymbol{r}' | \boldsymbol{k}'\sigma' \rangle a_{\boldsymbol{k}\sigma}^{\dagger} a_{\boldsymbol{k}'\sigma'} d^3r d^3r'$$

Use is then made of $\langle \boldsymbol{r} | H_{KE}^{v} | \boldsymbol{r}' \rangle = H_{KE}^{f} \delta(\boldsymbol{r} - \boldsymbol{r}')$ where

$$H_{KE}^{f} = -\frac{\hbar^2}{2m} \nabla^2$$

where $\nabla_r^2 = \frac{\partial^2}{\partial x^2} + \frac{\partial^2}{\partial y^2} + \frac{\partial^2}{\partial z^2}$. As the operator H_{KE}^{f} has no effect on the spin, it follows that

$$H_{KE} = \int \sum_{kk'\sigma\sigma'} a_{\boldsymbol{k}\sigma}^{\dagger} \phi_{\boldsymbol{k}}^{\dagger}(r) H_{KE}^{f} \delta(\boldsymbol{r} - \boldsymbol{r}') \delta_{\sigma\sigma'} a_{\boldsymbol{k}'\sigma'} \phi_{\boldsymbol{k}'}(\boldsymbol{r}') d^3r d^3r'$$

Note in the above that use has been made of

$$\langle \boldsymbol{r}' | \boldsymbol{k}'\sigma' \rangle = \phi_{\boldsymbol{k}'\sigma'}(\boldsymbol{r}') = \frac{1}{\sqrt{V}} e^{i\boldsymbol{k}'.\boldsymbol{r}'} |\sigma'\rangle$$

$$\langle \boldsymbol{k}\sigma | \boldsymbol{r} \rangle = \phi_{\boldsymbol{k}\sigma}^{\dagger}(\boldsymbol{r}) = \frac{1}{\sqrt{V}} e^{-i\boldsymbol{k}.\boldsymbol{r}} \langle \sigma |$$

Integrating with the Dirac delta function,

$$H_{KE} = \int \sum_{kk'\sigma} a^{\dagger}_{k\sigma}\phi^{\dagger}_{k}(\boldsymbol{r}) H^{f}_{KE} a_{k'\sigma}\phi_{k'}(\boldsymbol{r'}) d^3r$$

In field form, we have

$$H_{KE} = \int \sum_{\sigma} \psi^{\dagger}_{\sigma}(\boldsymbol{r}) \left(-\frac{\hbar^2}{2m}\nabla^2\right) \psi_{\sigma}(\boldsymbol{r}) d^3r$$

Now, by chain rule,

$$\psi^{\dagger}_{\sigma}(\boldsymbol{r})\nabla^2\psi_{\sigma}(\boldsymbol{r}) = \nabla.(\psi^{\dagger}_{\sigma}(\boldsymbol{r})\nabla\psi_{\sigma}(\boldsymbol{r})) - (\nabla\psi^{\dagger}_{\sigma}(\boldsymbol{r})).(\nabla\psi_{\sigma}(\boldsymbol{r}))$$

It thus follows that

$$H_{KE} = -\frac{\hbar^2}{2m}\int \sum_{\sigma} \nabla.(\psi^{\dagger}_{\sigma}(\boldsymbol{r})\nabla\psi_{\sigma}(\boldsymbol{r})) - (\nabla\psi^{\dagger}_{\sigma}(\boldsymbol{r})).(\nabla\psi_{\sigma}(\boldsymbol{r}))d^3r$$

But according to the divergence theorem,

$$\int \nabla.(\psi^{\dagger}_{\sigma}(\boldsymbol{r})\nabla\psi_{\sigma}(\boldsymbol{r}))d^3r = \int \psi^{\dagger}_{\sigma}(\boldsymbol{r})\nabla\psi_{\sigma}(\boldsymbol{r}).d\boldsymbol{S}$$

The LHS is a volume integral. The RHS is an integral over the surface enclosing the volume. As the field approaches zero in the case of infinite-sized volume, the integral vanishes. The integral thus vanishes for any volume size as the theorem holds true with no specification for volume size needed. Therefore,

$$H_{KE} = \frac{\hbar^2}{2m}\int \sum_{\sigma} (\nabla\psi^{\dagger}_{\sigma}(\boldsymbol{r})).(\nabla\psi_{\sigma}(\boldsymbol{r}))d^3r$$

Remarks and Reflections

An equivalence to the second-quantized formulation is observed in the first-quantized formalisms, e.g., an operator O written in terms of its projection over the space spanned by its basis eigenstates (where O_n are the eigenvalues) can be mapped to a second-quantized formulation as follows:

$$O = \sum_{n} P_n O_n = \sum_{n} |n\rangle O_n \langle n| \leftrightarrows \sum_{n} a^{\dagger}_{n} a_n O_n$$

where $|n\rangle \to a_n^\dagger$ is the creation operator and $\langle n| \to a_n$ is the annihilation operator. Note also that the density operator can be written in terms of its projection over the space spanned by its basis eigenstates

$$\rho = e^{-\beta H} = \sum_v |v\rangle\langle v| e^{-\beta E_v}$$

Problem 3.05 Second-Quantized Kinetic Energy for Discrete System

We have shown the expression for the kinetic energy in field form. Thus, in a 1D continuous form, it will be written as follows:

$$H_{KE} = \int \sum_\sigma \psi_\sigma^\dagger(x) \left(-\frac{\hbar^2}{2m}\nabla_x^2\right) \psi_\sigma(x) dx$$

Derive a discrete form of the above so that it can be used to keep track of electron propagation in a 1D nanoscale device, e.g., a carbon nanotube or any type of quantum wire. The discrete form is also used extensively in the non-equilibrium Green's function (NEGF) for the study of quantum transport in electronic devices.

Solution

In continuous spatial form, the kinetic energy is

$$H_{KE} = \int \sum_\sigma \psi_\sigma^\dagger(x) \left(-\frac{\hbar^2}{2m}\nabla_x^2\right) \psi_\sigma(x) dx$$

Discretization can be carried out for term $\nabla_x^2 \psi_\sigma(x)$. Apply the continuous form of $H^f = -\frac{\hbar}{2m}\frac{d^2}{dx^2}$ to a particular lattice point discretized with an inter-site distance a as shown in the following:

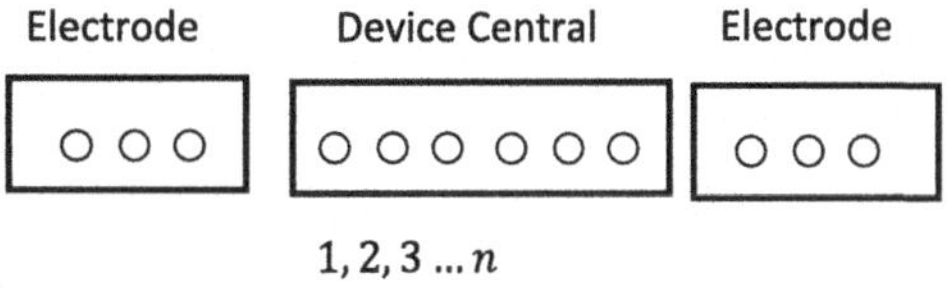

Fig. 1. A nanoscale device divided into three parts. Each part is discretized into spatial sites.

The spatial site within the central region is chosen and denoted by j. Now, with ψ_j representing $\psi(x = ja)$, and for small a,

$$\left(\frac{d\psi(x)}{dx}\right)_{j+\frac{1}{2}} \rightarrow \frac{1}{a}(\psi_{j+1} - \psi_j); \quad \left(\frac{d^2\psi(x)}{dx^2}\right)_j \rightarrow \frac{1}{a^2}(\psi_{j+1} - 2\psi_j + \psi_{j-1})$$

The approximation above leads, with straightforward substitution, to

$$(H^f\psi)_j = 2t\psi_j - t\psi_{j+1} - t\psi_{j-1}$$

where $t = \hbar^2/2ma^2$. Note that the finite difference method has clearly reflected the effects of local perturbation from neighboring sites. In fact, one could now define a H^r, which acts upon an orbital site ψ_j to yield the energy associated with the site and its nearest neighbors, i.e.,

$$H^r\psi_j = 2t\psi_j - t\psi_{j+1} - t\psi_{j-1}$$

Back to the problem, with the finite-difference method would thus yield the following:

$$H_{KE} = \sum_{\sigma j}(\psi_\sigma^\dagger)_j(2t_\sigma\psi_{\sigma j} - t_\sigma\psi_{\sigma(j-1)} - t_\sigma\psi_{\sigma(j+1)})$$

Finally,

$$H_{KE} = \sum_{j\sigma} 2t_\sigma\psi_{\sigma j}^\dagger\psi_{\sigma j} - t_\sigma\psi_{\sigma j}^\dagger\psi_{\sigma(j-1)} - t_\sigma\psi_{\sigma j}^\dagger\psi_{\sigma(j+1)}$$

Remarks and Reflections

If one represents a lattice site using the Hilbert space vector $|x_j\rangle$, and summing over all lattice sites, the Hamiltonian can be written as follows:

$$H_{KE} = \sum_i 2t|x_i\rangle\langle x_i| - t|x_i\rangle\langle x_{i-1}| - t|x_i\rangle\langle x_{i+1}|$$

In the second-quantized form, it is also commonly written as

$$H_{KE} = \sum_{i\sigma} 2t_\sigma c_{\sigma i}^\dagger c_{\sigma i} - t_\sigma c_{\sigma i}^\dagger c_{\sigma(i-1)} - t_\sigma c_{\sigma i}^\dagger c_{\sigma(i+1)}$$

Problem 3.06 Second-Quantized Electron Spin

Given that the spin operator in second-quantized form is expressed as follows:

$$\sigma_a^S = \sum_{kk'\sigma\sigma'} \langle k\sigma|\sigma_a|k'\sigma'\rangle a_{k\sigma}^\dagger a_{k'\sigma'}$$

Note that σ_a is the Pauli matrix in the Cartesian coordinates, while σ_a^S is the second-quantized spin operator. And k is the momentum and σ the spin quantum number. Show that the above leads to

(a) $\sigma_z^S = \sum_k (a_{k\uparrow}^\dagger a_{k\uparrow} - a_{k\downarrow}^\dagger a_{k\downarrow})$

(b) $\sigma_y^S = \sum_k -i(a_{k\uparrow}^\dagger a_{k\downarrow} + a_{k\downarrow}^\dagger a_{k\uparrow})$

(c) $\sigma_x^S = \sum_k (a_{k\uparrow}^\dagger a_{k\downarrow} + a_{k\downarrow}^\dagger a_{k\uparrow})$

Solution

Note that

$$\sigma_a^S = \sum_{kk'\sigma\sigma'} \langle k\sigma|\sigma_a|k'\sigma'\rangle a_{k\sigma}^\dagger a_{k'\sigma'} = \sum_{kk'\sigma\sigma'} \langle \sigma|\sigma_a|\sigma'\rangle \delta_{kk'} a_{k\sigma}^\dagger a_{k'\sigma'}$$

$$= \sum_{k\sigma\sigma'} \langle \sigma|\sigma_a|\sigma'\rangle a_{k\sigma}^\dagger a_{k\sigma'}$$

(a) For σ_z^S, one has

$$\sigma_z^S = \sum_{k\sigma\sigma'} \langle \sigma|\sigma_z|\sigma'\rangle a_{k\sigma}^\dagger a_{k\sigma'}$$

Take note of the following:

$$|\uparrow\rangle = |0\rangle = \begin{pmatrix} 1 \\ 0 \end{pmatrix} \quad |\downarrow\rangle = |1\rangle = \begin{pmatrix} 0 \\ 1 \end{pmatrix} \quad \text{and} \quad \sigma_z = \begin{pmatrix} 1 & 0 \\ 0 & -1 \end{pmatrix}$$

And one could work out that

$$\langle\uparrow|\sigma_z|\uparrow\rangle = 1, \quad \langle\downarrow|\sigma_z|\downarrow\rangle = -1, \quad \langle\uparrow|\sigma_z|\downarrow\rangle = 0, \quad \langle\downarrow|\sigma_z|\uparrow\rangle = 0$$

Therefore, summing over all possible spin pairing

$$\sigma_z^S = \sum_{k\sigma\sigma'} \langle \sigma|\sigma_z|\sigma'\rangle a_{k\sigma}^\dagger a_{k\sigma'} = \sum_k a_{k\uparrow}^\dagger a_{k\uparrow} - a_{k\downarrow}^\dagger a_{k\downarrow}$$

(b) For σ_y^S, one has

$$\sigma_y^S = \sum_{k\sigma\sigma'} \langle\sigma|\sigma_y|\sigma'\rangle a^\dagger_{k\sigma} a_{k\sigma'}$$

Take note of the following:

$$|\uparrow\rangle = |0\rangle = \begin{pmatrix}1\\0\end{pmatrix} \quad |\downarrow\rangle = |1\rangle = \begin{pmatrix}0\\1\end{pmatrix} \quad \text{and} \quad \sigma_y = \begin{pmatrix}0 & -i\\ i & 0\end{pmatrix}$$

And one could work out that

$$\langle\uparrow|\sigma_y|\uparrow\rangle = 0, \quad \langle\downarrow|\sigma_y|\downarrow\rangle = 0, \quad \langle\uparrow|\sigma_y|\downarrow\rangle = -i, \quad \langle\downarrow|\sigma_y|\uparrow\rangle = i$$

Therefore, summing over all possible spin pairing

$$\sigma_y^S = \sum_{k\sigma\sigma'} \langle\sigma|\sigma_z|\sigma'\rangle a^\dagger_{k\sigma} a_{k\sigma'} = \sum_k -i(a^\dagger_{k\uparrow} a_{k\downarrow} + a^\dagger_{k\downarrow} a_{k\uparrow})$$

(c) For σ_x^S, one has

$$\sigma_x^S = \sum_{kk'\sigma\sigma'} \langle\sigma|\sigma_x|\sigma'\rangle a^\dagger_{k\sigma} a_{k\sigma'}$$

Take note of the following:

$$|\uparrow\rangle = |0\rangle = \begin{pmatrix}1\\0\end{pmatrix} \quad |\downarrow\rangle = |1\rangle = \begin{pmatrix}0\\1\end{pmatrix} \quad \text{and} \quad \sigma_x = \begin{pmatrix}0 & 1\\ 1 & 0\end{pmatrix}$$

And one could work out that

$$\langle\uparrow|\sigma_x|\uparrow\rangle = 0, \quad \langle\downarrow|\sigma_x|\downarrow\rangle = 0, \quad \langle\uparrow|\sigma_x|\downarrow\rangle = 1, \quad \langle\downarrow|\sigma_x|\uparrow\rangle = 1$$

Therefore, summing over all possible spin pairing

$$\sigma_x^S = \sum_{k\sigma\sigma'} \langle\sigma|\sigma_x|\sigma'\rangle a^\dagger_{k\sigma} a_{k\sigma'} = \sum_k (a^\dagger_{k\uparrow} a_{k\downarrow} + a^\dagger_{k\downarrow} a_{k\uparrow})$$

Remarks and Reflections

Let's take a look at the σ_z with a more compact approach.

With $\sigma_z = \sum_{\boldsymbol{k}\boldsymbol{k}'\sigma\sigma'} \langle \boldsymbol{k}\sigma|\sigma_z|\boldsymbol{k}'\sigma'\rangle a^{\dagger}_{\boldsymbol{k}\sigma} a_{\boldsymbol{k}'\sigma'}$, one has

$$\sigma_z = \sum_{\boldsymbol{k}\boldsymbol{k}'\sigma\sigma'} \langle \sigma|\sigma_z|\sigma'\rangle \delta_{\boldsymbol{k}\boldsymbol{k}'} a^{\dagger}_{\boldsymbol{k}\sigma} a_{\boldsymbol{k}'\sigma'} = \sum_{\boldsymbol{k}\sigma\sigma'} \langle \sigma|\sigma_z|\sigma'\rangle a^{\dagger}_{\boldsymbol{k}\sigma} a_{\boldsymbol{k}\sigma'}$$

Note that σ, σ' both sum over $\uparrow, \downarrow$. Now,

$$\sum_{\sigma\sigma'} \langle \sigma|\sigma_z|\sigma'\rangle = \sum_{\sigma} \langle \sigma|\sigma'\rangle (\delta_{\sigma'\uparrow} - \delta_{\sigma'\downarrow}) = \sum_{\sigma} \delta_{\sigma\sigma'} (\delta_{\sigma'\uparrow} - \delta_{\sigma'\downarrow})$$

Therefore, Pauli z operator in second-quantized form is given by

$$\begin{aligned} \sigma_z &= \sum_{\boldsymbol{k}\sigma\sigma'} \langle \sigma|\sigma_z|\sigma'\rangle a^{\dagger}_{\boldsymbol{k}\sigma} a_{\boldsymbol{k}\sigma'} = \sum_{\boldsymbol{k}\sigma} \delta_{\sigma\sigma'} (\delta_{\sigma'\uparrow} - \delta_{\sigma'\downarrow}) a^{\dagger}_{\boldsymbol{k}\sigma} a_{\boldsymbol{k}\sigma'} \\ &= \sum_{\boldsymbol{k}} a^{\dagger}_{\boldsymbol{k}\uparrow} a_{\boldsymbol{k}\uparrow} - a^{\dagger}_{\boldsymbol{k}\downarrow} a_{\boldsymbol{k}\downarrow} \end{aligned}$$

Problem 3.07 Second-Quantized Spin–Orbit Coupling 1

The spin–orbit energy in a 2D X–Ysystem is written in the second-quantized form as follows:

$$H_{SO} = \sum_{\boldsymbol{k}\boldsymbol{k}'\sigma\sigma'} \alpha \langle \boldsymbol{k}\sigma|k_a\sigma_b|\boldsymbol{k}'\sigma'\rangle a^{\dagger}_{\boldsymbol{k}\sigma} a_{\boldsymbol{k}'\sigma'} \varepsilon_{abz}$$

where k is the momentum and σ is the spin quantum number. Show its explicit expression in

(a) momentum space,
(b) real space,
(c) real space discrete.

Note: (a, b) takes on indices (x, y, z).

Solution

(a) In momentum space,

$$H_{SO} = \sum_{\boldsymbol{k}\boldsymbol{k}'\sigma\sigma'} \alpha \langle \boldsymbol{k}\sigma|k_a\sigma_b|\boldsymbol{k}'\sigma'\rangle a^{\dagger}_{\boldsymbol{k}\sigma} a_{\boldsymbol{k}'\sigma'} \varepsilon_{abz}$$

Separate the spin and momentum operations,

$$
\begin{aligned}
H_{SO} &= \sum_{\boldsymbol{k}\boldsymbol{k}'\sigma\sigma'} \alpha\langle \boldsymbol{k}|k_a|\boldsymbol{k}'\rangle\langle\sigma|\sigma_b|\sigma'\rangle a^\dagger_{\boldsymbol{k}\sigma} a_{\boldsymbol{k}'\sigma'}\varepsilon_{abz} \\
&= \sum_{\boldsymbol{k}\boldsymbol{k}'\sigma\sigma'} \alpha k'_a \delta_{\boldsymbol{k}\boldsymbol{k}'} a^\dagger_{\boldsymbol{k}\sigma} a_{\boldsymbol{k}'\sigma'}\langle\sigma|\sigma_b|\sigma'\rangle\varepsilon_{abz} \\
&= \alpha\sum_{\sigma\sigma'} k_x\langle\sigma|\sigma_y|\sigma'\rangle a^\dagger_{\boldsymbol{k}\sigma} a_{\boldsymbol{k}\sigma'} - k_y\langle\sigma|\sigma_x|\sigma'\rangle a^\dagger_{\boldsymbol{k}\sigma} a_{\boldsymbol{k}\sigma'}
\end{aligned}
$$

(b) In a 2D X–Y real space, the resolution of identity is used,

$$
\begin{aligned}
H_{SO} &= \alpha\int\sum_{\boldsymbol{k}\sigma\sigma'}\langle\boldsymbol{k}\sigma|\boldsymbol{r}\rangle\langle\boldsymbol{r}|k_a\sigma_b|\boldsymbol{r}'\rangle\langle\boldsymbol{r}'|\boldsymbol{k}'\sigma'\rangle a^\dagger_{\boldsymbol{k}\sigma} a_{\boldsymbol{k}'\sigma'}\varepsilon_{abz} d^2r d^2r' \\
&= \alpha\left(\int\sum_{\boldsymbol{k}\sigma\sigma'}\langle\boldsymbol{k}\sigma|\boldsymbol{r}\rangle\langle\boldsymbol{r}|k_x\sigma_y|\boldsymbol{r}'\rangle\langle\boldsymbol{r}'|\boldsymbol{k}'\sigma'\rangle a^\dagger_{\boldsymbol{k}\sigma} a_{\boldsymbol{k}'\sigma'} d^2r d^2r'\right. \\
&\quad \left. - \int\sum_{\boldsymbol{k}\sigma\sigma'}\langle\boldsymbol{k}\sigma|\boldsymbol{r}\rangle\langle\boldsymbol{r}|k_y\sigma_x|\boldsymbol{r}'\rangle\langle\boldsymbol{r}'|\boldsymbol{k}'\sigma'\rangle a^\dagger_{\boldsymbol{k}\sigma} a_{\boldsymbol{k}'\sigma'} d^2r d^2r'\right)
\end{aligned}
$$

Separate the spin and momentum operations,

$$
\begin{aligned}
H_{SO} &= \alpha\left(\int\sum_{\boldsymbol{k}\boldsymbol{k}'\sigma\sigma'} a^\dagger_{\boldsymbol{k}\sigma}\phi^\dagger_{\boldsymbol{k}}(\boldsymbol{r})\langle\boldsymbol{r}|k_x|\boldsymbol{r}'\rangle\langle\sigma|\sigma_y|\sigma'\rangle a_{\boldsymbol{k}\sigma'}\phi_{\boldsymbol{k}'}(\boldsymbol{r}')d^2r d^2r'\right. \\
&\quad \left. - \int\sum_{\boldsymbol{k}\boldsymbol{k}'\sigma\sigma'} a^\dagger_{\boldsymbol{k}\sigma}\phi^\dagger_{\boldsymbol{k}}(\boldsymbol{r})\langle\boldsymbol{r}|k_y|\boldsymbol{r}'\rangle\langle\sigma|\sigma_x|\sigma'\rangle a_{\boldsymbol{k}\sigma'}\phi_{\boldsymbol{k}'}(\boldsymbol{r}')d^2r d^2r'\right)
\end{aligned}
$$

Finally,

$$
\begin{aligned}
H_{SO} &= \alpha\left(\int\sum_{\sigma\sigma'}\psi^\dagger_\sigma(\boldsymbol{r})(-i\partial_x)\psi_{\sigma'}(\boldsymbol{r})\langle\sigma|\sigma_y|\sigma'\rangle d^2r\right. \\
&\quad \left. - \int\sum_{\sigma\sigma'}\psi^\dagger_\sigma(\boldsymbol{r})(-i\partial_y)\psi_{\sigma'}(\boldsymbol{r})\langle\sigma|\sigma_x|\sigma'\rangle d^2r\right)
\end{aligned}
$$

(c) In real space discrete form, note that

$$
\left(\frac{d\psi(x)}{dx}\right)_j \to \frac{1}{2a}(\psi_{j+1} - \psi_{j-1})
$$

The k_x part:

$$H_{SO} = \alpha \sum_{\sigma\sigma'} k_x \langle\sigma|\sigma_y|\sigma'\rangle a^\dagger_{k\sigma} a_{k\sigma'}$$

$$\begin{aligned}
H_{SO} &= \alpha \int \sum_{\sigma\sigma'} \psi^\dagger_\sigma(x) \left(-i\frac{\partial}{\partial x}\right) \psi_{\sigma'}(x) \langle\sigma|\sigma_y|\sigma'\rangle dx \\
&= -\frac{i\alpha}{2a} \sum_{\sigma\sigma' m} \psi^\dagger_{\sigma m}[\psi_{\sigma'(m+1)} - \psi_{\sigma'(m-1)}]\langle\sigma|\sigma_y|\sigma'\rangle \\
&= -\frac{i\alpha}{2a} \sum_{\sigma\sigma' m} [\psi^\dagger_{\sigma m}\psi_{\sigma'(m+1)} - \psi^\dagger_{\sigma m}\psi_{\sigma'(m-1)}]\langle\sigma|\sigma_y|\sigma'\rangle
\end{aligned}$$

Note that m runs over the X coordinates
The k_y part:

$$H_{SO} = \alpha \sum_{\sigma\sigma'} -k_y \langle\sigma|\sigma_x|\sigma'\rangle a^\dagger_{k\sigma} a_{k\sigma'}$$

$$\begin{aligned}
H_{SO} &= -\alpha \int \sum_{\sigma\sigma'} \psi^\dagger_\sigma(y) \left(-i\frac{\partial}{\partial y}\right) \psi_{\sigma'}(y) \langle\sigma|\sigma_x|\sigma'\rangle dy \\
&= \frac{i\alpha}{2a} \sum_{\sigma\sigma' v} \psi^\dagger_{\sigma v}[\psi_{\sigma'(v+1)} - \psi_{\sigma'(v-1)}]\langle\sigma|\sigma_x|\sigma'\rangle \\
&= \frac{i\alpha}{2a} \sum_{\sigma\sigma' v} [\psi^\dagger_{\sigma v}\psi_{\sigma'(v+1)} - \psi^\dagger_{\sigma v}\psi_{\sigma'(v-1)}]\langle\sigma|\sigma_x|\sigma'\rangle
\end{aligned}$$

Note that v runs over the Y coordinates. Finally,

$$\begin{aligned}
H_{SO} = &-\frac{i\alpha}{2a} \sum_{\sigma\sigma' m} [\psi^\dagger_{\sigma m}\psi_{\sigma'(m+1)} - \psi^\dagger_{\sigma m}\psi_{\sigma'(m-1)}]\langle\sigma|\sigma_y|\sigma'\rangle \\
&+ \frac{i\alpha}{2a} \sum_{\sigma\sigma' v} [\psi^\dagger_{\sigma v}\psi_{\sigma'(v+1)} - \psi^\dagger_{\sigma v}\psi_{\sigma'(v-1)}]\langle\sigma|\sigma_x|\sigma'\rangle
\end{aligned}$$

Remarks and Reflections

The above is a general formulation for the linear spin–orbit coupling in a 2D X–Ysystem. This is because 2D linear spin–orbit coupling is found in many useful physical systems, e.g., semiconductor with Rashba and linear Dresselhaus as well as metal-based Rashba. In 2D systems like graphene, 2D topological surface states, and Weyl systems, the Hamiltonians are linear and mimic the linear spin–orbit coupling systems.

Problem 3.08 Second-Quantized Spin–Orbit Coupling 11

Show that the linear spin–orbit coupling can also be written as follows:

(a) Second-quantized form:

$$H_{SOC} = \frac{-i\alpha}{2a}\sum_{mv}(-i(c_{v,\uparrow}^{m\dagger}c_{v,\downarrow}^{m+1} - c_{v,\uparrow}^{m\dagger}c_{v,\downarrow}^{m-1}) + i(c_{v,\downarrow}^{m\dagger}c_{v,\uparrow}^{m+1} - c_{v,\downarrow}^{m\dagger}c_{v,\uparrow}^{m-1}))$$

$$+\frac{i\alpha}{2a}\sum_{mv}((c_{v,\uparrow}^{m\dagger}c_{v+1,\downarrow}^{m} - c_{v,\uparrow}^{m\dagger}c_{v-1,\downarrow}^{m}) + (c_{v,\downarrow}^{m\dagger}c_{v+1,\uparrow}^{m} - c_{v,\downarrow}^{m\dagger}c_{v-1,\uparrow}^{m}))$$

(b) Bra-Ket form:

$$H_{SOC} = -\frac{i\alpha}{2a}\sum_{m,v} -i(|x_v^m \uparrow\rangle\langle x_v^{m+1} \downarrow| - |x_v^m \uparrow\rangle\langle x_v^{m-1} \downarrow|)$$

$$+ i(|x_v^m \downarrow\rangle\langle x_v^{m+1} \uparrow| - |x_v^m \downarrow\rangle\langle x_v^{m-1} \uparrow|)$$

$$+\frac{i\alpha}{2a}\sum_{m,v}(|x_v^m \uparrow\rangle\langle x_{v+1}^m \downarrow| - |x_v^m \uparrow\rangle\langle x_{v-1}^m \downarrow|)$$

$$+ (|x_v^m \downarrow\rangle\langle x_{v+1}^m \uparrow| - |x_v^m \downarrow\rangle\langle x_{v-1}^m \uparrow|)$$

Solution

(a) In the last problem, spin–orbit coupling was shown as follows:

$$H_{SO} = -\frac{i\alpha}{2a}\sum_{\sigma\sigma' m}[\psi_{\sigma m}^\dagger\psi_{\sigma'(m+1)} - \psi_{\sigma m}^\dagger\psi_{\sigma'(m-1)}]\langle\sigma|\sigma_y|\sigma'\rangle$$

$$+\frac{i\alpha}{2a}\sum_{\sigma\sigma' v}[\psi_{\sigma v}^\dagger\psi_{\sigma'(v+1)} - \psi_{\sigma v}^\dagger\psi_{\sigma'(v-1)}]\langle\sigma|\sigma_x|\sigma'\rangle$$

Let's start with a 2D lattice site of $(x_i, y_i) \to (m, v)$. $\psi_{\sigma v}^\dagger$ and $c_{v\sigma}^{m\dagger}$ are dummy variables. Thus,

$$H_{SO} = -\frac{i\alpha}{2a}\sum_{\sigma\sigma' m}[c_{\sigma m}^\dagger c_{\sigma'(m+1)} - c_{\sigma m}^\dagger c_{\sigma'(m-1)}]\langle\sigma|\sigma_y|\sigma'\rangle$$

$$+\frac{i\alpha}{2a}\sum_{\sigma\sigma' v}[c_{\sigma v}^\dagger c_{\sigma'(v+1)} - c_{\sigma v}^\dagger c_{\sigma'(v-1)}]\langle\sigma|\sigma_x|\sigma'\rangle \tag{A}$$

Note that

$$\langle\uparrow|\sigma_y|\uparrow\rangle = \langle\downarrow|\sigma_y|\downarrow\rangle = 0 \quad \langle\uparrow|\sigma_y|\downarrow\rangle = -i \quad \langle\downarrow|\sigma_y|\uparrow\rangle = i$$

$$\langle\uparrow|\sigma_x|\uparrow\rangle = \langle\downarrow|\sigma_x|\downarrow\rangle = 0 \quad \langle\uparrow|\sigma_x|\downarrow\rangle = 1 \quad \langle\downarrow|\sigma_x|\uparrow\rangle = 1$$

Substituting the above into (A),

$$H_{SOC} = \frac{-i\alpha}{2a}\sum_{mv}(-i(c_{v,\uparrow}^{m\dagger}c_{v,\downarrow}^{m+1} - c_{v,\uparrow}^{m\dagger}c_{v,\downarrow}^{m-1}) + i(c_{v,\downarrow}^{m\dagger}c_{v,\uparrow}^{m+1} - c_{v,\downarrow}^{m\dagger}c_{v,\uparrow}^{m-1}))$$
$$+\frac{i\alpha}{2a}\sum_{mv}((c_{v,\uparrow}^{m\dagger}c_{v+1,\downarrow}^{m} - c_{v,\uparrow}^{m\dagger}c_{v-1,\downarrow}^{m}) + (c_{v,\downarrow}^{m\dagger}c_{v+1,\uparrow}^{m} - c_{v,\downarrow}^{m\dagger}c_{v-1,\uparrow}^{m}))$$

(b) Bra-Ket form,

$$c_{v,\uparrow}^{m\dagger} \to |x_v^m \uparrow\rangle$$
$$c_{v,\downarrow}^{m+1} \to \langle x_v^{m+1}\downarrow|$$

The same logic is applied to the other terms, and one has

$$H_{SOC} = -\frac{i\alpha}{2a}\sum_{m,v} -i(|x_v^m\uparrow\rangle\langle x_v^{m+1}\downarrow| - |x_v^m\uparrow\rangle\langle x_v^{m-1}\downarrow|)$$
$$+\,i(|x_v^m\downarrow\rangle\langle x_v^{m+1}\uparrow| - |x_v^m\downarrow\rangle\langle x_v^{m-1}\uparrow|)$$
$$+\frac{i\alpha}{2a}\sum_{m,v}(|x_v^m\uparrow\rangle\langle x_{v+1}^m\downarrow| - |x_v^m\uparrow\rangle\langle x_{v-1}^m\downarrow|)$$
$$+\,(|x_v^m\downarrow\rangle\langle x_{v+1}^m\uparrow| - |x_v^m\downarrow\rangle\langle x_{v-1}^m\uparrow|)$$

Remarks and Reflections

The second-quantized form was shown as follows:

$$H_{SOC} = \frac{-i\alpha}{2a}\sum_{mv}(-i(c_{v,\uparrow}^{m\dagger}c_{v,\downarrow}^{m+1} - c_{v,\uparrow}^{m\dagger}c_{v,\downarrow}^{m-1}) + i(c_{v,\downarrow}^{m\dagger}c_{v,\uparrow}^{m+1} - c_{v,\downarrow}^{m\dagger}c_{v,\uparrow}^{m-1}))$$
$$+\frac{i\alpha}{2a}\sum_{mv}((c_{v,\uparrow}^{m\dagger}c_{v+1,\downarrow}^{m} - c_{v,\uparrow}^{m\dagger}c_{v-1,\downarrow}^{m}) + (c_{v,\downarrow}^{m\dagger}c_{v+1,\uparrow}^{m} - c_{v,\downarrow}^{m\dagger}c_{v-1,\uparrow}^{m}))$$

Shuffling as follows:

$$\sum_j c_j^\dagger c_{j+1} - c_j^\dagger c_{j-1} \equiv \sum_j c_j^\dagger c_{j+1} - c_{j+1}^\dagger c_j$$

One can also write the above as

$$H_{SOC} = \frac{-i\alpha}{2a}\sum_{mv}(-i(c_{v,\uparrow}^{m\dagger}c_{v,\downarrow}^{m+1} - c_{v,\uparrow}^{m+1\dagger}c_{v,\downarrow}^{m}) + i(c_{v,\downarrow}^{m\dagger}c_{v,\uparrow}^{m+1} - c_{v,\downarrow}^{m+1\dagger}c_{v,\uparrow}^{m}))$$
$$+\frac{i\alpha}{2a}\sum_{mv}((c_{v,\uparrow}^{m\dagger}c_{v+1,\downarrow}^{m} - c_{v+1,\uparrow}^{m\dagger}c_{v,\downarrow}^{m}) + (c_{v,\downarrow}^{m\dagger}c_{v+1,\uparrow}^{m} - c_{v+1,\downarrow}^{m\dagger}c_{v,\uparrow}^{m}))$$

Dirac Delta Calculus

Problem 3.09 Dirac Delta and Fourier Transform

The application of calculus in physics goes as far back as the 16th century, when problems in classical dynamics become mathematically intractable with discrete summation. In the 19th century, when electrodynamic equations were formalized, vector calculus eventually becomes its language. One of the very common representations of the Dirac Delta is given by

$$\delta(x) = \frac{1}{2\pi}\int e^{ikx}dk$$

Using the techniques of Fourier transform (FT), show that

$$\delta(x - x_0) = \frac{1}{2\pi}\int e^{ik(x-x_0)}dk$$

Note: FT is the acronym for Fourier transform and IFT is the acronym for inverse Fourier transform.

Solution

One starts with the standard FT $g(k)$ and IFT $f(x)$ equation pair as shown in the following:

$$g(k) = \frac{1}{\sqrt{2\pi}}\int f(x)e^{-ikx}dx$$
$$f(x) = \frac{1}{\sqrt{2\pi}}\int g(k)e^{ikx}dk$$

Let $f(x)$ be $\delta(x)$, and one has

$$g(k) = \frac{1}{\sqrt{2\pi}}\int \delta(x)e^{-ikx}dx = \frac{1}{\sqrt{2\pi}}$$

which simply says that the FT of the $\delta(x)$ is $g(k)$ and its value is $\frac{1}{\sqrt{2\pi}}$. It thus follows that $\delta(x)$ can be found by performing an IFT of function $g(k)$, which goes according to the standard pair like

$$f(x) = \frac{1}{\sqrt{2\pi}} \int g(k) e^{ikx} dk$$

The LHS would be a Dirac delta function. The RHS merely makes use of the fact that $g(k) = \frac{1}{\sqrt{2\pi}}$. Upon substitution on the RHS, one has

$$\delta(x) = \left(\frac{1}{\sqrt{2\pi}}\right)\left(\frac{1}{\sqrt{2\pi}}\right) \int e^{ikx} dk$$

which implies

$$\delta(x - x_0) = \frac{1}{2\pi} \int e^{ik(x - x_0)} dk$$

Remarks and Reflections

Dirac delta function is an interesting mathematical tool. It is of great use to physicists and engineers alike. For example, Dirac delta function is important in electrodynamics, but how so? The simple but accurate answer is that the electric field of a point charge has the Math form of

$$\boldsymbol{E} \propto \frac{\boldsymbol{R}}{R^3}$$

as illustrated in Figure 2.

When divergence is performed on this function, a strange function which is zero anywhere in space except at the origin is generated. There are many explicit Math methods to represent such a function, but in abstract form, it is represented by the symbol $\delta(r)$.

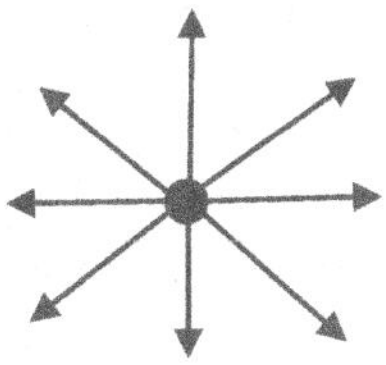

Fig. 2. A positive electrical charge with electric field radiating out from the source.

Problem 3.10 Dirac Delta and Calculus

At first glance, quantum mechanics does not seem to be heavy in calculus, which is true to some extent as the language of quantum mechanics is linear algebra. However, in more advanced topics of quantum physics which involves geometry, scattering, and interaction, calculus becomes very important again. In the following, let's start on Dirac delta and its calculus:

(a) Prove the identity

$$\delta(ax) = \frac{\delta(x)}{|a|}$$

(b) Show that

$$\int x\frac{d}{dx}\delta(x)dx = -\int \delta(x)dx$$

Solution

(a) Let $u = |a|x$, and one would have the following:

$$\int \delta(ax)dx = \int \delta(u)\frac{du}{|a|} = \frac{1}{|a|}$$

This would lead to $|a| \int \delta(ax)dx = 1$. Therefore, one can also write

$$|a| \int \delta(ax)dx = \int \delta(x)dx$$

By comparing the integrand of both sides, the following is obtained:

$$\delta(ax) = \frac{\delta(x)}{|a|}$$

(b) Using chain's rule,

$$\frac{d}{dx}(x\delta(x)) = \delta(x) + x\frac{d}{dx}\delta(x)$$

Since $\int \frac{d}{dx}(x\delta(x))dx = \frac{d}{dx}\int x\delta(x)dx = 0$, one arrives at

$$\frac{d}{dx}(x\delta(x)) = 0$$

It thus follows that

$$\delta(x) = -x\frac{d}{dx}\delta(x)$$

Remarks and Reflections

Strictly speaking, the above is an integrand identity. By quick inspection, it is clear that the Dirac delta is an even function where $\delta(x) = \delta(-x)$.

Problem 3.11 Dirac Delta Representations

As we progress, more calculus problems involving the Dirac delta will be reviewed. The Dirac delta function can have many representations. Show that the following expressions can be used to represent a Dirac delta function.

(a) $\delta(x) = \frac{1}{\pi}\lim\limits_{c\to 0}\frac{c}{x^2+c^2}$

(b) $\delta(x) = \frac{1}{\sqrt{\pi}}\lim\limits_{c\to 0}\frac{1}{c}e^{-x^2/c^2}$

Solution

(a) By inspection, one determines that the function $\lim\limits_{c\to 0}\frac{c}{x^2+c^2}$ plays the role of $\lim\limits_{c\to 0}\delta_c(x) = 0$ everywhere except at $x = 0$. How so?

Referring to Figure 3, one notes that the Dirac delta function has two important regions: $x = 0$ and $x \neq 0$. At $x = 0$,

$$\lim_{c\to 0}\frac{c}{x^2+c^2} = \lim_{c\to 0}\left(\frac{1}{c}\right) = +\infty$$

At $x \neq 0$,

$$\lim_{c\to 0}\frac{c}{x^2+c^2} = \left(\frac{0}{x^2+0}\right) = 0$$

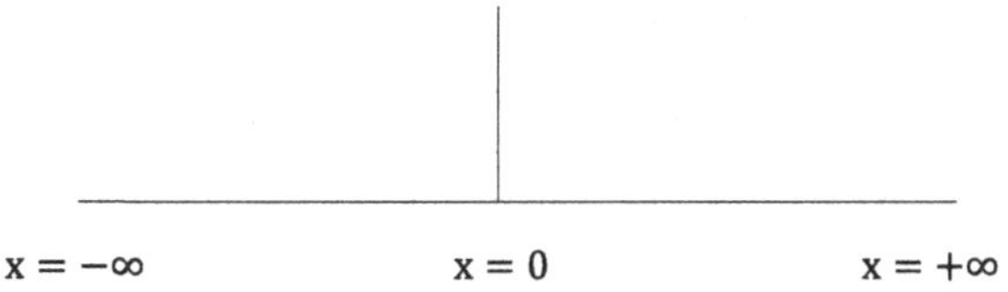

Fig. 3. A 1D Dirac delta function that is zero everywhere except at $x = 0$.

The above satisfies the most important feature of a Dirac delta function. It is thus plausible to consider representing the Dirac delta function with

$$\delta(x) = \lim_{c \to 0} \left(\frac{1}{A} \right) \frac{c}{x^2 + c^2}$$

where A is to be determined. To find A, one proceeds as follows:

$$\int_{-\infty}^{\infty} \lim_{c \to 0} \frac{c}{c^2 + x^2} dx = \int_{-\infty}^{\infty} A\delta(x) dx$$

which leads to

$$A = \left[\tan^{-1} \frac{x}{c} \right]_{-\infty}^{\infty} = \pi$$

Substituting A into $\lim_{c \to 0} \left(\frac{1}{A} \right) \frac{c}{x^2+c^2}$, one completes the proof that

$$\delta(x) = \frac{1}{\pi} \lim_{c \to 0} \frac{c}{x^2 + c^2}$$

(b) Similarly by inspection, one determines that the function $\lim_{c \to 0} \frac{1}{c}(e^{-x^2/c^2})$ plays the role of $\lim_{c \to 0} \delta_c(x) = 0$ everywhere except at $x = 0$. How so? Referring to Figure 4, one notes that the Dirac delta function has two important regions: $x = 0$ and $x \neq 0$.

At $x = 0$,

$$\lim_{c \to 0} \frac{1}{c}(e^{-x^2/c^2}) = \lim_{c \to 0} \frac{1}{c} = +\infty$$

At $x \neq 0$,

$$\lim_{c \to 0} \frac{1}{c}(e^{-x^2/c^2}) = \left(\frac{1}{0} \right) e^{-x^2/0} = 0$$

The above satisfies the most important feature of a Dirac delta function. It is thus plausible to consider representing the Dirac delta function with

$$\delta(x) = \left(\frac{1}{A}\right) \lim_{c\to 0} \frac{1}{c}(e^{-x^2/c^2})$$

where A is to be determined. To find A, one proceeds as follows:

$$\int_{-\infty}^{\infty} \lim_{c\to 0} \frac{1}{c}(e^{-x^2/c^2})dx = \int_{-\infty}^{\infty} A\delta(x)dx$$

$$\to A = \int_{-\infty}^{\infty} \lim_{c\to 0} \frac{1}{c}(e^{-x^2/c^2})dx$$

Using $v = \frac{x}{c}$, and with a change of variable, one has

$$A = \int_{-\infty}^{\infty} \lim_{c\to 0}(e^{-v^2})dv$$

Likewise, one can use a different variable and have

$$A = \int_{-\infty}^{\infty} \lim_{c\to 0}(e^{-u^2})du$$

Combining the two integral leads to the following:

$$A^2 = \int_{-\infty}^{\infty}\int_{-\infty}^{\infty} \lim_{c\to 0}(e^{-v^2-u^2})dv\,du$$

Now, it will be handy to use polar coordinates, i.e.,

$$A^2 = \int_0^{2\pi}\int_0^{\infty} e^{-R^2} R\,dR\,d\theta = 2\pi\left(-\frac{e^{-R^2}}{2}\right) = \pi$$

Therefore, substituting $A = \sqrt{\pi}$ into $\left(\frac{1}{A}\right) \lim\limits_{c\to 0} \frac{1}{c}(e^{-x^2/c^2})$, one completes the proof that

$$\delta(x) = \frac{1}{\sqrt{\pi}} \lim_{c\to 0} \frac{1}{c} e^{-x^2/c^2}$$

Problem 3.12 Vector Calculus and Dirac Delta

In geometry or topology-related topics, vector calculus is frequently used. In advanced quantum mechanics, there are many instances of applications that involve the use of vector calculus alongside Dirac functions. These problems will eventually be intertwined with Fourier transforms in second quantization and quantum field problems. In the following, prove that the following differential vector calculus identity is true

$$\boldsymbol{\nabla}\left(\frac{1}{R}\right)=\frac{-\boldsymbol{R}}{R^3}$$

With the above, where $\delta^3(\boldsymbol{R})=\delta(x)\delta(y)\delta(z)$, show that the divergence of the above is given by

$$\nabla^2\left(\frac{1}{R}\right)=-4\pi\delta^3(\boldsymbol{R})$$

Solution

Let's start in the spherical coordinates where $\boldsymbol{\nabla}U=\boldsymbol{e}'_{\boldsymbol{\theta}}\frac{1}{R}\frac{\partial}{\partial\theta}U+\boldsymbol{e}'_{\boldsymbol{\phi}}\frac{1}{R\sin\theta}\frac{\partial}{\partial\phi}U+\boldsymbol{e}'_{\boldsymbol{R}}\frac{\partial}{\partial R}U$. When U is only a function of R, e.g., $U=\frac{1}{R}$, one has

$$\begin{aligned}\boldsymbol{\nabla}\left(\frac{1}{R}\right)&=\boldsymbol{e}'_{\boldsymbol{R}}\frac{\partial}{\partial R}\left(\frac{1}{R}\right)=\boldsymbol{e}'_{\boldsymbol{R}}\left(-\frac{1}{R^2}\right)\\&=\frac{-\boldsymbol{R}}{R^3}\end{aligned}$$

Alternatively, in the Cartesian coordinates, use is made of $\boldsymbol{\nabla}=\frac{\partial}{\partial x}\boldsymbol{e_x}+\frac{\partial}{\partial y}\boldsymbol{e_y}+\frac{\partial}{\partial z}\boldsymbol{e_z}$. One would have

$$\begin{aligned}\boldsymbol{\nabla}\left(\frac{1}{R}\right)&=\left(\frac{\partial R}{\partial x}\frac{\partial}{\partial R}\boldsymbol{e_x}+\frac{\partial R}{\partial y}\frac{\partial}{\partial R}\boldsymbol{e_y}+\frac{\partial R}{\partial z}\frac{\partial}{\partial R}\boldsymbol{e_z}\right)\frac{1}{R}\\&=\left(\frac{x}{R}\boldsymbol{e_x}+\frac{y}{R}\boldsymbol{e_y}+\frac{z}{R}\boldsymbol{e_z}\right)\frac{-1}{R^2}\\\boldsymbol{\nabla}\left(\frac{1}{R}\right)&=\frac{-\boldsymbol{R}}{R^3}\end{aligned}$$

To show that $\nabla^2\left(\frac{1}{R}\right)=-4\pi\delta^3(R)$, one observes that

$$\nabla^2\left(\frac{1}{R}\right)=-\boldsymbol{\nabla}.\left(\frac{\boldsymbol{R}}{R^3}\right)$$

And

$$\begin{aligned}-\boldsymbol{\nabla}.\left(\frac{\boldsymbol{R}}{R^3}\right)&=-\left(\frac{\partial}{\partial x}\frac{x}{R^3}+\frac{\partial}{\partial y}\frac{y}{R^3}+\frac{\partial}{\partial z}\frac{z}{R^3}\right)\\&=\left(\frac{-R^3+3x^2R}{R^6}+\frac{-R^3+3y^2R}{R^6}+\frac{-R^3+3z^2R}{R^6}\right)\end{aligned}$$

Now, let $D^2 = R^2 + v^2$, and

$$-\boldsymbol{\nabla}.\left(\frac{\boldsymbol{R}}{R^3}\right) = \lim_{v\to 0} \boldsymbol{\nabla}.\left(\frac{\boldsymbol{D}}{D^3}\right)$$

where

$$-\lim_{v\to 0} \boldsymbol{\nabla}.\left(\frac{\boldsymbol{D}}{D^3}\right) = \lim_{v\to 0}\left(\frac{-D^3+3x^2D}{D^6} + \frac{-D^3+3y^2D}{D^6} + \frac{-D^3+3z^2D}{D^6}\right)$$

$$\to \lim_{v\to 0} \frac{-3v^2}{(R^2+v^2)^{\frac{5}{2}}} \equiv A\delta^3(R)$$

Index 5/2 ensures that dimension on the LHS is naturally consistent with the RHS. To find A, we need to perform integration on both sides as shown in the following:

$$\int_{-\infty}^{+\infty} \lim_{v\to 0} \frac{-3v^2}{(R^2+v^2)^{\frac{5}{2}}} d^3R = \int_{-\infty}^{+\infty} A\delta^3(R)d^3R$$

Considering that the limit taking and integration is interchangeable in priority as shown in

$$\int_{-\infty}^{+\infty} \lim_{v\to 0} \frac{-3v^2}{(R^2+v^2)^{\frac{5}{2}}} d^3R = \lim_{v\to 0} \int_{-\infty}^{+\infty} \frac{-3v^2}{(R^2+v^2)^{\frac{5}{2}}} d^3R$$

one has

$$\lim_{v\to 0} \int_{-\infty}^{+\infty} \frac{-3v^2}{(R^2+v^2)^{\frac{5}{2}}} d^3R = \int_{-\infty}^{+\infty} A\delta^3(R)d^3R$$

Note that in the above, $d^3R = dx\,dy\,dz$, therefore, limit is ordinarily taken from $-\infty$ to $+\infty$. Now, we will carry out integration using the spherical coordinates and their appropriate limits. Integral on the RHS is simple and can be visualized in the Cartesian coordinates to simply produce A. Working on the LHS, one is led to

$$\lim_{v\to 0} \int_{\phi=0}^{2\pi} \int_{\theta=0}^{\pi} \int_{0}^{+\infty} \frac{-3v^2}{(R^2+v^2)^{\frac{5}{2}}} dR\,d\theta\,d\phi R^2 \sin\theta = A$$

It follows that

$$4\pi \lim_{v\to 0} \int_{0}^{+\infty} \frac{-3v^2R^2}{(R^2+v^2)^{\frac{5}{2}}} dR = A$$

Now, we will use the substitution approach in the following, where $R = v\tan\alpha$. This leads to the following integral on the RHS (note the limit of α is such that R is taken from 0 to infinity),

$$A = 4\pi \lim_{v\to 0}\int_0^{\frac{\pi}{2}} \frac{-3v^5\tan^2\alpha\sec^2\alpha}{(v^2\tan^2\alpha + v^2)^{\frac{5}{2}}}d\alpha$$

The rest is a standard prescription to obtain A.

$$\begin{aligned} A &= 4\pi\int_0^{\frac{\pi}{2}} -3\sin^2\alpha\cos\alpha\, d\alpha \\ &= 4\pi\int_0^{\frac{\pi}{2}} -3\sin^2\alpha d\sin\alpha = -12\pi\frac{\sin^3\alpha}{3}\bigg|_0^{\frac{\pi}{2}} = -4\pi \end{aligned}$$

Therefore, it is finally proven that

$$\nabla^2\left(\frac{1}{R}\right) = A\delta^3(R) = -4\pi\delta^3(R)$$

Remarks and Reflections

Note that A can be found with an alternative approach. We know from $\nabla^2\left(\frac{1}{R}\right) = \lim\limits_{v\to 0}\frac{-3v^2}{(R^2+v^2)^{\frac{5}{2}}}$ that $\nabla^2\left(\frac{1}{R}\right)$ would have the following structure, i.e.,

$$\nabla^2\left(\frac{1}{R}\right) = A\delta^3(R)$$

Applying volume integral to both sides, one has

$$\iiint \nabla^2\left(\frac{1}{R}\right)d^3R = \iiint A\delta^3(R)d^3R$$

One then uses the divergence theorem, i.e.,

$$A = \int \boldsymbol{\nabla}.\boldsymbol{\nabla}\left(\frac{1}{R}\right)dV = \oint \boldsymbol{\nabla}\left(\frac{1}{R}\right).\boldsymbol{dS}$$

Recall that

$$\nabla\left(\frac{1}{R}\right) = \frac{-\boldsymbol{R}}{R^3}$$

And one has

$$A = \iint \frac{-\boldsymbol{R}}{R^3}.R^2\sin\theta\, d\theta\, d\phi \boldsymbol{e}'_{\boldsymbol{R}} = \iint -\sin\theta\, d\theta\, d\phi = -4\pi$$

Therefore,

$$\nabla^2\left(\frac{1}{R}\right) = -\boldsymbol{\nabla}.\left(\frac{\boldsymbol{R}}{R^3}\right) = -4\pi\delta^3(R)$$

For practice, perform the differential vector calculus of the above in the spherical coordinates. Note that

$$\nabla^2 U = \frac{1}{R^2}\frac{\partial}{\partial R}\left(R^2\frac{\partial U}{\partial R}\right) + \frac{1}{R^2\sin\theta}\frac{\partial}{\partial\theta}\left(\sin\theta\frac{\partial U}{\partial\theta}\right) + \frac{1}{R^2\sin^2\theta}\frac{\partial^2 U}{\partial\phi^2}$$

or

$$\boldsymbol{\nabla}.\boldsymbol{v} = \frac{1}{R^2}\frac{\partial}{\partial R}(R^2 v^R) + \frac{1}{R\sin\theta}\frac{\partial}{\partial\theta}(\sin\theta\, v^\theta) + \frac{1}{R\sin\theta}\frac{\partial}{\partial\phi}v^\phi$$

Condensed Matter Applications

Problem 3.13 Electron Background Energy

We have dealt with problems in second quantization and Dirac calculus. In the following, we will look into some applications in the condensed matter systems. In a condensed matter system, e.g., a bulky material, or the central region of an electronic device, electron interacts with the background ions. Interaction energies arising from ion–ion (H_b) and ion–electron (H_{be}) are given as follows:

$$H_b = \frac{1}{2}e^2\left(\frac{N}{V}\right)^2 \lim_{\mu\to 0}\iint \frac{e^{-\mu|\boldsymbol{x}-\boldsymbol{x}'|}}{|\boldsymbol{x}-\boldsymbol{x}'|}d^3x d^3x' = \lim_{\mu\to 0}\frac{1}{2}e^2\frac{N^2}{V}\frac{4\pi}{\mu^2}$$

$$H_{be} = -e^2\sum_i^N\left(\frac{N}{V}\right)\lim_{\mu\to 0}\int\frac{e^{-\mu|\boldsymbol{x}-\boldsymbol{r}'|}}{|\boldsymbol{x}-\boldsymbol{r}'|}d^3x = -\lim_{\mu\to 0}e^2\frac{N^2}{V}\frac{4\pi}{\mu^2}$$

where N is the total number of electrons and V is the volume of the material bulk. Prove that the positive ion–ion interaction H_b is given as above.

Solution

This problem is about the energy due to the positive background of the ions in a material. These are integration problems that require the change of variables. H_b is by the long range nature of Coulomb potential divergent, which means the integral tends to infinity. To avoid losing its physics to tricky infinity, the Yukawa term is introduced as follows:

$$e^{-\mu|\boldsymbol{x}-\boldsymbol{x}'|}$$

The background energy is

$$H_b = \frac{1}{2}e^2\left(\frac{N}{V}\right)^2 \lim_{\mu\to 0}\iint \frac{e^{-\mu|\boldsymbol{x}-\boldsymbol{x}'|}}{|\boldsymbol{x}-\boldsymbol{x}'|}d^3xd^3x' = \frac{1}{2}e^2\left(\frac{N}{V}\right)^2 \lim_{\mu\to 0} I$$

The integral is

$$I = \iint \frac{e^{-\mu|\boldsymbol{x}-\boldsymbol{x}'|}}{|\boldsymbol{x}-\boldsymbol{x}'|}d^3xd^3x' = \iint \frac{e^{-\mu y}}{y}d^3yd^3x$$

The x integral is a sphere, thus

$$I = V\int_0^{2\pi} d\phi \int_0^{\pi} \sin\theta d\theta \int_0^{\infty} \frac{y^2e^{-\mu y}}{y}dy$$

which leads to

$$I = 4\pi V\int_0^{\infty} ye^{-\mu y}\,dy = 4\pi V\int_0^{\infty} yd\left(\frac{e^{-\mu y}}{-\mu}\right)$$

$$= 4\pi V\left[\left(y\frac{e^{-\mu y}}{-\mu}\right)_0^{\propto} + \int_0^{\infty} dy\left(\frac{e^{-\mu y}}{\mu}\right)\right] = \frac{4\pi V}{\mu^2}$$

Therefore,

$$H_b = \lim_{\mu\to 0}\frac{1}{2}e^2\frac{N^2}{V}\frac{4\pi}{\mu^2}$$

Remarks and Reflections

In fact, one can see that the total energy related to ion interaction is

$$H_b + H_{be} = -\lim_{\mu\to 0}\frac{e^2}{2}\frac{N^2}{V}\frac{4\pi}{\mu^2}$$

This energy will be canceled exactly by another energy term arising from electron–electron interaction, known as Direct energy. These energies are relevant to quantum electronics, spintronics, and condensed matter. In the context of a quantum electronic device, this will be electron energy in the device central.

Problem 3.14 Electron Interaction Formulation

Besides electron–ion and ion–ion interactions, electrons interact with one another. The general expression for the electron–electron interaction is given by

$$H_{ee} = \frac{e^2}{2V} \sum_{\boldsymbol{kpq}} \sum_{\lambda_1 \lambda_2} M a^\dagger_{\boldsymbol{k_1}\lambda_1} a^\dagger_{\boldsymbol{k_2}\lambda_2} a_{\boldsymbol{k_4}\lambda_4} a_{\boldsymbol{k_3}\lambda_3}$$

where amplitude $M = \langle \boldsymbol{k_1}\lambda_1, \boldsymbol{k_2}\lambda_2 | V | \boldsymbol{k_3}\lambda_3, \boldsymbol{k_4}\lambda_4 \rangle$ is

$$M = \lim_{\mu \to 0} \frac{e^2}{V} \left(\frac{-4\pi}{q^2 + \mu^2} \right) \delta^{(\boldsymbol{k_1}+\boldsymbol{k_2}),(\boldsymbol{k_3}+\boldsymbol{k_4})}_{\lambda_1\lambda_3, \lambda_2\lambda_4}$$

Exact derivation for M will be carried out at a later stage. Note that $\boldsymbol{k}, \lambda$ represent momentum, spin, respectively. Show that the above can be written in the more intuitive form which depicts the process of momentum transfer q.

Note: We will be dealing with electrons which are fermionic pairs that abide by the anti-commutation rules: $\{a_k, a^\dagger_p\} = \delta_{kp}$; $\{a_k, a_p\} = 0$; $\{a^\dagger_k, a^\dagger_p\} = 0$. And, $a_{k\lambda}$ *is an operator that destroys a particle with momentum* k *and spin* λ; *whereas* $a^\dagger_{k\lambda}$ *is an operator that creates a particle with momentum* k *and spin* λ.

Solution

The important physics here comes from M which imposes — via $\delta^{(\boldsymbol{k_1}+\boldsymbol{k_2}),(\boldsymbol{k_3}+\boldsymbol{k_4})}_{\lambda_1\lambda_3, \lambda_2\lambda_4}$ aka δ — the physics of conservation of momentum (Figure 4), i.e.,

$$k_1 = k + q, \quad k_2 = p - q, \qquad k_3 = k, \quad k_4 = p$$

The physics is illustrated pictorially by

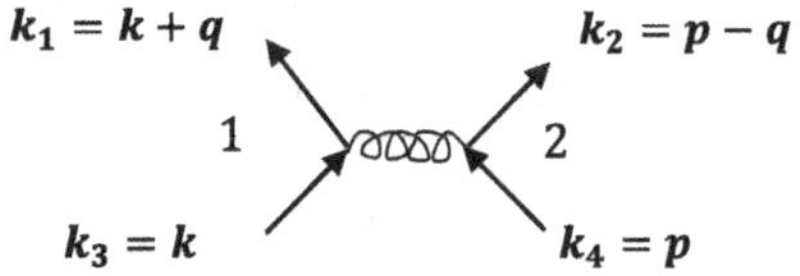

Fig. 4. A schematic illustration of two electrons interacting and exchanging momentum.

Note that (k_1, k_2, k_4, k_3) also corresponds to (K, L, M, N) in many textbooks. The physics is clear: electrons of momentum k and p interact and are destroyed with a momentum transfer q. They are re-created with momentum $k + q$ and $p - q$, respectively. The interaction energy can be classified based on the exchange of momentum, q. In the context of the Exchange energy where $q \neq 0$,

$$M = \lim_{\mu \to 0} \frac{e^2}{V} \left(\frac{-4\pi}{q^2 + \mu^2} \right) \delta$$

In the context of the Direct energy where $q = 0$,

$$M = \lim_{\mu \to 0} \frac{e^2}{V} \left(\frac{-4\pi}{\mu^2} \right) \delta$$

Therefore, the intuitive expressions are

$$H_{ee} = \lim_{\mu \to 0} \frac{e^2}{2V} \sum_{\substack{kpq \\ q \neq 0}} \sum_{\lambda_1 \lambda_2} \left(\frac{-4\pi}{q^2 + \mu^2} \right) a^\dagger_{(k+q)\lambda_1} a^\dagger_{(p-q)\lambda_2} a_{p,\lambda_2} a_{k,\lambda_1}$$

$$+ \lim_{\mu \to 0} \frac{e^2}{2V} \sum_{kp} \sum_{\lambda_1 \lambda_2} \left(\frac{-4\pi}{\mu^2} \right) a^\dagger_{k,\lambda_1} a^\dagger_{p,\lambda_2} a_{p,\lambda_2} a_{k,\lambda_1}$$

Remarks and Reflections

As seen above, the potential energy contains two terms, i.e., the finite momentum transfer and the zero momentum transfer terms. The former is also known as the Exchange energy, while the latter is called the Direct energy.

Problem 3.15 Electron–Electron Energy: Direct

The Direct energy, which is in fact the second term of the potential energy, is given by

$$H_D = \frac{e^2}{2V} \sum_{kp} \sum_{\lambda_1 \lambda_2} \frac{4\pi}{\mu^2} a^\dagger_{k\lambda_1} a^\dagger_{p\lambda_2} a_{p\lambda_2} a_{k\lambda_1}$$

Show that the background energy of an electronic system is given by

$$H_b + H_{be} = -\lim_{\mu \to 0} \frac{e^2}{2} \frac{N^2}{V} \frac{4\pi}{\mu^2}$$

will be exactly canceled by the Direct energy.

Solution

Method A

We will use the anti-commutation relations $\{a_{\boldsymbol{k}}, a_{\boldsymbol{p}}\} = 0$; $\{a_{\boldsymbol{k}}^\dagger, a_{\boldsymbol{p}}^\dagger\} = 0$ and arrive at

$$H_D = \frac{e^2}{2V} \sum_{\boldsymbol{kp}} \sum_{\lambda_1\lambda_2} \frac{-4\pi}{\mu^2} a_{\boldsymbol{k}\lambda_1}^\dagger a_{\boldsymbol{p}\lambda_2}^\dagger a_{\boldsymbol{k}\lambda_1} a_{\boldsymbol{p}\lambda_2}$$

Destroying a particle with the same quantum number twice yields a zero, i.e., $a_{\boldsymbol{k\lambda}} a_{\boldsymbol{k\lambda}} = 0$. Thus, the quantum numbers (k, λ_1) and (p, λ_2) must be different. We can now make use of $\{a_{\boldsymbol{k}\lambda_1}, a_{\boldsymbol{p}\lambda_2}^\dagger\} = \delta_{\boldsymbol{kp}}\delta_{\lambda_1\lambda_2} \to 0$. This leads to pairing as follows:

$$H_D = \frac{e^2}{2V} \sum_{\boldsymbol{kp}} \sum_{\lambda_1\lambda_2} \frac{4\pi}{\mu^2} a_{\boldsymbol{k}\lambda_1}^\dagger a_{\boldsymbol{k}\lambda_1} a_{\boldsymbol{p}\lambda_2}^\dagger a_{\boldsymbol{p}\lambda_2} \quad \text{where } (k, \lambda_1) \neq (p, \lambda_2)$$

Here, one recognizes that $a_{\boldsymbol{k},\lambda_1}^\dagger a_{\boldsymbol{k},\lambda_1} = \hat{n}_{k,\lambda_1}$ which is the number operator that gives the number of particles with quantum state $\boldsymbol{k}\lambda_1$. Hence,

$$H_D = \frac{e^2}{2V} \sum_{\boldsymbol{kp}\lambda_1\lambda_2}^{N} \frac{4\pi}{\mu^2} n_{\boldsymbol{k}\lambda_1} n_{\boldsymbol{p}\lambda_2} \quad \text{where } (k, \lambda_1) \neq (p, \lambda_2)$$

What can be observed or measured is the expectation or average value of this operator, which is just the number of particles with quantum state $\boldsymbol{k}\lambda_1$. The summation above is over momentum k and p which must be different. Every particle indexed by p should correspond to all other particles indexed by k except for the one with the same momentum as p. Therefore, each of p corresponds to $(N-1)$ of k. Since there will be N particles of p in the summation process, the Direct energy is

$$\begin{aligned} E_D = \langle G|H_D|G\rangle &= \lim_{\mu\to 0} \frac{e^2 4\pi}{2V\mu^2}(N^2 - N) \\ &= \lim_{\mu\to 0} \left(\frac{e^2 N^2 4\pi}{2V\mu^2} - \frac{e^2 N 4\pi}{2V\mu^2} \right) \end{aligned}$$

It can be seen that the first term of the Direct energy cancels the average of the background energy. The second term of the Direct energy goes to zero with proper limit taking. We will ignore the process of limit taking which makes the second term vanish. Therefore, background energy is exactly canceled by the Direct energy.

Method B

We begin with the same method by using the anti-commutation relations $\{a_k, a_p\} = 0$; $\{a_k^\dagger, a_p^\dagger\} = 0$ and arrive at

$$H_D = \frac{e^2}{2V} \sum_{kp} \sum_{\lambda_1 \lambda_2} \frac{-4\pi}{\mu^2} a_{k\lambda_1}^\dagger a_{p\lambda_2}^\dagger a_{k\lambda_1} a_{p\lambda_2}$$

Here goes the departure from the previous method. Without asserting that the quantum numbers (k, λ_1) and (p, λ_2) must be different so as to sidestep $a_{k\lambda} a_{k\lambda} = 0$, what we have is $\{a_{k\lambda_1}, a_{p\lambda_2}^\dagger\} = \delta_{kp}\delta_{\lambda_1\lambda_2}$ instead of $\{a_{k\lambda_1}, a_{p\lambda_2}^\dagger\} = \delta_{kp}\delta_{\lambda_1\lambda_2} \to 0$. This leads to pairing as follows:

$$\begin{aligned} H_D &= \frac{e^2}{2V} \sum_{kp} \sum_{\lambda_1 \lambda_2} \frac{-4\pi}{\mu^2} a_{k\lambda_1}^\dagger (\delta_{kp}\delta_{\lambda_1\lambda_2} - a_{k,\lambda_1} a_{p,\lambda_2}^\dagger) a_{p,\lambda_2} \\ &= \frac{e^2}{2V} \sum_{kp} \sum_{\lambda_1 \lambda_2} \frac{4\pi}{\mu^2} a_{k\lambda_1}^\dagger a_{k\lambda_1} a_{p\lambda_2}^\dagger a_{p\lambda_2} - \delta_{kp}\delta_{\lambda_1\lambda_2} a_{k\lambda_1}^\dagger a_{p\lambda_2} \\ &= \frac{e^2}{2V} \sum_{kp} \sum_{\lambda_1 \lambda_2} \frac{4\pi}{\mu^2} a_{k\lambda_1}^\dagger a_{k\lambda_1} a_{p\lambda_2}^\dagger a_{p\lambda_2} - a_{k\lambda_1}^\dagger a_{k\lambda_1} \\ &= \frac{e^2}{2V} \sum_{kp} \sum_{\lambda_1 \lambda_2} \frac{4\pi}{\mu^2} n_{k\lambda_1} n_{p\lambda_2} - n_{k\lambda_1} \end{aligned}$$

which leads once again to

$$E_D = \langle G|H_D|G\rangle = \lim_{\mu\to 0} \frac{e^2 4\pi}{2V\mu^2}(N^2 - N)$$

Problem 3.16 Electron–Electron Energy: Exchange

The formal expression of the Exchange energy in second-quantized operator form is given in the following:

$$\begin{aligned} H_{Ex} &= \lim_{\mu\to 0} \frac{e^2}{2V} \sum_{\substack{kpq \\ q\neq 0}} \sum_{\lambda_1 \lambda_2} \frac{4\pi}{q^2 + \mu^2} a_{(k+q)\lambda_1}^\dagger a_{(p-q)\lambda_2}^\dagger a_{p\lambda_2} a_{k\lambda_1} \\ &= \frac{e^2}{2V} \sum_{\substack{kpq \\ q\neq 0}} \sum_{\lambda_1 \lambda_2} \frac{4\pi}{q^2} a_{(k+q)\lambda_1}^\dagger a_{(p-q)\lambda_2}^\dagger a_{p\lambda_2} a_{k\lambda_1} \end{aligned}$$

Show that the expectation of the energy is given by the integral of two Heaviside functions

$$E_{Ex} = \frac{-4\pi e^2}{2Vq^2} 2 \iint_0^\infty \frac{d^3k d^3q}{(2\pi)^6} \theta(k_F - |\boldsymbol{k} + \boldsymbol{q}|)\theta(k_F - |\boldsymbol{k}|)$$

Note: (r, k) are vector quantities. They are sometimes written in boldface letters to emphasize their vector nature, e.g., when actual summation is to be performed over the Fermi sphere.

Solution

The total energy of the electronic system is given by

$$\begin{aligned} H &= H_K + (H_b + H_{be} + (H_{ee})_{q=0}) + (H_{ee})_{q\neq 0} \\ &= \sum_{k\lambda} \frac{\hbar^2 k^2}{2m} a^+_{k\lambda} a_{k\lambda} + (H_{ee})_{q\neq 0} \end{aligned}$$

Recalling that the Direct energy is $H_D = (H_{ee})_{q=0}$, we will label the exchange energy according to $H_{Ex} = (H_{ee})_{q\neq 0}$. The exchange energy can be written as

$$E_{Ex} = \langle G|H_{Ex}|G\rangle = \frac{e^2}{2V} \sum_{\substack{\boldsymbol{kpq} \\ q\neq 0}} \sum_{\lambda_1\lambda_2} \frac{4\pi}{q^2} \langle G|a^\dagger_{(\boldsymbol{k}+\boldsymbol{q})\lambda_1} a^\dagger_{(\boldsymbol{p}-\boldsymbol{q})\lambda_2} a_{\boldsymbol{p}\lambda_2} a_{\boldsymbol{k}\lambda_1}|G\rangle$$

Since $a^\dagger_{(\boldsymbol{p}-\boldsymbol{q})\lambda_2}$ and $a_{\boldsymbol{p}\lambda_2}$ are different because $\boldsymbol{q} \neq 0$, operator $a_{\boldsymbol{p}\lambda_2}$ can be shifted one notch (step) to the left by "jumping" over $a^\dagger_{(\boldsymbol{p}-\boldsymbol{q})\lambda_2}$ and adding a negative sign to the expression. This is simply satisfying $\{a_k, a^\dagger_p\} = \delta_{kp} = 0$. One is led to

$$E_{Ex} = \frac{e^2}{2V} \sum_{\substack{\boldsymbol{kpq} \\ q\neq 0}} \sum_{\lambda_1\lambda_2} \frac{-4\pi}{q^2} \langle G|a^\dagger_{(\boldsymbol{k}+\boldsymbol{q})\lambda_1} a_{\boldsymbol{p}\lambda_2} a^\dagger_{(\boldsymbol{p}-\boldsymbol{q})\lambda_2} a_{\boldsymbol{k}\lambda_1}|G\rangle$$

In Exchange, the momentum transfer is finite, i.e., $\boldsymbol{q} \neq 0$. One possible outcome of the above is that as $\boldsymbol{p}$ is destroyed, $\boldsymbol{k}$ acquires $\boldsymbol{q}$ to become $\boldsymbol{p}$. As $\boldsymbol{k}$ is destroyed, $\boldsymbol{p}$ loses $\boldsymbol{q}$ to become $\boldsymbol{k}$. In the momentum space (see Figure 5(a)), particle $\boldsymbol{p}$ has exchanged position with particle $\boldsymbol{k}$ by imparting momentum $\boldsymbol{q} = (\boldsymbol{p} - \boldsymbol{k})$ to $\boldsymbol{k}$. The interaction is considered for particles of the same spin. By contrast, in the case of Direct, the electron can replace

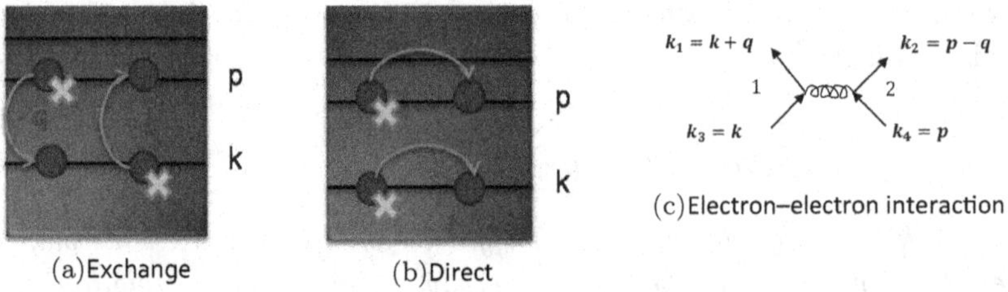

Fig. 5. Schematic illustration of electron–electron interaction with (a) a finite exchange of momentum, (b) no exchange of momentum. (c) Feynman diagram that illustrates the electron–electron interaction.

itself in the same position, implying zero momentum transfer as shown in Figure 5(b).

The anticipated process of Figure 5(a) is mathematically described by $\delta_{\lambda_1\lambda_2}\delta_{\boldsymbol{p},\boldsymbol{k}+\boldsymbol{q}}$. Therefore,

$$E_{Ex} = \sum_{\substack{\boldsymbol{k}\boldsymbol{q}\lambda_1 \\ \boldsymbol{q}\neq 0}} \frac{-4\pi e^2}{2Vq^2} \langle G | n_{\boldsymbol{k}+\boldsymbol{q},\lambda_1} n_{\boldsymbol{k},\lambda_1} | G \rangle$$

As electron fills up the Fermi sphere, one can figure out that the above would lead to

$$\begin{aligned} E_{Ex} &= 2 \sum_{\substack{\boldsymbol{k}\boldsymbol{q} \\ \boldsymbol{q}\neq 0}} \frac{-4\pi e^2}{2Vq^2} \theta(k_F - |\boldsymbol{k}+\boldsymbol{q}|)\theta(k_F - |\boldsymbol{k}|) \\ &= \frac{-4\pi e^2}{2Vq^2} 2 \iint_0^\infty \frac{d^3k d^3q}{(2\pi)^6} \theta(k_F - |\boldsymbol{k}+\boldsymbol{q}|)\theta(k_F - |\boldsymbol{k}|) \end{aligned}$$

Remarks and Reflections

From the second quantized to the integral expression, all operators and state vectors have disappeared. What emerges are a negative sign and the Heaviside functions. In other words, the second-quantized expression contains information that has now been translated to the negative sign as well as the limits for integration.

Problem 3.17 Scattering Strength of Electron Interaction

The central region of a device is where knowledge of electron energy is important. Interaction energy arising from electron–electron (e-e) interaction is

$$H_{ee} = \frac{e^2}{2V} \sum_{kpq} \sum_{\lambda_1\lambda_2} M a^{\dagger}_{k_1\lambda_1} a^{\dagger}_{k_2\lambda_2} a_{k_4\lambda_4} a_{k_3\lambda_3}$$

where $M = \langle \boldsymbol{k_1}\lambda_1, \boldsymbol{k_2}\lambda_2 | V | \boldsymbol{k_3}\lambda_3, \boldsymbol{k_4}\lambda_4 \rangle$ is the matrix element. It is the scalar strength of electron–electron interaction for both the Exchange and the Direct energies. Show the explicit derivation of M.

Solution

The matrix element can be written as follows:

$$M = \lim_{\mu\to 0} \frac{e^2}{V^2} \int dx_1^3\, dx_2^3 (e^{-i\boldsymbol{k_1}\cdot\boldsymbol{x_1}} \eta^{\dagger}_{\lambda_1})(e^{-i\boldsymbol{k_2}\cdot\boldsymbol{x_2}} \eta^{\dagger}_{\lambda_2})$$
$$\times \frac{e^{-\mu|\boldsymbol{x_1}-\boldsymbol{x_2}|}}{|\boldsymbol{x_1}-\boldsymbol{x_2}|} (e^{i\boldsymbol{k_3}\cdot\boldsymbol{x_1}} \eta_{\lambda_3})(e^{i\boldsymbol{k_4}\cdot\boldsymbol{x_2}} \eta_{\lambda_4})$$

With $\boldsymbol{x} = \boldsymbol{x_2}$; $\boldsymbol{y} = \boldsymbol{x_1} - \boldsymbol{x_2}$, the matrix is

$$M = \lim_{\mu\to 0} \frac{e^2}{V^2} \int dx^3\, dy^3 (e^{-i(\boldsymbol{k_1}+\boldsymbol{k_2}-\boldsymbol{k_3}-\boldsymbol{k_4})\cdot\boldsymbol{x}})(e^{i(\boldsymbol{k_3}-\boldsymbol{k_1})\cdot\boldsymbol{y}}) \frac{e^{-\mu|\boldsymbol{y}|}}{|\boldsymbol{y}|} \delta_{\lambda_1\lambda_3}\delta_{\lambda_2\lambda_4}$$
$$= \lim_{\mu\to 0} \frac{e^2}{V} \int dy^3 (e^{i(\boldsymbol{k_3}-\boldsymbol{k_1})\cdot\boldsymbol{y}}) \frac{e^{-\mu|\boldsymbol{y}|}}{|\boldsymbol{y}|} \delta_{\lambda_1\lambda_3}\delta_{\lambda_2\lambda_4}\delta_{\boldsymbol{k_1}+\boldsymbol{k_2},\boldsymbol{k_3}+\boldsymbol{k_4}}$$
$$= \lim_{\mu\to 0} \frac{e^2}{V} \int dy^3 (e^{i\boldsymbol{q}\cdot\boldsymbol{y}}) \frac{e^{-\mu|\boldsymbol{y}|}}{|\boldsymbol{y}|} \delta^{(\boldsymbol{k_1}+\boldsymbol{k_2}),(\boldsymbol{k_3}+\boldsymbol{k_4})}_{\lambda_1\lambda_3,\lambda_2\lambda_4}$$

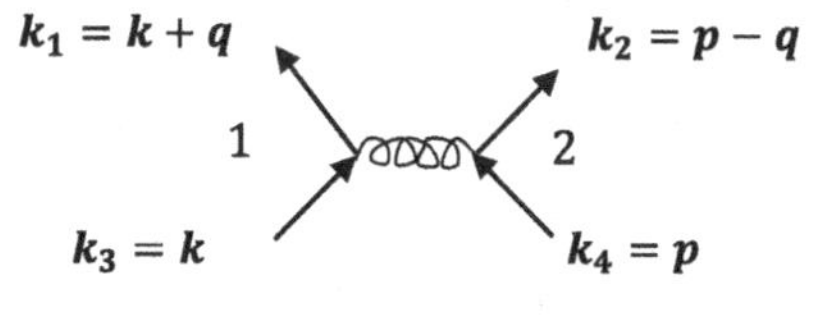

Fig. 6. Electron–electron interaction as illustrated by diagram.

In the above, use has been made of one of the following:

Volume summation (Continuous)	*Dirac delta function (Continuous)*
$\sum_k \to \int d^3k \frac{V}{(2\pi)^3}$	$\frac{1}{(2\pi)^3} \int e^{i(k-k').r} d^3r = \delta^3(k-k')$
Volume summation (Discrete)	*Dirac delta function (Discrete)*
$\sum_k$	$\int e^{i(k-k').r} d^3r = V\delta^3_{kk'}$

We will now represent $\delta^{(k_1+k_2),(k_3+k_4)}_{\lambda_1\lambda_3,\lambda_2\lambda_4}$ with just $\delta_{k\lambda}$ for brevity. Therefore, in the spherical coordinates,

$$M = \lim_{\mu\to 0} \frac{e^2}{V} \int y^2\, d\theta\, dy \sin\theta\, d\phi \left(\frac{e^{iqy\cos\theta - \mu|y|}}{|y|} \right) \delta_{k\lambda}$$

$$= \lim_{\mu\to 0} \frac{e^2}{V} 2\pi \int_0^\infty dy \int_{-1}^1 d\cos\theta \frac{e^{iqy\cos\theta - \mu y}}{y} \delta_{k\lambda}$$

And it follows that

$$M = \lim_{\mu\to 0} \frac{e^2}{V} \frac{2\pi}{iq} \int_0^\infty e^{-\mu y} dy [e^{iqy\cos\theta}]_{-1}^1 \delta_{k\lambda}$$

$$= \lim_{\mu\to 0} \frac{e^2}{V} \frac{2\pi}{iq} \int_0^\infty e^{-\mu y} dy (e^{iqy} - e^{-iqy}) \delta_{k\lambda}$$

Using standard trigonometry identities, one has

$$M = \lim_{\mu\to 0} \frac{e^2}{V} \frac{2\pi}{iq} \int_0^\infty e^{-\mu y} dy (2i \sin qy) \delta_{k\lambda}$$

$$= \lim_{\mu\to 0} \frac{e^2}{V} \frac{4\pi}{q} \int_0^\infty e^{-\mu y} dy (\sin qy) \delta_{k\lambda} \tag{A}$$

We will keep Eq. (A) in mind and perform an integration by parts that yields the following:

$$M = \lim_{\mu\to 0} \frac{e^2}{V} \left(\frac{-4\pi}{q^2} \right) \delta_{k\lambda} \int_0^\infty e^{-\mu y} d\cos qy$$

$$= \lim_{\mu\to 0} \frac{e^2}{V} \left(\frac{-4\pi}{q^2} \right) \left([e^{-\mu y} \cos qy]_0^\infty + \int_0^\infty \mu \cos qy e^{-\mu y} dy \right) \delta_{k\lambda}$$

Finally,

$$M = \lim_{\mu \to 0} \frac{e^2}{V} \left(\frac{-4\pi}{q^2} \right) \delta_{k\lambda} \left(1 + \frac{\mu^2}{q} \int_0^\infty \sin qy e^{-\mu y} dy \right) \tag{B}$$

Comparing Eqs. (A) and (B),

$$\lim_{\mu \to 0} \frac{e^2}{V} \left(\frac{4\pi}{q} \right) \delta_{k\lambda} \left(\int_0^\infty \sin qy e^{-\mu y} dy \right)$$
$$= \lim_{\mu \to 0} \frac{e^2}{V} \left(\frac{-4\pi}{q^2} \right) \delta_{k\lambda} \left(1 + \frac{\mu^2}{q} \int_0^\infty \sin qy e^{-\mu y} dy \right)$$

The above leads to

$$\int_0^\infty \sin qy e^{-\mu y} dy = \frac{\left(\frac{-4\pi}{q^2} \right)}{\left(\frac{4\pi}{q} + \frac{4\pi\mu^2}{q^3} \right)} = \frac{-1}{q + \frac{\mu^2}{q}}$$

Substituting this back into Eq. (A),

$$M = \lim_{\mu \to 0} \frac{e^2}{V} \left(\frac{-4\pi}{q^2 + \mu^2} \right) \delta_{k\lambda} = \frac{e^2}{V} \left(\frac{-4\pi}{q^2} \right) \delta_{k\lambda} \tag{C}$$

where again as a reminder, $\delta_{k\lambda} = \delta^{(\boldsymbol{k_1}+\boldsymbol{k_2}),(\boldsymbol{k_3}+\boldsymbol{k_4})}_{\lambda_1\lambda_3,\lambda_2\lambda_4}$.

Remarks and Reflections

In the context of the *Exchange* energy where $q \neq 0$,

$$M = \lim_{\mu \to 0} \frac{e^2}{V} \left(\frac{-4\pi}{q^2 + \mu^2} \right) \delta$$

In the context of the *Direct* energy where $q = 0$,

$$M = \lim_{\mu \to 0} \frac{e^2}{V} \left(\frac{-4\pi}{\mu^2} \right) \delta$$

Therefore, the electron–electron interaction energy, accounting for both *Exchange* and *Direct*, is given by

$$H_{ee} = \lim_{\mu \to 0} \frac{e^2}{2V} \sum_{\substack{\boldsymbol{kpq} \\ \boldsymbol{q} \neq 0}} \sum_{\lambda_1 \lambda_2} \left(\frac{4\pi}{q^2 + \mu^2} \right) a^\dagger_{(\boldsymbol{k}+\boldsymbol{q})\lambda_1} a^\dagger_{(\boldsymbol{p}-\boldsymbol{q})\lambda_2} a_{\boldsymbol{p}\lambda_2} a_{\boldsymbol{k}\lambda_1}$$
$$+ \lim_{\mu \to 0} \frac{e^2}{2V} \sum_{\boldsymbol{kp}} \sum_{\lambda_1 \lambda_2} \left(\frac{4\pi}{\mu^2} \right) a^\dagger_{\boldsymbol{k}\lambda_1} a^\dagger_{\boldsymbol{p}\lambda_2} a_{\boldsymbol{p}\lambda_2} a_{\boldsymbol{k}\lambda_1}$$

The term $\delta^{(k_1+k_2),(k_3+k_4)}_{\lambda_1\lambda_3,\lambda_2\lambda_4}$ has an important physical consequence: it captures the physics of momentum and spin conservation.

Problem 3.18 Quantifying the Exchange Energy

In an electron system, the total energy comprises only the kinetic energy and the correlation energy with finite momentum transfer, i.e., the exchange energy. By proper reasoning and pairing of the electron operators, the exchange energy can be derived as follows:

$$E_{exch} = \sum_{\substack{\boldsymbol{k}\boldsymbol{q}\lambda_1 \\ q\neq 0}} \frac{-4\pi e^2}{2Vq^2}\langle G|n_{\boldsymbol{k}+\boldsymbol{q},\lambda_1} n_{\boldsymbol{k},\lambda_1}|G\rangle$$

$$= 2\sum_{\substack{\boldsymbol{k}\boldsymbol{q} \\ q\neq 0}} \frac{-4\pi e^2}{2Vq^2}\theta(k_F - |\boldsymbol{k}+\boldsymbol{q}|)\theta(k_F - |\boldsymbol{k}|)$$

The above can be expressed in the form of an integral function

$$E_{exch} = \frac{-4\pi V e^2}{2q^2} 2 \iint \frac{d^3k d^3q}{(2\pi)^6}\theta(k_F - |\boldsymbol{k}+\boldsymbol{q}|)\theta(k_F - |\boldsymbol{k}|)$$

Show that this integral finally leads to the exchange energy per particle expression of

$$\frac{E_{exch}}{N} = -\frac{e^2}{2a_0}\frac{0.916}{r_S} = -\frac{0.916}{r_S} \quad \text{in Ryberg per particle}$$

Note: Integrals $\int d\boldsymbol{k}$ and $\int d^3k$ *mean the same.*

Solution

Let's start with one value of $\boldsymbol{q}$ in any direction; the strength (magnitude) of $\boldsymbol{q}$ must fall in the range of $0 < q < 2\mathrm{k_F}$ to ensure that for any $\boldsymbol{q}$, there exists a $\boldsymbol{k} < k_F$ such that $|\boldsymbol{k}+\boldsymbol{q}|$ is less than k_F. This is to ensure that the conditions imposed by $\theta(k_F - |\boldsymbol{k}+\boldsymbol{q}|)\theta(k_F - |\boldsymbol{k}|)$ can be simultaneously fulfilled. Let us check that when $|q| < 2k_F$: For all $\boldsymbol{k}$ that satisfies $\theta(k_F - |\boldsymbol{k}|)$, the condition of $\theta(k_F - |\boldsymbol{k} + \boldsymbol{q}|)$ CANNOT be fulfilled. In other words, $\theta(k_F - |\boldsymbol{k} + \boldsymbol{q}|)\theta(k_F - |\boldsymbol{k}|)$ CANNOT be simultaneously fulfilled.

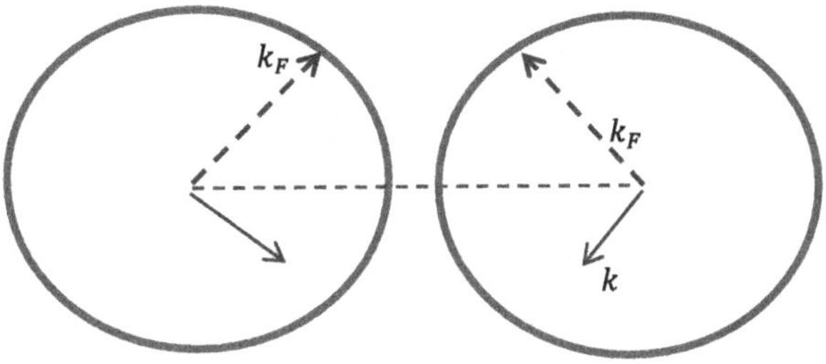

Let us check that when $|q| > 2k_F$: For all $\boldsymbol{k}$ that satisfies $\theta(k_F - |\boldsymbol{k}|)$, the condition of $\theta(k_F - |\boldsymbol{k} + \boldsymbol{q}|)$ CAN be fulfilled. In other words, $\theta(k_F - |\boldsymbol{k} + \boldsymbol{q}|)\theta(k_F - |\boldsymbol{k}|)$ CAN be simultaneously fulfilled. Integration is essentially over the overlapping region between the two circles of radius k_F each.

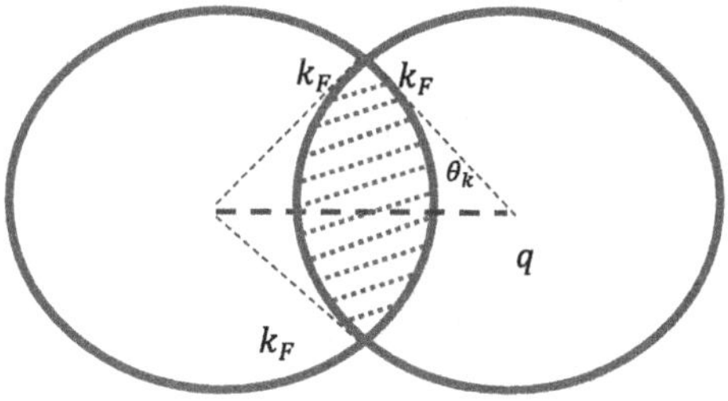

A 3D visualization is given in the following. It is clear there are two sets of parameters to sum over, i.e., $(\boldsymbol{q}, \boldsymbol{k})$. And $\boldsymbol{q}$ is free to move to inscribe ad sphere with a radius less than $2k_F$. Therefore,

$$d\theta_q \to (0, \pi) \to 2$$

$$d\phi_q \to (0, 2\pi) \to 2\pi$$

$$dq \to (0, 2k_F) \to$$

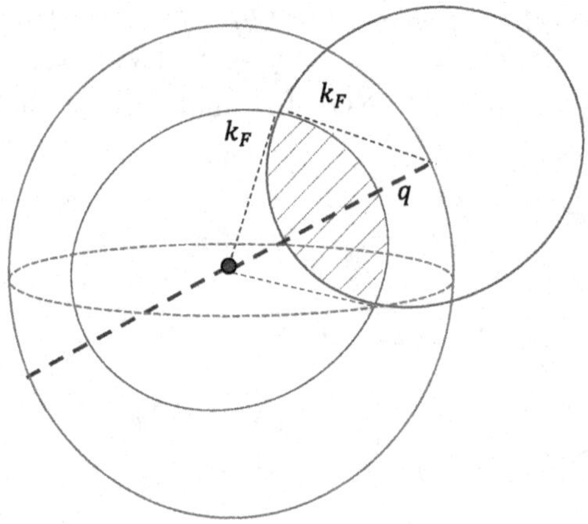

$$E_{exch} = \frac{-4\pi e^2}{2V} 2 \iint \frac{V^2}{(2\pi)^6} \theta(k_F - |\boldsymbol{k} + \boldsymbol{q}|)\theta(k_F - |\boldsymbol{k}|) d^3k d^3q$$
$$= \frac{-4\pi V e^2}{(2\pi)^6} (4\pi) \int_0^{2k_F} dq d^3k \theta(k_F - |\boldsymbol{k}|)$$

Now for every value of q, integration over k is carried out as follows:

	Steps to take for every q
1.	$d\theta_k \rightarrow \left(0, \cos^{-1} \frac{q^2}{2k_F}\right)$
2.	$\times 2$ is required because of region $1 + 2$
3.	$d\phi_k \rightarrow (0, 2\pi) \rightarrow 2\pi$
4.	$dk \rightarrow \left(\frac{q}{2\cos\theta_k}, 2k_F\right) \rightarrow$
5.	Pictorial indication of limits ●
6.	$\times 2$ is required because of region $(3 + 4) \times 2$

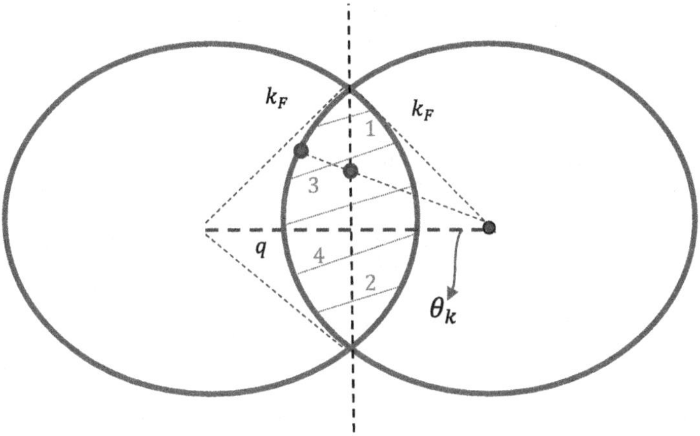

$$E_{exch} = \frac{-16\pi^2 V e^2}{(2\pi)^6} \int_0^{2k_F} dq d^3k \theta(k_F - |\boldsymbol{k}|)$$

$$= \frac{-e^2 V}{4\pi^4} 2 \int_0^{2k_F} dq \int_{\frac{q}{2k_F}}^{1} d\cos\theta_k \int_{\frac{q}{2\cos\theta_k}}^{k_F} k^2 dk \int_0^{2\pi} d\phi_k$$

$$= \frac{-e^2 V}{2\pi^3} 2 \int_0^{2k_F} dq \int_{\frac{q}{2k_F}}^{1} d\cos\theta_k \int_{\frac{q}{2\cos\theta_k}}^{k_F} k^2 dk$$

The total energy is

$$\frac{E_{exch}}{N} = -\frac{e^2}{2a_0}\frac{0.916}{r_S} = -\frac{0.916}{r_S} \quad \text{in Ryberg per particle}$$

Problem 3.19 Kinetic Energy per Particle

Using the spherical coordinates or other methods, deduce that the fermionic gas is characterized by physical properties as follows:

(a) Fermi wavevector of a bulk electron gas is given by

$$k_F = (3\pi^2 n)^{1/3}$$

(b) Kinetic energy per particle in the ground state is

$$E_{avg} = \frac{3}{5}\frac{\hbar^2 k_F^2}{2m}$$

(c) Density of states per bulk volume is

$$D(E) = \frac{1}{2\pi^2}\left(\frac{2m}{\hbar^2}\right)^{\frac{3}{2}} E^{\frac{1}{2}}\theta(E)$$

where n is carrier density.

Solution

(a) The number of particles in a non-interacting system is formally described by the following expression:

$$N = \sum_{\boldsymbol{k}\lambda} \langle G|n_{\boldsymbol{k}\lambda}|G\rangle = \sum_{\boldsymbol{k}\lambda} \theta(k_F - \boldsymbol{k}) = \frac{V}{(2\pi)^3}\int_{-\propto}^{\propto} d^3k \sum_{\lambda} \theta(k_F - |\boldsymbol{k}|)$$

where $n_{\boldsymbol{k}\lambda}$ is the number operator, V is the volume of the sample, $\theta(k_F - |\boldsymbol{k}|)$ is a Heaviside function, and $|k| = \sqrt{k_x^2 + k_y^2 + k_z^2}$. Summing over spin leads to a factor of 2. The Fermi wavevector expressed is thus

$$N = \frac{V}{8\pi^3}\left(\frac{4}{3}\pi k_F^3\right) \rightarrow k_F = \left(3\pi^2 \frac{N}{V}\right)^{\frac{1}{3}}$$

where use has been made of

Volume summation (Continuous)	*Dirac delta function (Continuous)*
$\sum_{\boldsymbol{k}} \rightarrow \int d^3k \frac{V}{(2\pi)^3}$	$\frac{1}{(2\pi)^3}\int e^{i(\boldsymbol{k}-\boldsymbol{k}').\boldsymbol{r}} d^3r = \delta^3(k - k')$
Volume summation (Discrete)	*Dirac delta function (Discrete)*
$\sum_{\boldsymbol{k}}$	$\int e^{i(\boldsymbol{k}-\boldsymbol{k}').\boldsymbol{r}} d^3r = V\delta^3_{kk'}$

And

$$\sum_{\boldsymbol{k}} \rightarrow \frac{V}{(2\pi)^3} \int d^3k = \frac{V}{(2\pi)^3} \int dk_x\, dk_y\, dk_z$$

Note that the k-space volume is given by $\left(\frac{2\pi}{L_x}\right)\left(\frac{2\pi}{L_y}\right)\left(\frac{2\pi}{L_z}\right) = \frac{(2\pi)^3}{V}$.

(b) Averaging the kinetic energy over the ground state, we have

$$E = \langle G|H_{KE}|G\rangle = \frac{\hbar^2}{2m} \sum_{\boldsymbol{k}\lambda} k^2 \langle G|n_{\boldsymbol{k}\lambda}|G\rangle = \frac{\hbar^2}{2m} \sum_{\boldsymbol{k}\lambda} k^2 \theta(k_F - \boldsymbol{k})$$

$$= \frac{\hbar^2}{2m} \frac{V}{(2\pi)^3} \int_0^\infty d\boldsymbol{k} \sum_{\lambda} k^2 \theta(k_F - \boldsymbol{k})$$

$$\Rightarrow \frac{E_k}{N} = \frac{3}{5} \frac{\hbar^2 k_F^2}{2m}$$

(c) The kinetic energy can now be expressed in terms of the Fermi wavevector and electron density. It follows that the density of states per bulk volume can be deduced by proper substitution to be

$$D(E) = \frac{dN}{dE} = \frac{V}{2\pi^2} \left(\frac{2m}{\hbar^2}\right)^{\frac{3}{2}} E^{\frac{1}{2}} \theta(E)$$

Remarks and Reflections

Note that $n_{\boldsymbol{k}\lambda} = a_{\boldsymbol{k}\lambda}^\dagger a_{\boldsymbol{k}\lambda}$, and $\langle G|n_{\boldsymbol{k}\lambda}|G\rangle$ is also represented by the Heaviside function of $\theta(k_F - k)$.

$$\theta(k_F - k) = \begin{cases} 1 & \textit{for } k < k_F \\ 0 & \textit{for } k > k_F \end{cases}$$

The number of particles as well as the total energy of the system will be given by

$$N = \int D(E) dE; \quad E_T = \int E D(E) dE$$

Appendix 3A. Dirac Delta Identities

Dirac delta functions are one of the most commonly used functions in physics, both classical and quantum. This table summarizes some identities for quick and off-the-cuff uses. Students can apply them directly to their problems or resort to proofs as quick exercises.

	Dirac Delta Identities
1.	$\int f(x)\delta(x-x_0)dx = f(x_0)$
2.	$\int \delta(x-x_1)\delta(x-x_2)dx = \delta(x_1-x_2)$
3.	$\delta(x-x') = \frac{1}{2\pi}\int e^{ik(x-x')}dk$
	$\delta(k-k') = \frac{1}{2\pi}\int e^{i(k-k')x}dx$
4.	$\delta(x) = \frac{d}{dx}\theta(x)$ where $\theta(x) = 1$ for $x > 0$; 0 for $x < 0$; $\frac{1}{2}$ for $x = 0$
5.	$\delta(ax) = \frac{\delta(x)}{\lvert a\rvert}$
6.	Sokhotski–Plemelj theorem: $\lim_{\eta\to 0}\frac{1}{x\pm i\eta} = P\left(\frac{1}{x}\right)\mp i\pi\delta(x)$
7.	$\delta(x) = \lim_{c\to 0}\frac{1}{\pi}\frac{c}{c^2+x^2}dx$
8.	$\delta(x) = \frac{1}{\sqrt{\pi}}\lim_{c\to 0}\frac{1}{c}e^{-x^2/c^2}$
9.	$\theta(x) = \frac{1}{2\pi i}\int_{-\infty}^{\infty}\lim_{c\to 0}\frac{e^{-i\omega x}}{\omega+ic}d\omega$

Appendix 3B. Vector Calculus Identities

Vector calculus is used most extensively in electromagnetic theories and classical mechanics. In quantum mechanics, vector calculus is heavily used in fields that involve magnetic fields, spin, and radiation. In advanced quantum mechanics, vector calculus is needed in the studies of gauge, topology, geometric phase, and curve space.

	Vector Calculus Identities
	General
1.	$\boldsymbol{A}.(\boldsymbol{B}\times \boldsymbol{C}) = \boldsymbol{B}.(\boldsymbol{C}\times \boldsymbol{A}) = \boldsymbol{C}.(\boldsymbol{A}\times \boldsymbol{B})$
2.	$\boldsymbol{A}\times(\boldsymbol{B}\times \boldsymbol{C}) = (\boldsymbol{A}.\boldsymbol{C})\boldsymbol{B} - (\boldsymbol{A}.\boldsymbol{B})\boldsymbol{C}$
3.	$(\boldsymbol{A}\times \boldsymbol{B})\times \boldsymbol{C} = (\boldsymbol{A}.\boldsymbol{C})\boldsymbol{B} - (\boldsymbol{B}.\boldsymbol{C})\boldsymbol{A}$
4.	$(\boldsymbol{A}\times \boldsymbol{B})\times(\boldsymbol{C}\times \boldsymbol{D}) = [(\boldsymbol{A}\times \boldsymbol{B}).\boldsymbol{D}]\boldsymbol{C} - [(\boldsymbol{A}\times \boldsymbol{B}).\boldsymbol{C}]\boldsymbol{D}$
5.	$\boldsymbol{\nabla}(\boldsymbol{A}.\boldsymbol{B}) = (\boldsymbol{B}.\boldsymbol{\nabla})\boldsymbol{A} + (\boldsymbol{A}.\boldsymbol{\nabla})\boldsymbol{B} + \boldsymbol{B}\times(\boldsymbol{\nabla}\times \boldsymbol{A}) + \boldsymbol{A}\times(\boldsymbol{\nabla}\times \boldsymbol{B})$
	Divergence Function
6.	$\boldsymbol{\nabla}.(\boldsymbol{A}\times \boldsymbol{B}) = \boldsymbol{B}.(\boldsymbol{\nabla}\times \boldsymbol{A}) - \boldsymbol{A}.(\boldsymbol{\nabla}\times \boldsymbol{B})$
7.	$\boldsymbol{\nabla}.(\phi \boldsymbol{A}) = (\boldsymbol{\nabla}\phi).\boldsymbol{A} + \phi(\boldsymbol{\nabla}.\boldsymbol{A})$
8.	$\boldsymbol{\nabla}.\boldsymbol{\nabla}\phi = \nabla^2\phi$
9.	$\boldsymbol{\nabla}.(\boldsymbol{\nabla}\times \boldsymbol{A}) = \boldsymbol{0}$
	Curl Function
10.	$\boldsymbol{\nabla}\times(\boldsymbol{A}\times \boldsymbol{B}) = (\boldsymbol{B}.\boldsymbol{\nabla})\boldsymbol{A} - \boldsymbol{B}(\boldsymbol{\nabla}.\boldsymbol{A}) - (\boldsymbol{A}.\boldsymbol{\nabla})\boldsymbol{B} + \boldsymbol{A}(\boldsymbol{\nabla}.\boldsymbol{B})$
11.	$\boldsymbol{\nabla}\times(\phi \boldsymbol{A}) = (\boldsymbol{\nabla}\phi)\times \boldsymbol{A} + \phi(\boldsymbol{\nabla}\times A)$
12.	$\boldsymbol{\nabla}\times(\boldsymbol{\nabla}\times \boldsymbol{A}) = \boldsymbol{\nabla}(\boldsymbol{\nabla}.\boldsymbol{A}) - \boldsymbol{\nabla}^{\boldsymbol{2}}\boldsymbol{A}$
13.	$\boldsymbol{\nabla}\times(\boldsymbol{\nabla}\phi) = 0$
14.	$\boldsymbol{\nabla}\times(\boldsymbol{A}+\boldsymbol{B}) = \boldsymbol{\nabla}\times(\boldsymbol{A}) + \boldsymbol{\nabla}\times(\boldsymbol{B})$
15.	$\boldsymbol{A}\times(\boldsymbol{\nabla}\times \boldsymbol{A}) = \boldsymbol{\nabla}\left(\frac{1}{2}A^2\right) - (\boldsymbol{A}.\boldsymbol{\nabla})\boldsymbol{A}$
16.	$\boldsymbol{\nabla}\times(\boldsymbol{\nabla}\times \boldsymbol{A}) = \boldsymbol{\nabla}(\boldsymbol{\nabla}.\boldsymbol{A}) - \boldsymbol{\nabla}^{\boldsymbol{2}}\boldsymbol{A}$

Appendix 3C. Fourier Transform Identities

Fourier Transform Identities

$f(x) = \sum_{-\infty}^{+\infty} a_n e^{ik_n x}$, where $k_n = \frac{2\pi n}{L}$

and $a_n = \frac{1}{L}\int_0^L f(x)e^{-ik_n x}dx$

In the case where $f(x)$ is not periodic, but a function, one can find its spectrum $g(k)$ using the Fourier transform with which one has

$$g(k) = \frac{1}{\sqrt{2\pi}} \int f(x)e^{-ikx}dx$$

$$f(x) = \frac{1}{\sqrt{2\pi}} \int g(k)e^{ikx}dk$$

where $g(k) = FT\{f(x)\}$.

	Function $f(x)$	Unitary, angular FT $g(k)$
1.	$\partial_x^n f(x)$	$(ik)^n g(k)$
2.	$f(x-a)$	$e^{-iak}g(k)$
3.	$f(ax)$	$\frac{1}{\|a\|} g\left(\frac{k}{a}\right)$
4.	$(f * h)x$	$\sqrt{2\pi} g_f(k) g_h(k)$
5.	$f(x)h(x)$	$\frac{(g_f * g_h)k}{\sqrt{2\pi}}$

Chapter 4

Non-equilibrium Green's Function

This chapter deals with the non-equilibrium Green's Function (NEGF) — a modern topic with problems designed to suit applications in nanoscale electronic and spintronic devices. As Green's functions in quantum mechanical context are built on second quantization, requisite knowledge of second quantization is needed and Chapter 3 would come in handy. In this chapter, Green's functions are first introduced in the context of general applications in condensed matter systems. It is followed by a full-fledged non-equilibrium formulation for the study of quantum transport in electronic as well as spinor-based devices. The latter part of this chapter is particularly specialized and problems are formulated for research purposes. In general, problems here could be used to augment the numerous courses of postgraduate quantum mechanics.

Green's Function for Quantum Transport

Problem 4.01	Green's Function for Quantum Electronics
Problem 4.02	Retarded and Advanced Green's Function — Fourier Transform
Problem 4.03	Green's Function: Density of States
Problem 4.04	Green's Function in Electronic Charge Distribution I
Problem 4.05	Green's Function in Electronic Charge Distribution II
Problem 4.06	Mathematical Methods: Trigonometric Integral

Non-equilibrium Green's Function

Problem 4.07	Non-equilibrium Green's Function (NEGF): Electronic Charge Current
Problem 4.08	Non-equilibrium Green's Function (NEGF): Kinetic Spin Current

Problem 4.09 Non-equilibrium Green's Function (NEGF): Magnetic Spin Current
Problem 4.10 Non-equilibrium Green's Function (NEGF): Spin–Orbit Spin Current

Green's Function for Quantum Transport

Problem 4.01 Green's Function for Quantum Electronics

The retarded Green's function is by definition

$$G^R(\boldsymbol{x}\boldsymbol{x}', tt') = -i\langle \boldsymbol{x}|e^{-iH^v\tau}|\boldsymbol{x}'\rangle\theta(\tau)$$

where $\tau = t - t'$. This mathematical expression is the probability amplitude (chance) that an electron found at an earlier time t' in location $\boldsymbol{x}'$ will be found at a later time t in location $\boldsymbol{x}$. Therefore, the retarded Green's function has the physical meaning of electron propagation. It is widely used in the physics of quantum electronics, spintronics, or pseudo-spintronics (e.g., graphene valley spin). With the eigenstate set of $(|v\rangle, |v'\rangle)$ as basis states, show that the non-interacting retarded Green's function can be written in a simple form as shown in the following:

$$iG^R(\boldsymbol{x}\boldsymbol{x}', tt') = \theta(\tau)\sum_v \phi_v(\boldsymbol{x})\phi_v^\dagger(\boldsymbol{x}')e^{-iE_v\tau}$$

Note: *r, k stand for position and momentum coordinates, respectively. Integration is carried out for range $(-\infty, +\infty)$ and V is the volume containing the particles.*

Note: *$(\boldsymbol{r}, \boldsymbol{k})$ are vector quantities. The integrals of $\int d^3r = \int dx\,dy\,dz = \int d\boldsymbol{r}$ have the same meaning. Sometimes, they are written like $\int d^3x = \int d\boldsymbol{x}$, $\int d^3y = \int d\boldsymbol{y}$. Likewise, the integrals of $\int d^3k = \int dk_x dk_y dk_z = \int d\boldsymbol{k}$ have the same meaning.*

Solution

With the resolution of identity, one carries out a change of basis as follows:

$$\begin{aligned} G^R(\boldsymbol{x}\boldsymbol{x}', tt') &= -i\langle \boldsymbol{x}|e^{-iH\tau}|\boldsymbol{x}'\rangle\theta(\tau) \\ &= -i\sum_{vv'}\langle \boldsymbol{x}|v\rangle\langle v|e^{-iH\tau}|v'\rangle\langle v'|\boldsymbol{x}'\rangle\theta(\tau) \end{aligned}$$

Under the non-interacting circumstances, when $|v\rangle$ and $|v'\rangle$ are the *eigenstates* of the Hamiltonian, one would have

$$\begin{aligned}
G^R(xx', tt') &= -i\sum_{vv'} \phi_v(\boldsymbol{x})\langle v|v'\rangle e^{-iE_{v'}\tau}\phi^\dagger_{v'}(\boldsymbol{x}')\theta(\tau) \\
&= -i\sum_{vv'} \phi_v(\boldsymbol{x})\delta_{vv'}e^{-iE_{v'}\tau}\phi^\dagger_{v'}(\boldsymbol{x}')\theta(\tau) \\
&= -i\theta(\tau)\sum_{v} \phi_v(\boldsymbol{x})\phi^\dagger_{v}(\boldsymbol{x}')e^{-iE_v\tau}
\end{aligned}$$

Therefore,

$$iG^R(\boldsymbol{x}\boldsymbol{x}', tt') = \theta(\tau)\sum_{v} \phi_v(\boldsymbol{x})\phi^\dagger_{v}(\boldsymbol{x}')e^{-iE_v\tau}$$

The above can also be written as

$$iG^R(\boldsymbol{x}\boldsymbol{x}', tt') = \frac{1}{V}\theta(\tau)\sum_{k} e^{i\boldsymbol{k}.(\boldsymbol{x}-\boldsymbol{x}')}e^{-iE_k\tau}$$

Note that

$$\langle\boldsymbol{x}|\boldsymbol{k}\rangle = \phi_{\boldsymbol{k}}(\boldsymbol{x}) = \frac{1}{\sqrt{V}}e^{i\boldsymbol{k}.\boldsymbol{x}} \quad \langle\boldsymbol{k}|\boldsymbol{x}\rangle = \phi^\dagger_{\boldsymbol{k}}(\boldsymbol{x}) = \frac{1}{\sqrt{V}}e^{-i\boldsymbol{k}.\boldsymbol{x}}$$

Remarks and Reflections

What about the case when $|n\rangle$ and $|n'\rangle$ are *not the eigenstates* of the Hamiltonian? Let's start once again with

$$G^R(nn', tt') = -i\langle n|e^{-iH^v\tau}|n'\rangle\theta(\tau)$$

where $|n_i\rangle$ belongs to a complete set of basis states. One can then carry out a change of basis and write

$$\begin{aligned}
iG^R(xx', tt') &= \langle x|e^{-iH^v\tau}|x'\rangle\theta(\tau) \\
&= \sum_{nn'}\langle x|n\rangle\langle n|e^{-iH^v\tau}|n'\rangle\langle n'|x'\rangle\theta(\tau)
\end{aligned}$$

Finally, one keeps Green's function in a general form as follows:

$$iG^R(xx', tt') = \sum_{nn'} \phi_n(x)G^R(nn', tt')\phi^\dagger_{n'}(x')$$

Points to Ponder

Many body physics began with the many body wavefunction which gives the probability amplitude of finding N number of particles in a system. But to find the explicit solution for the many body wavefunction is an immense task. This is not surprising as the wavefunction means the probability amplitude of the whereabouts of all the particles!

Thus, Green's function which means the probability of finding a particle at place r' at time t', given it was found at r at time t is probably easier to find. In fact, Green's function for a particle even in the presence of interaction due to many other particles is more useful than the wavefunction, especially with respect to electron dynamics. This is the major motivation behind the elaborate use of Green's function to understand electron propagation and scattering in nanoscale devices.

Problem 4.02 Retarded and Advanced Green's Function — Fourier Transform

The retarded Green's function has the physical meaning of electron propagation. It is widely used in quantum electronics and spintronics. The retarded and the advanced Green's function are by definition

$$G^R(\boldsymbol{x}\boldsymbol{x}', tt') = -i\langle \boldsymbol{x}|e^{-iH\tau}|\boldsymbol{x}'\rangle\theta(\tau)$$
$$G^A(\boldsymbol{x}\boldsymbol{x}', tt') = i\langle \boldsymbol{x}|e^{-iH\tau}|\boldsymbol{x}'\rangle\theta(-\tau)$$

respectively, where $\tau = t - t'$. Perform a Fourier transform of the retarded Green's function above in both space and time.

Solution

Let's focus on the retarded Green's function first. Fourier transform of $G^R(\boldsymbol{x}, \boldsymbol{x}', \tau)$ in real space is carried out as follows:

$$\begin{aligned} iG^R(K,\tau) &= \int \sum_{\boldsymbol{k}} \phi_{\boldsymbol{k}}(\boldsymbol{x})\phi^\dagger_{\boldsymbol{k}}(\boldsymbol{x}')e^{-iE_k\tau}\theta(\tau)e^{-i\boldsymbol{K}.(\boldsymbol{x}-\boldsymbol{x}')}d(\boldsymbol{x}-\boldsymbol{x}') \\ &= \int \sum_{\boldsymbol{k}} e^{-i(\boldsymbol{K}-\boldsymbol{k}).(\boldsymbol{x}-\boldsymbol{x}')}e^{-iE_k\tau}\theta(\tau)d(\boldsymbol{x}-\boldsymbol{x}') \end{aligned}$$

Note that

$$\langle \boldsymbol{x}|\boldsymbol{k}\rangle = \phi_{\boldsymbol{k}}(\boldsymbol{x}) = \frac{1}{\sqrt{V}}e^{i\boldsymbol{k}.\boldsymbol{x}} \quad \langle \boldsymbol{k}|\boldsymbol{x}\rangle = \phi^\dagger_{\boldsymbol{k}}(\boldsymbol{x}) = \frac{1}{\sqrt{V}}e^{-i\boldsymbol{k}.\boldsymbol{x}}$$

Note that use is made of either the continuous or the discrete approaches shown in the following table (to be adapted to one dimension).

Volume summation (Continuous)	*Dirac delta function (Continuous)*
$\sum_k \to \int d^3k \frac{V}{(2\pi)^3}$	$\frac{1}{(2\pi)^3}\int e^{i(\boldsymbol{k}-\boldsymbol{k}').\boldsymbol{r}} d^3r = \delta^3(k-k')$
Volume summation (Discrete)	*Dirac delta function (Discrete)*
$\sum_k$	$\int e^{i(\boldsymbol{k}-\boldsymbol{k}').\boldsymbol{r}} d^3r = V\delta^3_{kk'}$

Continuous summation:

$$iG^R(\boldsymbol{K},\tau) = \frac{1}{V}\sum_{\boldsymbol{k}} 2\pi\delta(\boldsymbol{K}-\boldsymbol{k})e^{-iE_k\tau}\theta(\tau) = e^{-iE_K\tau}\theta(\tau)$$

Discrete summation:

$$iG^R(\boldsymbol{K},\tau) = \frac{1}{V}\sum_{\boldsymbol{k}} V\delta_{\boldsymbol{Kk}}e^{-iE_k\tau}\theta(\tau) = e^{-iE_K\tau}\theta(\tau)$$

In the commonly used k notation, $iG^R(\boldsymbol{k},\tau) = e^{-iE_k\tau}\theta(\tau)$. Fourier transform of $G^R(\boldsymbol{x},\boldsymbol{x}',\tau)$ in time-space only is carried out as follows:

$$\begin{aligned} iG^R(\boldsymbol{x},\boldsymbol{x}',E) &= \int \sum_{\boldsymbol{k}} \phi_{\boldsymbol{k}}(\boldsymbol{x})\phi^\dagger_{\boldsymbol{k}}(\boldsymbol{x}')e^{-iE_k\tau}e^{iE\tau}\theta(\tau)d\tau \\ &= \int \sum_{\boldsymbol{k}} \phi_{\boldsymbol{k}}(\boldsymbol{x})\phi^\dagger_{\boldsymbol{k}}(\boldsymbol{x}')e^{i(E-E_k)\tau}\theta(\tau)d\tau \end{aligned}$$

The integration of a Heaviside function (more details in the following time-space transform) leads to

$$iG^R(\boldsymbol{x}\boldsymbol{x}',E) = \sum_{\boldsymbol{k}} \frac{\phi_{\boldsymbol{k}}(\boldsymbol{x})\phi^\dagger_{\boldsymbol{k}}(\boldsymbol{x}')}{(E-E_{\boldsymbol{k}}+i\eta)}$$

Now, Fourier transform of $G^R(\boldsymbol{x},\boldsymbol{x}',\tau)$ in both real and time-space is carried out. Let us start with a Green's function already transformed in real space, i.e., $iG^R(\boldsymbol{k},\tau) = e^{-iE_k\tau}\theta(\tau)$. Its *time-space transform* is

$$iG^R(\boldsymbol{k},E) = \lim_{n\to 0}\int_{-\propto}^{+\propto} e^{iE\tau}.e^{-iE_k\tau}\theta(\tau)e^{-\eta\tau}\,d\tau$$

$$= \lim_{n\to 0}\int_{0}^{\propto} e^{i(E-E_k)\tau}e^{-\eta\tau}\,d\tau$$

$$= \frac{1}{i(E-E_{\boldsymbol{k}}+i\eta)}[e^{i(E-E_k)\tau}e^{-\eta\tau}]_0^{\propto}$$

$$= \frac{-1}{i(E-E_{\boldsymbol{k}}+i\eta)}$$

Therefore,

$$G^R(\boldsymbol{k},E) = \frac{1}{(E-E_{\boldsymbol{k}}+i\eta)}$$

Fourier transform of $G^A(\boldsymbol{x},\boldsymbol{x}',\tau)$ in both real and time-space will be

$$iG^A(\boldsymbol{k},\tau) = e^{-iE_k\tau}\theta(-\tau)$$

$$iG^A(\boldsymbol{k},E) = \lim_{\eta\to 0}\int_{-\propto}^{+\propto} e^{-iE(-\tau)}.e^{+iE_k(-\tau)}\theta(-\tau)e^{-\eta(-\tau)}d(-\tau)$$

$$= \lim_{\eta\to 0}\int_{0}^{\propto} e^{-i(E-E_k)(-\tau)}e^{-\eta(-\tau)}d(-\tau)$$

$$= \frac{1}{-i(E-E_{\boldsymbol{k}}-i\eta)}[e^{-i(E-E_k)(-\tau)}e^{-\eta(-\tau)}]_0^{\propto}$$

$$= \frac{1}{i(E-E_{\boldsymbol{k}}-i\eta)}$$

Therefore,

$$G^A(\boldsymbol{k},E) = \frac{1}{(E-E_{\boldsymbol{k}}-i\eta)}$$

Problem 4.03 Green's Function: Density of States

The spectral function has the physical significance of the density of states via

$$D(E) = -\frac{1}{\pi}ImG^R(\boldsymbol{k},E) = \frac{1}{2\pi}A(\boldsymbol{k},E)$$

Broadening of the spectral function occurs in the presence of interaction. For

$$G^R(\boldsymbol{k},\tau) = -i\theta(\tau)e^{-iE_k\tau}e^{-\frac{\tau}{t_d}}$$

prove that the spectral function is broadened as given by

$$A(\boldsymbol{k},E) = \frac{2/t_d}{(E-E_{\boldsymbol{k}})^2 + (1/t_d)^2}$$

Solution

The spectral function is a delta function which suggests that injecting an electron into the system can generate excitation only when the energy of the electron is E_k. It is thus related to the density of states at a given energy.

$$\int A\frac{dE}{2\pi} = 1$$

Now,

$$\begin{aligned}A(\boldsymbol{k},E) &= -2\,Im\,G^R(\boldsymbol{k},E)\\ &= -2\,Im\int_{-\infty}^{\infty} d\tau e^{iE\tau}(-i\theta(\tau))e^{-iE_k\tau}e^{\frac{-\tau}{t_d}}\,d\tau\end{aligned}$$

It follows that

$$\begin{aligned}A(\boldsymbol{k},E) &= 2\,Im\left[i\int_0^{\infty} d\tau e^{i(E-E_k)\tau}e^{\frac{-\tau}{t_d}}\,d\tau\right]\\ &= 2\,Im\left[\frac{1}{(E-E_{\boldsymbol{k}})-\frac{i}{t_d}}\right]\end{aligned}$$

Therefore,

$$A(\boldsymbol{k},E) = \frac{\frac{2}{t_d}}{(E-E_{\boldsymbol{k}})^2 + \left(\frac{1}{t_d}\right)^2}$$

Remarks and Reflections

The explicit expression of the spectral function is also given by

$$A(\boldsymbol{k},E) = 2\pi\delta(E-E_{\boldsymbol{k}})$$

Note that Green's function's formalism of density of states is generally given by

$$G^R = \lim_{\eta \to 0} \frac{1}{E - E_{\boldsymbol{k}} + i\eta}$$

$$G^A = \lim_{\eta \to 0} \frac{1}{E - E_{\boldsymbol{k}} - i\eta}$$

Therefore,

$$D(E) = \frac{1}{2\pi} i(G^R - G^A) = \frac{1}{2\pi}\left(\lim_{\eta \to 0} \frac{1}{E - E_{\boldsymbol{k}} + i\eta} - \lim_{\eta \to 0} \frac{1}{E - E_{\boldsymbol{k}} - i\eta}\right)$$

$$= \frac{1}{2\pi}\left(\lim_{\eta \to 0} \frac{2\eta}{[(E - E_{\boldsymbol{k}})^2 + \eta^2]}\right)$$

Referring to the formulations for Dirac delta functions,

$$D(E) = \delta(E - E_{\boldsymbol{k}})$$

All in all,

$$D(E) = \frac{1}{2\pi} i(G^R - G^A) = \delta(E - E_{\boldsymbol{k}})$$

Therefore, an important function that can be derived from the studies of the electronic systems using the NEGF method is the spectral function, which by definition is

$$A(k, E) = -2\, Im\, G^R(\boldsymbol{k}, E)$$

Therefore, the spectral function has the physical significance of the density of states via

$$D(E) = \frac{1}{2\pi} i(G^R - G^A) = -\frac{1}{\pi}\, Im\, G^R(\boldsymbol{k}, E)$$

Therefore, the spectral function is related to the density of states by

$$D(E) = \frac{1}{2\pi} A(\boldsymbol{k}, E)$$

Problem 4.04 Green's Function in Electronic Charge Distribution 1

The electrical potential for a fixed charge distribution is given by a differential equation:

$$\nabla_r^2 \phi(\boldsymbol{r}) = -\frac{1}{\varepsilon} \rho(\boldsymbol{r})$$

Express the potential function in the form of an integral equation using Green's function.

Solution

Define Green's function as

$$\nabla_r^2 G(\boldsymbol{r}-\boldsymbol{r}') = \delta(\boldsymbol{r}-\boldsymbol{r}')$$

Making use of the Dirac delta function, one can write

$$\begin{aligned}\nabla_r^2\phi(\boldsymbol{r}) &= -\frac{1}{\varepsilon}\int \delta(\boldsymbol{r}-\boldsymbol{r}')\rho(\boldsymbol{r}')d\boldsymbol{r}' \\ &= -\frac{1}{\varepsilon}\int \nabla_r^2 G(\boldsymbol{r}-\boldsymbol{r}')\rho(r')d\boldsymbol{r}'\end{aligned}$$

It thus follows that

$$\phi(\boldsymbol{r}) = -\frac{1}{\varepsilon}\int G(\boldsymbol{r}-\boldsymbol{r}')\rho(\boldsymbol{r}')d\boldsymbol{r}'$$

The electrical potential distribution can be found by solving

$$\nabla_r^2 G(\boldsymbol{r}-\boldsymbol{r}') = \delta(\boldsymbol{r}-\boldsymbol{r}')$$

Problem 4.05 Green's Function in Electronic Charge Distribution 11

For a point charge located at $r = 0$, the electric potential is $\phi(\boldsymbol{r}) = \frac{1}{4\pi\varepsilon r}$. From the previous expression of $\phi(\boldsymbol{r}) = -\frac{1}{\varepsilon}\int G(\boldsymbol{r}-\boldsymbol{r}')\rho(\boldsymbol{r}')d\boldsymbol{r}'$, one could infer that $G(\boldsymbol{r}) = -\frac{1}{4\pi r}$. Once again, making use of previous Green's function expression in the form of $\nabla^2 G(\boldsymbol{r}) = \delta(\boldsymbol{r})$, show that the explicit expression of Green's function is indeed

$$G(\boldsymbol{r}) = -\frac{1}{4\pi r}$$

Solution

Method A

The Fourier transform (FT) of the above is

$$FT\{\nabla^2 G(\boldsymbol{r})\} = (ik)^2 G(\boldsymbol{k})$$

leading to

$$A\int \delta(\boldsymbol{r})e^{-i\boldsymbol{k}.\boldsymbol{r}}d\boldsymbol{r} = (ik)^2 G(\boldsymbol{k})$$

which gives $G(\boldsymbol{k}) = A\frac{-1}{k^2}$. Thus, $G(\boldsymbol{r})$ can be found by inverse Fourier transform of $G(\boldsymbol{k})$, i.e.,

$$G(\boldsymbol{r}) = B\int d^3ke^{i\boldsymbol{k}.\boldsymbol{r}}G(\boldsymbol{k}) = AB\int d\boldsymbol{k}\frac{-e^{i\boldsymbol{k}.\boldsymbol{r}}}{k^2}$$
$$= \left(\frac{1}{2\pi}\right)^3\int d\boldsymbol{k}\frac{-e^{i\boldsymbol{k}.\boldsymbol{r}}}{k^2}$$

Note that $AB = \left(\frac{1}{2\pi}\right)^3$ relates to one of the following:

$$A = B = \left(\frac{1}{\sqrt{2\pi}}\right)^3; \quad A = 1, B = \left(\frac{1}{2\pi}\right)^3; \quad B = 1, A = \left(\frac{1}{2\pi}\right)^3$$

Now, one needs to derive

$$G(\boldsymbol{r}) = \left(\frac{1}{2\pi}\right)^3\int d\boldsymbol{k}\frac{-e^{i\boldsymbol{k}.\boldsymbol{r}}}{k^2}$$

Using theorem $\int_{-\infty}^{+\infty} f(q)e^{iq.r}d\boldsymbol{q} = \frac{4\pi}{r}\int_0^{+\infty} qf(q)\sin(rq)d\boldsymbol{q}$, it can be found that

$$G(\boldsymbol{r}) = \left(\frac{1}{2\pi}\right)^3\int d\boldsymbol{k}\frac{-e^{i\boldsymbol{k}.\boldsymbol{r}}}{k^2} = -\left(\frac{1}{2\pi}\right)^3\frac{4\pi}{r}\int_0^{+\infty}\frac{\sin(rk)}{k}d\boldsymbol{k}$$

Since $\int\frac{\sin y}{y}dy = \frac{\pi}{2}$,

$$G(\boldsymbol{r}) = \frac{-1}{2r\pi^2}\int_0^{\infty}\frac{siny}{y}dy = \frac{-1}{4\pi r}$$

Method B

Similarly, one needs to derive $G(\boldsymbol{r}) = \left(\frac{1}{2\pi}\right)^3\int d\boldsymbol{k}\frac{-e^{i\boldsymbol{k}.\boldsymbol{r}}}{k^2}$. In spherical coordinates,

$$G(\boldsymbol{r}) = \left(\frac{1}{2\pi}\right)^3\iiint\frac{-e^{ikr\cos\theta}}{k^2}k^2\sin\theta\, d\theta\, d\phi\, dk$$
$$= \frac{-1}{4\pi^2}\int\frac{e^{ikr} - e^{-ikr}}{ikr}dk$$

Since $\int\frac{\sin y}{y}dy = \frac{\pi}{2}$,

$$G(r) = \frac{-1}{2r\pi^2}\int\frac{\sin y}{y}dy = \frac{-1}{4\pi r}$$

Remarks and Reflections

In all the methods above, use has been repeatedly made of the integral

$$\int \frac{\sin y}{y} dy = \frac{\pi}{2}$$

Problem 4.06 Mathematical Methods: Trigonometric Integral

In the previous Greens' function problems, use is invariably made of the integral expression

$$\int_0^\infty \frac{\sin y}{y} dy = \frac{\pi}{2}$$

Prove the expression above.

Solution

Method A

One starts with

$$I(a,b) = \int_0^\infty e^{-ay} \frac{\sin by}{y} dy$$

where

$$I(0,1) = \int_0^\infty \frac{\sin y}{y} dy$$

It follows that $\frac{dI}{da} = -\int_0^\infty e^{-ay} \sin by \, dy$. The above can continue straightforwardly using integration by parts. Alternatively, one can perform

$$\frac{dI}{da} = -Im \int_0^\infty e^{(ib-a)y} dy = Im \, \frac{1}{ib-a} = \frac{-b}{a^2+b^2}$$

Now,

$$I(a,b) = \int_\infty^a \frac{dI}{da'} da' = \int_\infty^a \frac{-b}{a'^2+b^2} da' = -tan^{-1}\frac{a}{b} + \frac{\pi}{2}\, sgn[b]$$

where $sgn[b]$ is a scalar *1* that takes on the sign of constant b. Thus, the integral

$$I(0,1) = \frac{\pi}{2}$$

Method B

Use is made of $\int_0^\infty e^{-ay}\sin y\, da = \frac{\sin y}{y}$, and switching integration sequence, one can write the integral in the form of a double integral as follows:

$$I = \int_0^\infty \frac{\sin y}{y} dy = \int_0^\infty \left(\int_0^\infty e^{-ay} \sin y\, dy \right) da$$

With integration by parts,

$$I = \int_0^\infty - \left[\frac{e^{-ay}}{1+a^2} (a \sin y + \cos y) \right]_0^\infty da = \int_0^\infty \frac{1}{1+a^2} da = \frac{\pi}{2}$$

Method C

In Laplace transform,

$$F(s) = \int_0^\infty f(t) e^{-st}\, dt$$

$$f(t) = \frac{g(t)}{t} \Rightarrow F(s) = \int_s^\infty G(s) ds$$

Now, we choose $g(t) = \sin t$,

$$F(s) = \int_0^\infty \frac{g(t)}{t} e^{-st}\, dt = \int_0^\infty \frac{\sin t}{t} e^{-st}\, dt$$

There's a look-up table for $g(t)$ and $G(s)$.

$$\begin{aligned} F(s) &= \int_s^\infty G(s) ds = \int_s^\infty \frac{1}{1+s^2} ds \\ &= A\, tan^{-1}\, s|\frac{\infty}{s} \\ &= \frac{\pi}{2} - A \tan s \end{aligned}$$

Lastly,

$$\begin{aligned} \int_0^\infty \frac{\sin t}{t} dt &= \lim_{s \to 0} \int_0^\infty \frac{\sin t}{t} e^{-st}\, dt \\ &= \lim_{s \to 0} F(s) = \lim_{s \to 0} \left(\frac{\pi}{2} - A \tan s \right) \\ &= \frac{\pi}{2} \end{aligned}$$

Non-equilibrium Green's Function (NEGF)

Problem 4.07 Non-equilibrium Green's Function (NEGF): Electronic Charge Current

In electronic devices, electrical bias sets up a potential difference across the device and drives the electron channel into a state of non-equilibrium. The Keldysh forms of Green's functions are developed to study the quantum transport of an electron or a hole in these devices. In quantum electronic transport, this form of Green's function is also known as the non-equilibrium Green's function or the NEGF in short form. The kinetic energy of an electronic device system is given by

$$H_{KE} = \sum_{\sigma''} 2a^{\dagger}_{i\sigma''}a_{i\sigma''} - a^{\dagger}_{i+1,\sigma''}a_{i\sigma''} - a^{\dagger}_{i,\sigma''}a_{i+1,\sigma''}$$

where i is the index that runs through the space that spans the device. The charge current of the device is given by

$$\frac{dN_C}{dt} = J^C_{m,m+1} = \left\langle \left[H_{KE}, \sum_{\sigma} a^{\dagger}_{m\sigma}a_{m\sigma} \right] \right\rangle$$

Find the explicit expression of the charge current in Green's function.

Solution

Let's focus on the on-site kinetic energy term: $\sum_{\sigma''} 2a^{\dagger}_{i\sigma''}a_{i\sigma''}$

$$\begin{aligned}
J^C_{m,m+1} &= \langle [H_{KE}, a^{\dagger}_m a_m] \rangle \\
&= \left\langle 2\sum_{i\sigma''\sigma} \left(a^{\dagger}_{i\sigma''}a_{i\sigma''}a^{\dagger}_{m\sigma}a_{m\sigma} - a^{\dagger}_{m\sigma}a_{m\sigma}a^{\dagger}_{i\sigma''}a_{i\sigma''} \right) \right\rangle \\
&= 2\sum_{i\sigma''\sigma} \Big(\langle a^{\dagger}_{i\sigma''}(\delta_{im}\delta_{\sigma''\sigma} - a^{\dagger}_{m\sigma}a_{i\sigma''})a_{m\sigma} \rangle \\
&\quad - \langle a^{\dagger}_{m\sigma}(\delta_{mi}\delta_{\sigma\sigma''} - a^{\dagger}_{i\sigma''}a_{m\sigma})a_{i\sigma''} \rangle \Big)
\end{aligned}$$

In the above, use has been made of the fermionic physics of $\{a_m, a^{\dagger}_{m'}\} = \delta_{mm'}$. Moving on,

$$\begin{aligned}
J^C_{m,m+1} = \sum_{i\sigma''\sigma} \Big(&\langle a^{\dagger}_{m\sigma}a_{m\sigma} \rangle - \langle a^{\dagger}_{i\sigma''}a^{\dagger}_{m\sigma}a_{i\sigma''}a_{m\sigma} \rangle \\
&- \langle a^{\dagger}_{m\sigma}a_{m\sigma} \rangle + \langle a^{\dagger}_{m\sigma}a^{\dagger}_{i\sigma''}a_{m\sigma}a_{i\sigma''} \rangle \Big) = 0 \qquad \text{(A)}
\end{aligned}$$

where use is now made of $\{a_k, a_{k'}\} = 0$. We shall turn our attention to the first overlapping term: $\sum_{\sigma''} -a^\dagger_{i+1,\sigma''} a_{i\sigma''}$, where

$$J^C_{m,m+1} = \langle [H_{KE}, a^\dagger_m a_m] \rangle = \Big\langle - \sum\nolimits_{i\sigma''\sigma} \Big(a^\dagger_{i+1,\sigma''} a_{i\sigma''} a^\dagger_{m\sigma} a_{m\sigma} - a^\dagger_{m\sigma} a_{m\sigma} a^\dagger_{i+1,\sigma''} a_{i\sigma''} \Big) \Big\rangle$$

$$= - \sum_{i\sigma''\sigma} \Big(\langle a^\dagger_{i+1,\sigma''} (\delta_{im}\delta_{\sigma''\sigma} - a^\dagger_{m\sigma} a_{i\sigma''}) a_{m\sigma} \rangle - \langle a^\dagger_{m\sigma} (\delta_{m,i+1}\delta_{\sigma\sigma''} - a^\dagger_{i+1,\sigma''} a_{m\sigma}) a_{i\sigma''} \rangle \Big)$$

In the above, use has been made of the fermionic physics of $\{a_m, a^\dagger_{m'}\} = \delta_{mm'}$. Moving on,

$$J^C_{m,m+1} = - \sum_{i\sigma''\sigma} (\langle a^\dagger_{m+1,\sigma} a_{m\sigma} \rangle - \langle a^\dagger_{i+1,\sigma''} a^\dagger_{m\sigma} a_{i\sigma''} a_{m\sigma} \rangle - \langle a^\dagger_{m\sigma} a_{m-1,\sigma} \rangle + \langle a^\dagger_{m\sigma} a^\dagger_{i+1,\sigma''} a_{m\sigma} a_{i\sigma''} \rangle)$$

$$= - \sum_{\sigma} (\langle a^\dagger_{m+1,\sigma} a_{m\sigma} \rangle - \langle a^\dagger_{m\sigma} a_{m-1,\sigma} \rangle)$$

where use has been of $\{a^\dagger_m, a^\dagger_{m'}\} = 0$ and $\{a_m, a_{m'}\} = 0$. The above is none other than

$$J^C_{m,m+1} = - \sum_{\sigma} \int \left(G^<_{\substack{m+1,\sigma \\ m\sigma}} - G^<_{\substack{m\sigma \\ m-1,\sigma}} \right) dE \tag{B}$$

where the first term can be interpreted as the *forward* current while the second term as the *backward* current. The current can be roughly understood by referring to the illustration shownin Figure 1.

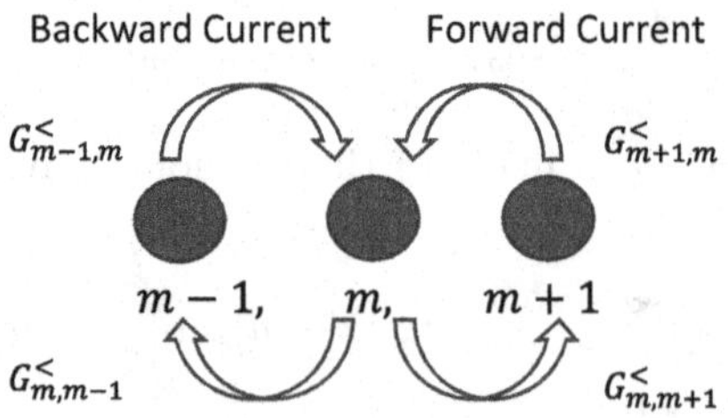

Fig. 1. A rough illustration of the physical significance of Green's function.

We will now turn our attention to the second overlapping term: $\sum_{\sigma''} -a^{\dagger}_{i,\sigma''}a_{i+1,\sigma''}$, where

$$J^{C}_{m,m+1} = \langle [H_{KE}, a^{\dagger}_{m}a_{m}]\rangle = \Big\langle -\sum\nolimits_{i\sigma''\sigma}(a^{\dagger}_{i\sigma''}a_{i+1,\sigma''}a^{\dagger}_{m\sigma}a_{m\sigma}$$
$$- a^{\dagger}_{m\sigma}a_{m\sigma}a^{\dagger}_{i\sigma''}a_{i+1,\sigma''})\Big\rangle$$
$$= -\sum_{i\sigma''\sigma}(\langle a^{\dagger}_{i\sigma''}(\delta_{i+1,m}\delta_{\sigma''\sigma} - a^{\dagger}_{m\sigma}a_{i+1,\sigma''})a_{m\sigma}\rangle$$
$$- \langle a^{\dagger}_{m\sigma}(\delta_{mi}\delta_{\sigma\sigma''} - a^{\dagger}_{i\sigma''}a_{m\sigma})a_{i+1,\sigma''}\rangle)$$

In the above, use has been made of the fermionic physics of $\{a_m, a^{\dagger}_{m'}\} = \delta_{mm'}$. Moving on,

$$J^{C}_{m,m+1} = -\sum_{i\sigma''\sigma}(\langle a^{\dagger}_{m-1,\sigma}a_{m\sigma}\rangle - \langle a^{\dagger}_{i\sigma''}a^{\dagger}_{m\sigma}a_{i+1,\sigma''}a_{m\sigma}\rangle$$
$$- \langle a^{\dagger}_{m\sigma}a_{m+1,\sigma}\rangle + \langle a^{\dagger}_{m\sigma}a^{\dagger}_{i\sigma''}a_{m\sigma}a_{i+1,\sigma''}\rangle)$$
$$= -\sum_{\sigma}(\langle a^{\dagger}_{m-1,\sigma}a_{m\sigma}\rangle - \langle a^{\dagger}_{m\sigma}a_{m+1,\sigma}\rangle)$$

where use has been made of $\{a_m, a_{m'}\} = 0$. The above is none other than

$$J^{C}_{m,m+1} = -\sum_{\sigma}\int\left(G^{<}_{\substack{m-1,\sigma\\ m\sigma}} - G^{<}_{\substack{m\sigma\\ m+1,\sigma}}\right)dE \qquad \text{(C)}$$

where the first term can be interpreted as the *backward* current while the second term as the *forward* current. Combining (A), (B), and (C), the total charge current is given by

$$J^{C}_{m,m+1} = -\sum_{\sigma}\int\left(G^{<}_{\substack{m+1,\sigma\\ m\sigma}} - G^{<}_{\substack{m\sigma\\ m-1,\sigma}}\right) + \left(G^{<}_{\substack{m-1,\sigma\\ m\sigma}} - G^{<}_{\substack{m\sigma\\ m+1,\sigma}}\right)dE$$

The final expression is therefore given by

$$J^{S_n}_{m,m+1} = -\sum_{\sigma\sigma'}\langle\sigma|S_n|\sigma'\rangle\int\left(G^{<}_{\substack{m+1,\sigma\\ m\sigma}} - G^{<}_{\substack{m\sigma\\ m+1,\sigma}}\right)$$
$$+\left(G^{<}_{\substack{m-1,\sigma\\ m\sigma}} - G^{<}_{\substack{m\sigma\\ m-1,\sigma}}\right)dE$$

where the first term is now the *forward* current while the second term is the *backward* current. Refer to Figure 1 for a rough understanding.

Remarks and Reflections

Points to Ponder

In field theoretic condensed matter physics, Green's function has been used to study the effect of interactions on electron energy, transport and so forth. Thus, the NEGF is also a natural tool to include many body physics in nanoelectronics.

The incorporation of the effects of electron–electron and electron–phonon scatterings is particularly useful. In the more recent developments where additional degrees of freedom are studied, e.g., spintronics and graphene electronics, the NEGF provides a platform to include spin–orbit coupling and spin–spin interaction.

Problem 4.08 Non-equilibrium Green's Function (NEGF): Kinetic Spin Current

Spin current consists of the kinetic, magnetic, and spin–orbit. This problem deals with the kinetic spin current. The kinetic energy of an electronic device system is given by

$$H_{KE} = \sum_{\sigma''} 2a^\dagger_{i\sigma''} a_{i\sigma''} - a^\dagger_{i+1,\sigma''} a_{i\sigma''} - a^\dagger_{i\sigma''} a_{i+1,\sigma''}$$

where σ'' is the index for the spin degree of freedom and i is the index that runs through the space that spans the device. The spin current (kinetic) of the device is given by

$$\frac{dN_S}{dt} = J^{S_n}_{m,m+1} = \left\langle \left[H_{KE}, \sum\nolimits_{\sigma\sigma'} \langle\sigma|S_n|\sigma'\rangle a^\dagger_{m\sigma} a_{m\sigma'} \right] \right\rangle$$

where σ, σ' are the spin indices. Find the explicit expression of the kinetic spin current in Green's function.

Solution

Let's focus on the first overlapping term of the kinetic energy $-a^\dagger_{i+1,\sigma''} a_{i\sigma''}$,

$$\begin{aligned} J^{S_n(1)}_{m,m+1} &= \left\langle \left[H_{KE}, \sum\nolimits_{\sigma\sigma'} \langle\sigma|S_n|\sigma'\rangle a^\dagger_{m\sigma} a_{m\sigma'} \right] \right\rangle \\ &= -\left\langle \sum_{i\sigma''\sigma\sigma'} \langle\sigma|S_n|\sigma'\rangle (a^\dagger_{i+1,\sigma''} a_{i\sigma''} a^\dagger_{m\sigma} a_{m\sigma'} - a^\dagger_{m\sigma} a_{m\sigma'} a^\dagger_{i+1,\sigma''} a_{i\sigma''}) \right\rangle \end{aligned}$$

$$= -\sum_{i\sigma''\sigma\sigma'} \langle\sigma|S_n|\sigma'\rangle(\langle a^{\dagger}_{i+1,\sigma''}(\delta_{im}\delta_{\sigma''\sigma} - a^{\dagger}_{m\sigma}a_{i\sigma''})a_{m\sigma'}\rangle$$

$$- \langle a^{\dagger}_{m\sigma}(\delta_{m,i+1}\delta_{\sigma'\sigma''} - a^{\dagger}_{i+1,\sigma''}a_{m\sigma'})a_{i\sigma''}\rangle)$$

In the above, use has been made of the fermionic physics of $\{a_m, a^{\dagger}_{m'}\} = \delta_{mm'}$. Moving on,

$$J^{S_n(1)}_{m,m+1} = -\sum_{i\sigma''\sigma\sigma'} \langle\sigma|S_n|\sigma'\rangle(\langle a^{\dagger}_{m+1,\sigma}a_{m\sigma'}\rangle$$

$$- \langle a^{\dagger}_{i+1,\sigma''}a^{\dagger}_{m\sigma}a_{i\sigma''}a_{m\sigma'}\rangle - \langle a^{\dagger}_{m\sigma}a_{m-1,\sigma'}\rangle$$

$$+ \langle a^{\dagger}_{m\sigma}a^{\dagger}_{i+1,\sigma''}a_{m\sigma'}a_{i\sigma''}\rangle)$$

which leads, with $\{a^{\dagger}_m, a^{\dagger}_{m'}\} = 0$ and $\{a_m, a_{m'}\} = 0$, to the simple expression of

$$J^{S_n(1)}_{m,m+1} = -\sum_{\sigma\sigma'}\langle\sigma|S_n|\sigma'\rangle(\langle a^{\dagger}_{m+1,\sigma}a_{m\sigma'}\rangle - \langle a^{\dagger}_{m\sigma}a_{m-1,\sigma'}\rangle)$$

where use has been made of $\{a_m, a_{m'}\} = 0$. The above is none other than

$$J^{S_n(1)}_{m,m+1} = -\sum_{\sigma\sigma'}\langle\sigma|S_n|\sigma'\rangle \int \left(G^{<}_{\substack{m+1,\sigma\\ m\sigma'}} - G^{<}_{\substack{m\sigma\\ m-1,\sigma'}}\right) dE \qquad \text{(A)}$$

where the first term can be interpreted as the *forward* current while the second term as the *backward* current. We will now focus our attention on the second overlapping term of $-a^{\dagger}_{i\sigma''}a_{i+1,\sigma''}$, where

$$J^{S_n(2)}_{m,m+1} = \left\langle -\sum_{\sigma\sigma'}\langle\sigma|S_n|\sigma'\rangle(a^{\dagger}_{i\sigma''}a_{i+1,\sigma''}a^{\dagger}_{m\sigma}a_{m\sigma'} - a^{\dagger}_{m\sigma}a_{m\sigma'}a^{\dagger}_{i\sigma''}a_{i+1,\sigma''})\right\rangle$$

$$= -\sum_{\sigma\sigma'}\langle\sigma|S_n|\sigma'\rangle\langle a^{\dagger}_{i\sigma''}(\delta_{i+1,m}\delta_{\sigma''\sigma} - a^{\dagger}_{m\sigma}a_{i+1,\sigma''})a_{m\sigma'}$$

$$- a^{\dagger}_{m\sigma}(\delta_{mi}\delta_{\sigma'\sigma''} - a^{\dagger}_{i\sigma''}a_{m\sigma'})a_{i+1,\sigma''}\rangle$$

In the above, use has been made of the fermionic physics of $\{a_m, a^{\dagger}_{m'}\} = \delta_{mm'}$. Moving on,

$$J^{S_n(2)}_{m,m+1} = -\sum_{\sigma\sigma'}\langle\sigma|S_n|\sigma'\rangle(\langle a^{\dagger}_{m-1,\sigma}a_{m\sigma'}\rangle$$

$$- \langle a^{\dagger}_{i\sigma''}a^{\dagger}_{m\sigma}a_{i+1,\sigma''}a_{m\sigma'}\rangle - \langle a^{\dagger}_{m\sigma}a_{m+1,\sigma'}\rangle$$

$$+ \langle a^{\dagger}_{m\sigma}a^{\dagger}_{i\sigma''}a_{m\sigma'}a_{i+1,\sigma''}\rangle)$$

which leads, with $\{a_m^\dagger, a_{m'}^\dagger\} = 0$ and $\{a_m, a_{m'}\} = 0$, to the simple expression of

$$J_{m,m+1}^{S_n(2)} = -\sum_{\sigma\sigma'} \langle\sigma|S_n|\sigma'\rangle(\langle a_{m-1,\sigma}^\dagger a_{m\sigma'}\rangle - \langle a_{m\sigma}^\dagger a_{m+1,\sigma'}\rangle)$$

where use has been made of $\{a_m, a_{m'}\} = 0$. The above is none other than

$$J_{m,m+1}^{S_n(2)} = -\sum_{\sigma\sigma'} \langle\sigma|S_n|\sigma'\rangle \int \left(G^{<}_{\substack{m-1,\sigma\\ m\sigma'}} - G^{<}_{\substack{m\sigma\\ m+1,\sigma'}}\right) dE \qquad \text{(B)}$$

where the first term can be interpreted as the *backward* current while the second term as the *forward* current. Combining (A) and (B), the total spin current is given by

$$\begin{aligned} J_{m,m+1}^{S_n} &= J_{m,m+1}^{S_n(1)} + J_{m,m+1}^{S_n(2)} \\ &= -\sum_{\sigma\sigma'} \langle\sigma|S_n|\sigma'\rangle \int \left(G^{<}_{\substack{m+1,\sigma\\ m\sigma'}} - G^{<}_{\substack{m\sigma\\ m-1,\sigma'}}\right) \\ &\quad + \left(G^{<}_{\substack{m-1,\sigma\\ m\sigma'}} - G^{<}_{\substack{m\sigma\\ m+1,\sigma'}}\right) dE \end{aligned}$$

The final expression is given by

$$\begin{aligned} J_{m,m+1}^{S_n} &= -\sum_{\sigma\sigma'} \langle\sigma|S_n|\sigma'\rangle \int \left(G^{<}_{\substack{m+1,\sigma\\ m\sigma'}} - G^{<}_{\substack{m\sigma\\ m+1,\sigma'}}\right) \\ &\quad + \left(G^{<}_{\substack{m-1,\sigma\\ m\sigma'}} - G^{<}_{\substack{m\sigma\\ m-1,\sigma'}}\right) dE \end{aligned}$$

where the first term is now the *forward* current while the second term is the *backward* current.

Remarks and Reflections

What about the on-site kinetic energy $2a_{i\sigma''}^\dagger a_{i\sigma''}$? Show that this term generates zero spin current.

Problem 4.09 Non-equilibrium Green's Function (NEGF): Magnetic Spin Current

Spin current consists of the kinetic, magnetic, and spin–orbit. This problem deals with the magnetic spin current. The magnetic energy of an electronic device system is given by

$$H_{mag} = \sum_{i\sigma''\sigma'''} \langle \sigma''|S^a|\sigma'''\rangle M_a a^\dagger_{i\sigma''} a_{i\sigma'''}$$

The spin current (magnetic) of the device is given by

$$\frac{dN_S}{dt} = J^{S_n}_{m,m} = \left\langle \left[H_{mag}, \sum_{\sigma\sigma'} \langle\sigma|S_n|\sigma'\rangle a^\dagger_{m\sigma} a_{m\sigma'} \right] \right\rangle$$

Find the explicit expression of the magnetic spin current in Green's function.

Solution

$$\frac{dN_S}{dt} = J^{S_n}_{m,m} = \left\langle \sum_{\substack{i\sigma\sigma' \\ \sigma''\sigma'''}} \langle\sigma''|S^a|\sigma'''\rangle\langle\sigma|S_n|\sigma'\rangle M_a (a^\dagger_{i\sigma''} a_{i\sigma'''} a^\dagger_{m\sigma} a_{m\sigma'} \right.$$
$$\left. - a^\dagger_{m\sigma} a_{m\sigma'} a^\dagger_{i\sigma''} a_{i\sigma'''}) \right\rangle$$
$$= \left\langle \sum_{\substack{i\sigma\sigma' \\ \sigma''\sigma'''}} \langle\sigma''|S^a|\sigma'''\rangle\langle\sigma|S_n|\sigma'\rangle M_a (a^\dagger_{i\sigma''} (\delta_{im}\delta_{\sigma\sigma'''} \right.$$
$$\left. - a^\dagger_{m\sigma} a_{i\sigma'''}) a_{m\sigma'} \rangle - a^\dagger_{m\sigma}(\delta_{im}\delta_{\sigma'\sigma''} - a^\dagger_{i\sigma''} a_{m\sigma'}) a_{i\sigma'''}) \right\rangle$$

In the above, use has been made of the fermionic physics of $\{a_m, a^\dagger_{m'}\} = \delta_{mm'}$. Moving on,

$$J^{S_n}_{m,m} = \sum_{\substack{\sigma\sigma' \\ \sigma''\sigma'''}} \langle\sigma''|S^a|\sigma'''\rangle\langle\sigma|S_n|\sigma'\rangle M_a (\langle a^\dagger_{m\sigma''} a_{m\sigma'}\rangle \delta_{\sigma\sigma'''}$$
$$- \langle a^\dagger_{i\sigma''} a^\dagger_{m\sigma} a_{i\sigma'''} a_{m\sigma'}\rangle - \langle a^\dagger_{m\sigma} a_{i\sigma'''}\rangle \delta_{im}\delta_{\sigma'\sigma''}$$
$$+ \langle a^\dagger_{m\sigma} a^\dagger_{i\sigma''} a_{m\sigma'} a_{i\sigma'''}\rangle)$$

With $\{a^\dagger_m, a^\dagger_{m'}\} = 0$ and $\{a_m, a_{m'}\} = 0$, one has

$$J^{S_n}_{m,m} = \sum_{\substack{\sigma\sigma' \\ \sigma''\sigma'''}} \langle\sigma''|S^a|\sigma'''\rangle\langle\sigma|S_n|\sigma'\rangle M_a (\langle a^\dagger_{m\sigma''} a_{m\sigma'}\rangle \delta_{\sigma\sigma'''} - \langle a^\dagger_{m\sigma} a_{m\sigma'''}\rangle \delta_{\sigma'\sigma''})$$

Therefore, as $[\langle\sigma''|S^a|\sigma'''\rangle, \langle\sigma|S_n|\sigma'\rangle] = 0$, one has

$$J_{m,m}^{S_n} = \sum_{\sigma\sigma'\sigma''\sigma'''} \langle\sigma''|S^a|\sigma'''\rangle\langle\sigma|S_n|\sigma'\rangle M_a\langle a_{m\sigma''}^\dagger a_{m\sigma'}\rangle\delta_{\sigma\sigma'''}$$
$$- \langle\sigma|S_n|\sigma'\rangle\langle\sigma''|S^a|\sigma'''\rangle M_a\langle a_{m\sigma}^\dagger a_{m\sigma'''}\rangle\delta_{\sigma'\sigma''}$$

Applying the resolution of identity,

$$J_{m,m}^{S_n} = \sum_{\sigma'\sigma''}\langle\sigma''|S^a S_n|\sigma'\rangle M_a\langle a_{m\sigma''}^\dagger a_{m\sigma'}\rangle - \sum_{\sigma\sigma'''}\langle\sigma|S_n S^a|\sigma'''\rangle M_a\langle a_{m\sigma}^\dagger a_{m\sigma'''}\rangle$$
$$= \sum_{\sigma'\sigma''}\langle\sigma''|S^a S_n - S_n S^a|\sigma'\rangle M_a\langle a_{m\sigma''}^\dagger a_{m\sigma'}\rangle \qquad \text{(A)}$$

In flat space, contra- and co-variance can be neglected. S_a and S_n are Pauli matrices,

$$S_n S_a = i\varepsilon_{nak}S_k + \delta_{na}I, \quad S_a S_n = i\varepsilon_{ank}S_k + \delta_{na}I$$

Therefore,

$$S_a S_n - S_n S_a = 2iS_k\varepsilon_{ank}$$

This leads to

$$J_{m,m}^{S_n} = \sum_{\sigma'\sigma''}\langle\sigma''|2iS_k M_a\varepsilon_{ank}|\sigma'\rangle M_a\langle a_{m\sigma''}^\dagger a_{m\sigma'}\rangle \qquad \text{(B)}$$

Remarks and Reflections

Note that Eqs. (A) and (B) have the following classical correspondence:

$$J_{m,m}^{S_n} = \sum_{\sigma'\sigma''}\langle\sigma''|S^a S_n - S_n S^a|\sigma'\rangle M_a\langle a_{m\sigma''}^\dagger a_{m\sigma'}\rangle \to [\boldsymbol{S.M,S}]$$
$$J_{m,m}^{S_n} = \sum_{\sigma'\sigma''}\langle\sigma''|2iS_k M_a\varepsilon_{ank}|\sigma'\rangle M_a\langle a_{m\sigma''}^\dagger a_{m\sigma'}\rangle \to 2i\boldsymbol{S}\times\boldsymbol{M}$$
$$\frac{d\boldsymbol{S}}{dt} = [\boldsymbol{S.M,S}] = 2i\boldsymbol{S}\times\boldsymbol{M}$$

The magnetic spin current is none other than spin precession in classical physics.

Problem 4.10 Non-equilibrium Green's Function: Spin–Orbit Spin Current

Spin current consists of the kinetic, magnetic, and spin–orbit. This problem deals with the spin–orbit spin current. The spin–orbit energy of an electronic device system is given by

$$H_{SOC} = \sum_{i,\sigma\sigma'} (a_{i\sigma}^{\dagger} a_{i+1,\sigma'} - a_{i+1,\sigma}^{\dagger} a_{i\sigma'}) \langle \sigma | S_a | \sigma' \rangle \varepsilon_{iaz}$$

The spin current (spin–orbit) of the device is given by

$$\frac{dN_S}{dt} = J_m^{S_n} = \left\langle \left[H_{SOC}, \sum_{\sigma\sigma'} \langle \sigma | S_n | \sigma' \rangle a_{m\sigma}^{\dagger} a_{m\sigma'} \right] \right\rangle$$

(a) Find the explicit expression of the spin current in Green's function.
(b) The spin current composes of two components: the adiabatic and the precession. Explain the physical significance of the adiabatic Green's function current.
(c) Show that the precession Green's function current is consistent with classical precession.

Solution

(a) Green's function spin current

$$J_{m,m+1}^{S_n} = (-1) \left\langle \left[\sum_{i,\sigma''\sigma'''} (a_{i\sigma''}^{\dagger} a_{i+1,\sigma'''} - a_{i+1,\sigma''}^{\dagger} a_{i\sigma'''}) \langle \sigma'' | S_a | \sigma''' \rangle \varepsilon_{iaz}, \right. \right.$$

$$\left. \left. \times \sum_{\sigma\sigma'} \langle \sigma | S_n | \sigma' \rangle a_{m\sigma}^{\dagger} a_{m\sigma'} \right] \right\rangle$$

Let's examine the FIRST term on the RHS. This current is based on energy term $a_{i\sigma''}^{\dagger} a_{i+1,\sigma'''}$.

$$J_m^{S_n 1} = (-1) \left\langle \sum_{\substack{i,\sigma''\sigma''' \\ \sigma\sigma'}} (a_{i\sigma''}^{\dagger} a_{i+1,\sigma'''} a_{m\sigma}^{\dagger} a_{m\sigma'} \right.$$

$$\left. - a_{m\sigma}^{\dagger} a_{m\sigma'} a_{i\sigma''}^{\dagger} a_{i+1,\sigma'''}) [S_a]_{\sigma'''}^{\sigma''} [S_n]_{\sigma'}^{\sigma} \varepsilon_{iaz} \right\rangle$$

$$= (-1)\left\langle \sum_{\substack{i,\sigma''\sigma'''\\ \sigma\sigma'}} (a^\dagger_{i\sigma''}(\delta_{\sigma'''\sigma}\delta_{m,i+1} - a^\dagger_{m\sigma}a_{i+1,\sigma'''})a_{m\sigma'} \right.$$

$$\left. - a^\dagger_{m\sigma}(\delta_{\sigma'\sigma''}\delta_{mi} - a^\dagger_{i\sigma''}a_{m\sigma'})a_{i+1,\sigma'''})[S_a]^{\sigma''}_{\sigma'''}[S_n]^{\sigma}_{\sigma'}\varepsilon_{iaz}\right\rangle$$

where $[S_a]^{\sigma''}_{\sigma'''} = \langle\sigma''|S_a|\sigma'''\rangle$, and $[S_n]^{\sigma}_{\sigma'} = \langle\sigma|S_n|\sigma'\rangle$

$$J^{S_n 1}_{m,} = (-1)\sum_{\substack{i,\sigma''\sigma'''\\ \sigma\sigma'}} (\langle a^\dagger_{i\sigma''}a_{m\sigma'}\rangle\delta_{\sigma'''\sigma}\delta_{m,i+1} - \langle a^\dagger_{i\sigma''}a^\dagger_{m\sigma}a_{i+1,\sigma'''}a_{m\sigma'}\rangle$$

$$- \langle a^\dagger_{m\sigma}a_{i+1,\sigma'''}\rangle\delta_{\sigma'\sigma''}\delta_{mi} + \langle a^\dagger_{m\sigma}a^\dagger_{i\sigma''}a_{m\sigma'}a_{i+1,\sigma'''}\rangle)[S_a]^{\sigma''}_{\sigma'''}[S_n]^{\sigma}_{\sigma'}\varepsilon_{iaz}$$

With $\{a^\dagger_m, a^\dagger_{m'}\} = 0$ and $\{a_m, a_{m'}\} = 0$, one has

$$J^{S_n 1}_{m} = (-1)\sum_{\substack{\sigma''\sigma'''\\ \sigma\sigma'}} (\langle a^\dagger_{m-1,\sigma''}a_{m\sigma'}\rangle\delta_{\sigma'''\sigma}[S_a]^{\sigma''}_{\sigma'''}[S_n]^{\sigma}_{\sigma'}\varepsilon_{m-1,az}$$

$$- \langle a^\dagger_{m\sigma}a_{m+1,\sigma'''}\rangle\delta_{\sigma'\sigma''}[S_n]^{\sigma}_{\sigma'}[S_a]^{\sigma''}_{\sigma'''}\varepsilon_{maz})$$

Note that $\varepsilon_{m-1,az} = \varepsilon_{maz}$ because $m, m-1$ are of the same dimension. Applying the resolution of identity,

$$J^{S_n 1}_{m} = (-1)\varepsilon_{maz}\left(\sum_{\sigma'\sigma''}\langle a^\dagger_{m-1,\sigma''}a_{m\sigma'}\rangle\langle\sigma''|S_aS_n|\sigma'\rangle\right.$$

$$\left. - \sum_{\sigma\sigma'''}\langle a^\dagger_{m\sigma}a_{m+1,\sigma'''}\rangle\langle\sigma|S_nS_a|\sigma'''\rangle\right)$$

In terms of Green's function,

$$J^{S_n 1}_{m} = -\varepsilon_{maz}\sum_{\sigma'\sigma''}\left(\langle\sigma''|S_aS_n|\sigma'\rangle\int G^<_{\substack{m-1,\sigma''\\ m\sigma'}}dE\right)$$

$$- \sum_{\sigma''\sigma'}\left(\langle\sigma''|S_nS_a|\sigma'\rangle\int G^<_{\substack{m\sigma''\\ m+1,\sigma'}}dE\right)$$

Let's examine the SECOND term on the RHS. This current is based on energy term $-a^\dagger_{i+1,\sigma} a_{i\sigma'}$.

$$J_m^{S_n 2} = (-1) \left\langle \sum_{\substack{i,\sigma''\sigma''' \\ \sigma\sigma'}} -(a^\dagger_{i+1,\sigma''} a_{i\sigma'''} a^\dagger_{m\sigma} a_{m\sigma'} \right.$$

$$\left. - a^\dagger_{m\sigma} a_{m\sigma'} a^\dagger_{i+1,\sigma''} a_{i\sigma'''})[S_a]^{\sigma''}_{\sigma'''}[S_n]^{\sigma}_{\sigma'}\varepsilon_{iaz} \right\rangle$$

$$= (-1) \left\langle \sum_{\substack{i,\sigma''\sigma''' \\ \sigma\sigma'}} -(a^\dagger_{i+1,\sigma''}(\delta_{\sigma'''\sigma}\delta_{mi} - a^\dagger_{m\sigma} a_{i\sigma'''}) a_{m\sigma'} \right.$$

$$\left. + a^\dagger_{m\sigma}(\delta_{\sigma'\sigma''}\delta_{m,i+1} - a^\dagger_{i+1,\sigma''} a_{m\sigma'}) a_{i\sigma'''})[S_a]^{\sigma''}_{\sigma'''}[S_n]^{\sigma}_{\sigma'}\varepsilon_{iaz} \right\rangle$$

where $[S_a]^{\sigma''}_{\sigma'''} = \langle \sigma'' | S_a | \sigma''' \rangle$, and $[S_n]^{\sigma}_{\sigma'} = \langle \sigma | S_n | \sigma' \rangle$

$$J_{m,}^{S_n 2} = (-1) \sum_{\substack{i,\sigma''\sigma''' \\ \sigma\sigma'}} (-\langle a^\dagger_{i+1,\sigma''} a_{m\sigma'} \delta_{\sigma'''\sigma}\delta_{mi}$$

$$+ \langle a^\dagger_{i+1,\sigma''} a^\dagger_{m\sigma} a_{i\sigma'''} a_{m\sigma'} \rangle + \langle a^\dagger_{m\sigma} a_{i\sigma'''} \rangle \delta_{\sigma'\sigma''}\delta_{m,i+1}$$

$$- \langle a^\dagger_{m\sigma} a^\dagger_{i+1,\sigma''} a_{m\sigma'} a_{i\sigma'''} \rangle)[S_a]^{\sigma''}_{\sigma'''}[S_n]^{\sigma}_{\sigma'}\varepsilon_{iaz}$$

With $\{a^\dagger_m, a^\dagger_{m'}\} = 0$ and $\{a_m, a_{m'}\} = 0$, one has

$$J_m^{S_n 1} = (-1) \sum_{\substack{\sigma''\sigma''' \\ \sigma\sigma'}} \big(- \langle a^\dagger_{m+1,\sigma''} a_{m\sigma'} \rangle \delta_{\sigma'''\sigma}[S_a]^{\sigma''}_{\sigma'''}[S_n]^{\sigma}_{\sigma'}\varepsilon_{maz}$$

$$+ \langle a^\dagger_{m\sigma} a_{m-1,\sigma'''} \rangle \delta_{\sigma'\sigma''}[S_n]^{\sigma}_{\sigma'}[S_a]^{\sigma''}_{\sigma'''}\varepsilon_{m-1,az}\big)$$

Note that $\varepsilon_{m-1,az} = \varepsilon_{maz}$ because $m, m-1$ are of the same dimension. Applying the resolution of identity,

$$J_m^{S_n 2} = (-1)\varepsilon_{maz} \left(\sum_{\sigma'\sigma''} -\langle a^\dagger_{m+1,\sigma''} a_{m\sigma'} \rangle \langle \sigma'' | S_a S_n | \sigma' \rangle \right.$$

$$\left. + \sum_{\sigma\sigma'''} \langle a^\dagger_{m\sigma} a_{m-1,\sigma'''} \rangle \langle \sigma | S_n S_a | \sigma''' \rangle \right)$$

In terms of Green's function,

$$J_m^{S_n 2} = -\varepsilon_{maz} \sum_{\sigma'\sigma''} - \left(\langle \sigma'' | S_a S_n | \sigma' \rangle \int G^<_{\substack{m+1,\sigma'' \\ m\sigma'}} dE \right) + \sum_{\sigma''\sigma'} \left(\langle \sigma'' | S_n S_a | \sigma' \rangle \int G^<_{\substack{m\sigma'' \\ m-1,\sigma'}} dE \right)$$

Putting the two current parts together,

$$J_m^{S_n 1} = -\varepsilon_{maz} \sum_{\sigma'\sigma''} \left(\langle \sigma'' | S_a S_n | \sigma' \rangle \int G^<_{\substack{m-1,\sigma'' \\ m\sigma'}} dE \right) - \sum_{\sigma''\sigma'} \left(\langle \sigma'' | S_n S_a | \sigma' \rangle \int G^<_{\substack{m\sigma'' \\ m+1,\sigma'}} dE \right)$$

$$J_m^{S_n 2} = -\varepsilon_{maz} \sum_{\sigma'\sigma''} - \left(\langle \sigma'' | S_a S_n | \sigma' \rangle \int G^<_{\substack{m+1,\sigma'' \\ m\sigma'}} dE \right) + \sum_{\sigma''\sigma'} \left(\langle \sigma'' | S_n S_a | \sigma' \rangle \int G^<_{\substack{m\sigma'' \\ m-1,\sigma'}} dE \right)$$

Note that $J_m^{S_n 1}$ is based on $a^\dagger_{i\sigma} a_{i+1,\sigma'}$, while $J_m^{S_n 2}$ is based on $-a^\dagger_{i+1,\sigma} a_{i\sigma'}$. Therefore, referring to the Hamiltonian $H_{SOC} = \sum_{i,\sigma\sigma'} (a^\dagger_{i\sigma} a_{i+1,\sigma'} - a^\dagger_{i+1,\sigma} a_{i\sigma'}) \langle \sigma | S_a | \sigma' \rangle \varepsilon_{iaz}$, the total current is

$$J_m^{Sn} = J_m^{S_n 1} + J_m^{S_n 2}$$

Backward current:

$$J_m^{S_n B} = -\varepsilon_{maz} \sum_{\sigma'\sigma''} \left(\langle \sigma'' | S_a S_n | \sigma' \rangle \int G^<_{\substack{m-1,\sigma'' \\ m\sigma'}} dE \right) + \sum_{\sigma''\sigma'} \left(\langle \sigma'' | S_n S_a | \sigma' \rangle \int G^<_{\substack{m\sigma'' \\ m-1,\sigma'}} dE \right)$$

Forward current:

$$J_m^{S_n F} = -\varepsilon_{maz} \sum_{\sigma''\sigma'} - \left(\langle \sigma'' | S_n S_a | \sigma' \rangle \int G^<_{\substack{m\sigma'' \\ m+1,\sigma'}} dE \right) - \sum_{\sigma'\sigma''} \left(\langle \sigma'' | S_a S_n | \sigma' \rangle \int G^<_{\substack{m+1,\sigma'' \\ m\sigma'}} dE \right)$$

Let's examine the forward current only:

$$J_m^{S_nF} = -\varepsilon_{maz} \sum_{\sigma''\sigma'} - \left(\langle \sigma'' | (\delta_{na} I + S_k \varepsilon_{nak}) | \sigma' \rangle \int G^{<}_{\substack{m\sigma'' \\ m+1,\sigma'}} dE \right)$$
$$- \sum_{\sigma'\sigma''} \left(\langle \sigma'' | (\delta_{na} I + S_k \varepsilon_{ank}) | \sigma' \rangle \int G^{<}_{\substack{m+1,\sigma'' \\ m\sigma'}} dE \right)$$

Shuffling,

$$J_m^{S_nF} = -\sum_{\sigma''\sigma'} - \left(\delta_{na}\varepsilon_{maz} \langle \sigma'' | I | \sigma' \rangle \int G^{<}_{\substack{m\sigma'' \\ m+1,\sigma'}} + G^{<}_{\substack{m+1,\sigma'' \\ m\sigma'}} dE \right)^A$$
$$- \sum_{\sigma'\sigma''} \left(\langle \sigma'' | (S_k \varepsilon_{nak} \varepsilon_{maz}) | \sigma' \rangle \int G^{<}_{\substack{m\sigma'' \\ m+1,\sigma'}} - G^{<}_{\substack{m+1,\sigma'' \\ m\sigma'}} dE \right)^P$$

Note that the forward current consists of two terms indicated by superscripts A (adiabatic) and P (precession). Take note that m is the only spatial index; n, a, k are all spin indices; z is the index of the electric field, which is permanently along the Z axis in 2D linear spin–orbit system.

(b) The Adiabatic spin current

$$(J_m^n)^A = -\sum_{\sigma''\sigma'} - \left(\delta_{na}\varepsilon_{maz} \langle \sigma'' | I | \sigma' \rangle \int G^{<}_{\substack{m\sigma'' \\ m+1,\sigma'}} + G^{<}_{\substack{m+1,\sigma'' \\ m\sigma'}} dE \right)^A$$

Note that in a 2D system, m takes on (x, y) only. $(J_m^{S_nF})^A$ allows two permutations of its indices as follows:

$$(J_x^y)^A = -\sum_{\sigma} - \left(\int G^{<}_{\substack{x\sigma \\ x+1,\sigma}} + G^{<}_{\substack{x+1,\sigma \\ x\sigma}} dE \right)$$
$$(J_y^x)^A = -\sum_{\sigma} \left(\int G^{<}_{\substack{y\sigma \\ y+1,\sigma}} + G^{<}_{\substack{y+1,\sigma \\ y\sigma}} dE \right)$$

The expression for $(J_x^y)^A$ shows that the spin Y current depends on the momentum X. Likewise, $(J_y^x)^A$ shows that the spin X current depends on the momentum Y. This is consistent with the energy landscape of the 2D spin–orbit system that the electron would settle down to when it is moving slowly enough

$$\boldsymbol{\sigma}.(\boldsymbol{p}\times\boldsymbol{e_z})\rightarrow\sigma_x p_y-\sigma_y p_x$$

(c) The Precession spincurrent

$$(J_m^n)^P=-\sum_{\sigma'\sigma''}\left(\langle\sigma''|(S_k\varepsilon_{nak}\varepsilon_{maz})|\sigma'\rangle\int G^<_{\substack{m\sigma''\\ m+1,\sigma'}}-G^<_{\substack{m+1,\sigma''\\ m\sigma'}}dE\right)^P$$

$$\begin{aligned}(J_m^n)^P&=-\sum_{\sigma'\sigma''}\left(\langle\sigma''|(S_k\varepsilon_{nak}\varepsilon_{maz})|\sigma'\rangle\int G^<_{\substack{m\sigma''\\ m+1,\sigma'}}-G^<_{\substack{m+1,\sigma''\\ m\sigma'}}dE\right)\\&=-\sum_{\sigma'\sigma''}\left(\langle\sigma''|(S_k\varepsilon_{kna}\varepsilon_{zma})|\sigma'\rangle\int G^<_{\substack{m\sigma''\\ m+1,\sigma'}}-G^<_{\substack{m+1,\sigma''\\ m\sigma'}}dE\right)\\&=-\sum_{\sigma'\sigma''}\left(\langle\sigma''|S_k(\delta_{kz}\delta_{nm}-\delta_{km}\delta_{nz})|\sigma'\rangle\int G^<_{\substack{m\sigma''\\ m+1,\sigma'}}-G^<_{\substack{m+1,\sigma''\\ m\sigma'}}dE\right)\\&=-\sum_{\sigma'\sigma''}\left(\langle\sigma''|(S_z\delta_{nm}-S_m\delta_{nz})|\sigma'\rangle\int G^<_{\substack{m\sigma''\\ m+1,\sigma'}}-G^<_{\substack{m+1,\sigma''\\ m\sigma'}}dE\right)\end{aligned}\tag{A}$$

Note that in a 2D system, m takes on (x,y) only. $(J_m^n)^P$ allows four permutations of its indices as follows:

$$(J_x^x)^P=-\sum_{\sigma'\sigma''}\left(\langle\sigma''|(S_z)|\sigma'\rangle\int G^<_{\substack{x\sigma''\\ x+1,\sigma'}}-G^<_{\substack{x+1,\sigma''\\ x\sigma'}}dE\right)$$

$$(J_y^y)^P=-\sum_{\sigma'\sigma''}\left(\langle\sigma''|(S_z)|\sigma'\rangle\int G^<_{\substack{y\sigma''\\ y+1,\sigma'}}-G^<_{\substack{y+1,\sigma''\\ y\sigma'}}dE\right)$$

$$(J_x^z)^P=-\sum_{\sigma'\sigma''}\left(\langle\sigma''|(-S_x)|\sigma'\rangle\int G^<_{\substack{x\sigma''\\ x+1,\sigma'}}-G^<_{\substack{x+1,\sigma''\\ x\sigma'}}dE\right)$$

$$(J_y^z)^P=-\sum_{\sigma'\sigma''}\left(\langle\sigma''|(-S_y)|\sigma'\rangle\int G^<_{\substack{y\sigma''\\ y+1,\sigma'}}-G^<_{\substack{y+1,\sigma''\\ y\sigma'}}dE\right)$$

Let's check with the semi-classical physics of spin current. For simplicity, we drop all constants and focus on the mathematical structure

$$\frac{d\boldsymbol{S}}{dt}=\boldsymbol{S}\times(\boldsymbol{p}\times\boldsymbol{e_z})\rightarrow\frac{dS_n}{dt}=S_k(\boldsymbol{p}\times\boldsymbol{e_z})_b\varepsilon_{kbn}$$

It follows that

$$\frac{dS_n}{dt} = S_k(p_m|e_z|\varepsilon_{mzb})\varepsilon_{kbn}$$
$$= S_k(p_m|e_z|\varepsilon_{mzb})\varepsilon_{nkb}$$
$$= S_k p_m(\delta_{mn}\delta_{zk} - \delta_{mk}\delta_{nz})$$
$$= S_z p_m \delta_{mn} - S_m p_m \delta_{nz} \qquad \text{(B)}$$

It's worth noting that Equations (A) and (B) are equivalent. We would now provide an explicit relationship in Eqs. (C), (D), and (E):

The spin-X current is

$$\frac{dS_x}{dt} = S_z p_m \delta_{mx} - S_m p_m \delta_{xz} = S_z p_x$$
$$\rightarrow (J_x^x)^P = -\sum_{\sigma'\sigma''}\left(\langle\sigma''|(S_z)|\sigma'\rangle \int G^{<}_{\substack{x\sigma'' \\ x+1,\sigma'}} - G^{<}_{\substack{x+1,\sigma'' \\ x\sigma'}} dE\right) \qquad \text{(C)}$$

The spin-Y current is

$$\frac{dS_y}{dt} = S_z p_m \delta_{my} - S_m p_m \delta_{yz} = S_z p_y$$
$$\rightarrow (J_y^y)^P = -\sum_{\sigma'\sigma''}\left(\langle\sigma''|(S_z)|\sigma'\rangle \int G^{<}_{\substack{y\sigma'' \\ y+1,\sigma'}} - G^{<}_{\substack{y+1,\sigma'' \\ y\sigma'}} dE\right) \qquad \text{(D)}$$

The spin-Z current is

$$\frac{dS_z}{dt} = S_z p_m \delta_{mz} - S_m p_m \delta_{zz} = -S_x p_x - S_y P_y$$
$$\rightarrow (J_x^z)^P + (J_y^z)^P = -\sum_{\sigma'\sigma''}\left(\langle\sigma''|(-S_x)|\sigma'\rangle \int G^{<}_{\substack{x\sigma'' \\ x+1,\sigma'}} - G^{<}_{\substack{x+1,\sigma'' \\ x\sigma'}} dE\right)$$
$$-\sum_{\sigma'\sigma''}\left(\langle\sigma''|(-S_y)|\sigma'\rangle \int G^{<}_{\substack{y\sigma'' \\ y+1,\sigma'}} - G^{<}_{\substack{y+1,\sigma'' \\ y\sigma'}} dE\right)$$
$$\text{(E)}$$

Chapter 5

Gauge and Topology

This chapter is about gauge and topology with modern emphasis on applications in new materials like graphene and topological systems. Coordinate transformation is introduced to familiarize readers with the frame of references and their physical consequences. Metric of space is implied but not explicitly dealt with. The Berry–Pancharatnam phase is introduced in applications like graphene and semiconductor systems. Towards the end, the topological nature of phase and curvature are introduced with special attention paid to numerous models in condensed matter systems, e.g., the Su–Schrieffer–Heeger (SSH) model and polarization in the 2D Chern insulator. This is a specialized topic suitable for research purposes. Problems here could also be used for postgraduate quantum mechanics.

Coordinate Transformation

Problem 5.01 Coordinate Change and Basis Vectors
Problem 5.02 Coordinate Change and Component Vectors
Problem 5.03 Pauli Matrices and Coordinate Change

Berry-Pancharatnam Gauge

Problem 5.04 Frame Rotation Gauge Field
Problem 5.05 Path Integral and Gauge
Problem 5.06 Magnetic Monopole
Problem 5.07 Berry Pancharatnam Phase in 2D systems
Problem 5.08 Graphene Gauge Field
Problem 5.09 Graphene Berry Pancharatnam Curvature
Problem 5.10 Spin Hall Gauge in a 2D System 1
Problem 5.11 Spin Hall Gauge in a 2D System 11
Problem 5.12 Spin Hall Gauge: Integral Calculus

Non-Abelian Gauge

Problem 5.13 Non-Abelian Gauge Transformation 1
Problem 5.14 Non-Abelian Gauge Transformation 11
Problem 5.15 Non-Abelian Gauge Curvature

Topology and Gauge

Problem 5.16 Topological Models
Problem 5.17 Exact Solution to the Su-Schrieffer-Heeger (SSH) Model
Problem 5.18 Topological Characterization of the SSH model
Problem 5.19 Topological Polarization of the 2D Chern Insulator
Problem 5.20 Berry Curvature of the Dirac Model

Appendix

Appendix 5A. Spherical Coordinates in Different Basis

Coordinate Transformation

Problem 5.01 Coordinate Change and Basis Vectors

One of the most popular coordinate systems besides the Cartesian coordinates is the spherical coordinate system. The spherical coordinates can be expressed on different bases. Note that in the spherical system, the basis ($\boldsymbol{e}_\theta\ \boldsymbol{e}_\phi\ \boldsymbol{e}_R$) that has the dimension of length is known as the holonomic basis. Conversely, the non-holonomic basis ($\boldsymbol{e}'_\theta\ \boldsymbol{e}'_\phi\ \boldsymbol{e}'_R$) would have no dimension. Both bases describe the same length element $\boldsymbol{dl}$. Show that the basis vectors in the spherical basis are related to the Cartesian basis as follows:

$$\begin{pmatrix} \boldsymbol{e}_\theta \\ \boldsymbol{e}_\phi \\ \boldsymbol{e}_R \end{pmatrix} = \begin{pmatrix} R\cos\theta\cos\phi & R\cos\theta\sin\phi & -R\sin\theta \\ -R\sin\theta\sin\phi & R\sin\theta\cos\phi & \boldsymbol{0} \\ \sin\theta\cos\phi & \sin\theta\sin\phi & \cos\theta \end{pmatrix} \begin{pmatrix} \boldsymbol{e}_x \\ \boldsymbol{e}_y \\ \boldsymbol{e}_z \end{pmatrix}$$

Hence, show that

$$\begin{pmatrix} \boldsymbol{e}_x \\ \boldsymbol{e}_y \\ \boldsymbol{e}_z \end{pmatrix} = \begin{pmatrix} \frac{\cos\phi\cos\theta}{R} & \frac{\sin\phi}{-R\sin\theta} & \sin\theta\cos\phi \\ \frac{\sin\phi\cos\theta}{R} & \frac{\cos\phi}{R\sin\theta} & \sin\theta\sin\phi \\ \frac{-\sin\theta}{R} & 0 & \cos\theta \end{pmatrix} \begin{pmatrix} \boldsymbol{e}_\theta \\ \boldsymbol{e}_\phi \\ \boldsymbol{e}_R \end{pmatrix}$$

Solution

Note that $x = R\sin\theta\cos\phi, y = R\sin\theta\sin\phi, z = R\cos\theta$ and

$$\boldsymbol{R} = x\boldsymbol{e_x} + y\boldsymbol{e_y} + z\boldsymbol{e_z}$$

Therefore,

$$\begin{aligned}
\boldsymbol{e_\theta} &= \frac{\partial \boldsymbol{R}}{\partial\theta} = \frac{\partial x}{\partial\theta}\boldsymbol{e_x} + \frac{\partial y}{\partial\theta}\boldsymbol{e_y} + \frac{\partial z}{\partial\theta}\boldsymbol{e_z} \\
&= R\cos\theta\cos\phi\boldsymbol{e_x} + R\cos\theta\sin\phi\boldsymbol{e_y} - R\sin\theta\boldsymbol{e_z} \\
\boldsymbol{e_\phi} &= \frac{\partial \boldsymbol{R}}{\partial\phi} = \frac{\partial x}{\partial\phi}\boldsymbol{e_x} + \frac{\partial y}{\partial\phi}\boldsymbol{e_y} + \frac{\partial z}{\partial\phi}\boldsymbol{e_z} = -R\sin\theta\sin\phi\boldsymbol{e_x} + R\sin\theta\cos\phi\boldsymbol{e_y} \\
\boldsymbol{e_R} &= \frac{\partial \boldsymbol{R}}{\partial R} = \frac{\partial x}{\partial R}\boldsymbol{e_x} + \frac{\partial y}{\partial R}\boldsymbol{e_y} + \frac{\partial z}{\partial R}\boldsymbol{e_z} \\
&= \sin\theta\cos\phi\boldsymbol{e_x} + \sin\theta\sin\phi\boldsymbol{e_y} + \cos\theta\,\boldsymbol{e_z}
\end{aligned}$$

In summary,

$$\begin{aligned}
\begin{pmatrix} \boldsymbol{e_\theta} \\ \boldsymbol{e_\phi} \\ \boldsymbol{e_R} \end{pmatrix} &= \begin{pmatrix} \frac{\partial x}{\partial\theta} & \frac{\partial y}{\partial\theta} & \frac{\partial z}{\partial\theta} \\ \frac{\partial x}{\partial\phi} & \frac{\partial y}{\partial\phi} & \frac{\partial z}{\partial\phi} \\ \frac{\partial x}{\partial R} & \frac{\partial y}{\partial R} & \frac{\partial z}{\partial R} \end{pmatrix} \begin{pmatrix} \boldsymbol{e_x} \\ \boldsymbol{e_y} \\ \boldsymbol{e_z} \end{pmatrix} \\
&= \begin{pmatrix} R\cos\theta\cos\phi & R\cos\theta\sin\phi & -R\sin\theta \\ -R\sin\theta\sin\phi & R\sin\theta\cos\phi & \boldsymbol{0} \\ \sin\theta\cos\phi & \sin\theta\sin\phi & \cos\theta \end{pmatrix} \begin{pmatrix} \boldsymbol{e_x} \\ \boldsymbol{e_y} \\ \boldsymbol{e_z} \end{pmatrix} \qquad \text{(A)}
\end{aligned}$$

The holonomic basis has its origin in

$$d\boldsymbol{l} = \frac{\partial \boldsymbol{l}}{\partial\theta}d\theta + \frac{\partial \boldsymbol{l}}{\partial\phi}d\phi + \frac{\partial \boldsymbol{l}}{\partial r}dr \quad \to \quad d\boldsymbol{l} = \boldsymbol{e_\theta}\,d\theta + \boldsymbol{e_\phi}\,d\phi + \boldsymbol{e_R}\,dr$$

By contrast, the non-holonomic basis has its origin in

$$d\boldsymbol{l} = \frac{\partial \boldsymbol{l}}{\partial l_\theta}dl_\theta + \frac{\partial \boldsymbol{l}}{\partial l_\phi}dl_\phi + \frac{\partial \boldsymbol{l}}{\partial l_r}dl_r \quad \to \quad d\boldsymbol{l} = \boldsymbol{e'_\theta}\,dl_\theta + \boldsymbol{e'_\phi}\,dl_\phi + \boldsymbol{e'_R}\,dl_r$$

Therefore, Eq. (A) has its non-holonomic counterpart in

$$\begin{pmatrix} \boldsymbol{e'_\theta} \\ \boldsymbol{e'_\phi} \\ \boldsymbol{e'_R} \end{pmatrix} = \begin{pmatrix} \cos\theta\cos\phi & \cos\theta\sin\phi & -\sin\theta \\ -\sin\phi & \cos\phi & \boldsymbol{0} \\ \sin\theta\cos\phi & \sin\theta\sin\phi & \cos\theta \end{pmatrix} \begin{pmatrix} \boldsymbol{e_x} \\ \boldsymbol{e_y} \\ \boldsymbol{e_z} \end{pmatrix} \qquad \text{(B)}$$

It's worth noting that

$$\boldsymbol{e_R} = \boldsymbol{e'_R}$$

For more details on the holonomic (non-holonomic) basis, refer to Table 1. Now, one can take the inverse matrix or find M

$$\begin{pmatrix} \boldsymbol{e_x} \\ \boldsymbol{e_y} \\ \boldsymbol{e_z} \end{pmatrix} = M \begin{pmatrix} \boldsymbol{e_\theta} \\ \boldsymbol{e_\phi} \\ \boldsymbol{e_R} \end{pmatrix} = \begin{pmatrix} \frac{\partial\theta}{\partial x} & \frac{\partial\phi}{\partial x} & \frac{\partial R}{\partial x} \\ \frac{\partial\theta}{\partial y} & \frac{\partial\phi}{\partial y} & \frac{\partial R}{\partial y} \\ \frac{\partial\theta}{\partial z} & \frac{\partial\phi}{\partial z} & \frac{\partial R}{\partial z} \end{pmatrix} \begin{pmatrix} \boldsymbol{e_\theta} \\ \boldsymbol{e_\phi} \\ \boldsymbol{e_R} \end{pmatrix}$$

Referring to $x = R\sin\theta\cos\phi, y = R\sin\theta\sin\phi, z = R\cos\theta$, one can express (θ, ϕ, R) in terms of (x, y, z), e.g., $\tan\phi = \frac{y}{x}$ and $\left(\frac{\partial\phi}{\partial x}, \frac{\partial\phi}{\partial y}, \frac{\partial\phi}{\partial z}\right)$ can be found. Likewise, $R^2 = x^2 + y^2 + z^2$, and $\left(\frac{\partial R}{\partial x}, \frac{\partial R}{\partial y}, \frac{\partial R}{\partial z}\right)$ can be found. As for $\left(\frac{\partial\theta}{\partial x}, \frac{\partial\theta}{\partial y}, \frac{\partial\theta}{\partial z}\right)$, the relation $\tan^2\theta = \left(\frac{x}{z}\right)^2 + \left(\frac{y}{z}\right)^2$ is used.

With $\tan^2\theta = \left(\frac{x}{z}\right)^2 + \left(\frac{y}{z}\right)^2$, one has $2\tan\theta\sec^2\theta\left(\frac{\partial\theta}{\partial x}\right) = \frac{2x}{z^2}$, which leads to

$$\left(\frac{\partial\theta}{\partial x}\right) = \frac{\cos\theta\cos\phi}{R}, \left(\frac{\partial\theta}{\partial y}\right) = \frac{\cos\theta\sin\phi}{R},$$

$$\left(\frac{\partial\theta}{\partial z}\right) = -\frac{\cos^3\theta}{\sin\theta}\left(\frac{x^2+y^2}{z^3}\right) = -\frac{\sin\theta}{R}$$

With $\tan\phi = \frac{y}{x}$, one has $\sec^2\phi\left(\frac{\partial\phi}{\partial x}\right) = -\frac{y}{x^2}$, which leads to

$$\left(\frac{\partial\phi}{\partial x}\right) = -\frac{y}{x^2}\cos^2\phi = \frac{\sin\phi}{-R\sin\theta}, \quad \left(\frac{\partial\phi}{\partial y}\right) = \frac{\cos\phi}{R\sin\theta}, \quad \left(\frac{\partial\phi}{\partial z}\right) = 0$$

With $R^2 = x^2 + y^2 + z^2$, one has $\left(\frac{\partial R}{\partial x}\right) = \frac{x}{R}$, which leads to

$$\frac{\partial R}{\partial x} = \frac{x}{R} = \sin\theta\cos\phi, \quad \frac{\partial R}{\partial y} = \frac{y}{R} = \sin\theta\sin\phi, \quad \frac{\partial R}{\partial x} = \frac{z}{R} = \cos\theta$$

Therefore,

$$\begin{pmatrix} \frac{\partial\theta}{\partial x} & \frac{\partial\phi}{\partial x} & \frac{\partial R}{\partial x} \\ \frac{\partial\theta}{\partial y} & \frac{\partial\phi}{\partial y} & \frac{\partial R}{\partial y} \\ \frac{\partial\theta}{\partial z} & \frac{\partial\phi}{\partial z} & \frac{\partial R}{\partial z} \end{pmatrix} = \begin{pmatrix} \frac{\cos\phi\cos\theta}{R} & \frac{\sin\phi}{-R\sin\theta} & \sin\theta\cos\phi \\ \frac{\sin\phi\cos\theta}{R} & \frac{\cos\phi}{R\sin\theta} & \sin\theta\sin\phi \\ \frac{-\sin\theta}{R} & 0 & \cos\theta \end{pmatrix}$$

The above leads to

$$\begin{pmatrix} e_x \\ e_y \\ e_z \end{pmatrix} = \begin{pmatrix} \dfrac{\cos\phi\cos\theta}{R} & \dfrac{\sin\phi}{-R\sin\theta} & \sin\theta\cos\phi \\ \dfrac{\sin\phi\cos\theta}{R} & \dfrac{\cos\phi}{R\sin\theta} & \sin\theta\sin\phi \\ \dfrac{-\sin\theta}{R} & 0 & \cos\theta \end{pmatrix} \begin{pmatrix} e_\theta \\ e_\phi \\ e_R \end{pmatrix}$$

Remarks and Reflections

In 2D, we use $\boldsymbol{r}$ as the 2D version of R, where

$$\begin{pmatrix} e_\phi \\ e_r \end{pmatrix} = \begin{pmatrix} -r\sin\phi & r\cos\phi \\ \cos\phi & \sin\phi \end{pmatrix} \begin{pmatrix} e_x \\ e_y \end{pmatrix}$$

Inversely,

$$\begin{pmatrix} e_x \\ e_y \end{pmatrix} = -\frac{1}{r} \begin{pmatrix} \sin\phi & -r\cos\phi \\ -cos\phi & -r\,sin\phi \end{pmatrix} \begin{pmatrix} e_\phi \\ e_r \end{pmatrix}$$

In Cartesian coordinates,

$$\boldsymbol{r} = x\boldsymbol{e_x} + y\boldsymbol{e_y}$$

In spherical coordinates,

$$\begin{aligned} \boldsymbol{r} &= r\cos\phi\,\boldsymbol{e_x} + r\sin\phi\,\boldsymbol{e_y} \\ &= r\cos\phi\left(-\frac{1}{r}\sin\phi\,\boldsymbol{e_\phi} + \cos\phi\,\boldsymbol{e_r}\right) + r\sin\phi\left(\frac{1}{r}\cos\phi\boldsymbol{e_\phi} + \sin\phi\,\boldsymbol{e_r}\right) \\ &= r\,\boldsymbol{e_r} \end{aligned}$$

Note that $\boldsymbol{r} = r\boldsymbol{e_r}\ (\phi)$, as one can see $\boldsymbol{r}$ as

$$\boldsymbol{r} = r\left(\cos\phi\,\boldsymbol{e_x} + \sin\phi\,\boldsymbol{e_y}\right)$$

Problem 5.02 Coordinate Change and Component Vectors

In a general expression, $\boldsymbol{R} = R^a\boldsymbol{e_a}$. In Cartesian coordinates,

$$\boldsymbol{R} = R^x\boldsymbol{e_x} + R^y\boldsymbol{e_y} + R^z\boldsymbol{e_z} = x\boldsymbol{e_x} + y\boldsymbol{e_y} + z\boldsymbol{e_z}$$

Reverting the general expression of $R^a\boldsymbol{e_a} = R^a\,\Lambda_a^c\boldsymbol{e_c} = \left(R^b\Lambda_b^a\right)\left(\Lambda_a^c\boldsymbol{e_c}\right)$, where b, c run over θ, ϕ, R, one has $R^a\boldsymbol{e_a} = R^b\boldsymbol{e_c}\delta_b^c = R^b\boldsymbol{e_b}$. Therefore, in spherical coordinates, one writes

$$\boldsymbol{R} = R^R \boldsymbol{e_R} + R^\theta \boldsymbol{e_\theta} + R^\phi \boldsymbol{e_\phi}$$

Find the explicit expressions of R^R, R^θ, R^ϕ.

Solution

With the above, one can establish that $\boldsymbol{e_a} = (\Lambda_a^c \boldsymbol{e_c})$ and $R^a = (R^b \Lambda_b^a)$, which lead to

$$\begin{pmatrix} \boldsymbol{e_x} \\ \boldsymbol{e_y} \\ \boldsymbol{e_z} \end{pmatrix} = \begin{pmatrix} \dfrac{\partial\theta}{\partial x} & \dfrac{\partial\phi}{\partial x} & \dfrac{\partial R}{\partial x} \\ \dfrac{\partial\theta}{\partial y} & \dfrac{\partial\phi}{\partial y} & \dfrac{\partial R}{\partial y} \\ \dfrac{\partial\theta}{\partial z} & \dfrac{\partial\phi}{\partial z} & \dfrac{\partial R}{\partial z} \end{pmatrix} \begin{pmatrix} \boldsymbol{e_\theta} \\ \boldsymbol{e_\phi} \\ \boldsymbol{e_R} \end{pmatrix}, \begin{pmatrix} R^x \\ R^y \\ R^z \end{pmatrix} = \begin{pmatrix} \dfrac{\partial x}{\partial\theta} & \dfrac{\partial x}{\partial\phi} & \dfrac{\partial x}{\partial R} \\ \dfrac{\partial y}{\partial\theta} & \dfrac{\partial y}{\partial\phi} & \dfrac{\partial y}{\partial R} \\ \dfrac{\partial z}{\partial\theta} & \dfrac{\partial z}{\partial\phi} & \dfrac{\partial z}{\partial R} \end{pmatrix} \begin{pmatrix} R^\theta \\ R^\phi \\ R^R \end{pmatrix}$$

As $R^x = x = R\sin\theta\cos\phi, R^y = y = R\sin\theta\sin\phi, R^z = z = R\cos\theta$ one has

$$\begin{pmatrix} R^x \\ R^y \\ R^z \end{pmatrix} = \begin{pmatrix} R\cos\theta\cos\phi & -R\sin\theta\sin\phi & \sin\theta\cos\phi \\ R\cos\theta\sin\phi & R\sin\theta\cos\phi & \sin\theta\sin\phi \\ -R\sin\theta & 0 & \cos\theta \end{pmatrix} \begin{pmatrix} R^\theta \\ R^\phi \\ R^R \end{pmatrix}$$

To solve for R^R, R^θ, R^ϕ one needs to find a useful pair out of the equations above. Referring to R^x, R^y, a useful pair is found as follows:

$$R\cos\theta\cos\phi\, R^\theta - R\sin\theta\sin\phi\, R^\phi + \sin\theta\cos\phi\, R^R = R\sin\theta\cos\phi \qquad \text{(A)}$$

$$R\cos\theta\sin\phi\, R^\theta + R\sin\theta\cos\phi\, R^\phi + \sin\theta\sin\phi\, R^R = R\sin\theta\sin\phi \qquad \text{(B)}$$

Multiply (A) with $\sin\phi$ and (B) with $\cos\phi$ and perform (B)–(A)

$$R\sin\theta\cos^2\phi\, R^\phi + R\sin\theta\sin^2\phi\, R^\phi = 0$$

$$\rightarrow R^\phi = 0$$

Referring to R^z and the result of $R^\phi = 0$ for Eq. (A), one forms a second useful pair as follows:

$$-R\sin\theta R^\theta + \cos\theta R^R = R\cos\theta$$

$$R\cos\theta\, R^\theta + \sin\theta\, R^R = R\sin\theta$$

Using a method similar to that in the first useful pair, one finds

$$R^\theta = 0, \; R^R = R$$

Therefore,

$$\boldsymbol{R} = R^R \boldsymbol{e_R} + R^\theta \boldsymbol{e_\theta} + R^\phi \boldsymbol{e_\phi} = R\boldsymbol{e_R}(\theta, \phi)$$

where $\boldsymbol{e_R}(\theta,\phi) = \sin\theta\cos\phi\boldsymbol{e_x} + \sin\theta\sin\phi\boldsymbol{e_y} + \cos\theta\,\boldsymbol{e_z}$ was derived earlier.

Remarks and Reflections

Note in the last problem and this one that

$$\begin{pmatrix} \frac{\partial\theta}{\partial x} & \frac{\partial\phi}{\partial x} & \frac{\partial R}{\partial x} \\ \frac{\partial\theta}{\partial y} & \frac{\partial\phi}{\partial y} & \frac{\partial R}{\partial y} \\ \frac{\partial\theta}{\partial z} & \frac{\partial\phi}{\partial z} & \frac{\partial R}{\partial z} \end{pmatrix} = \begin{pmatrix} \frac{\cos\phi\cos\theta}{R} & \frac{\sin\phi}{-R\sin\theta} & \sin\theta\cos\phi \\ \frac{\sin\phi\cos\theta}{R} & \frac{\cos\phi}{R\sin\theta} & \sin\theta\sin\phi \\ \frac{-\sin\theta}{R} & 0 & \cos\theta \end{pmatrix}$$

$$\begin{pmatrix} \frac{\partial x}{\partial\theta} & \frac{\partial x}{\partial\phi} & \frac{\partial x}{\partial R} \\ \frac{\partial y}{\partial\theta} & \frac{\partial y}{\partial\phi} & \frac{\partial y}{\partial R} \\ \frac{\partial z}{\partial\theta} & \frac{\partial z}{\partial\phi} & \frac{\partial z}{\partial R} \end{pmatrix} = \begin{pmatrix} R\cos\theta\cos\phi & -R\sin\theta\sin\phi & \sin\theta\cos\phi \\ R\cos\theta\sin\phi & R\sin\theta\cos\phi & \sin\theta\sin\phi \\ -R\sin\theta & 0 & \cos\theta \end{pmatrix}$$

Observe that

$$\begin{aligned} &\frac{\partial x}{\partial\theta}\frac{\partial\theta}{\partial x} + \frac{\partial x}{\partial\phi}\frac{\partial\phi}{\partial x} + \frac{\partial x}{\partial R}\frac{\partial R}{\partial x} \\ &\quad = R\cos\theta\cos\phi\left(\frac{\cos\phi\cos\theta}{R}\right) + R\sin\theta\sin\phi\left(\frac{\sin\phi}{R\sin\theta}\right) \\ &\qquad + \sin\theta\cos\phi\sin\theta\cos\phi = 1 \\ &\frac{\partial\theta}{\partial x}\frac{\partial x}{\partial\theta} + \frac{\partial\theta}{\partial y}\frac{\partial y}{\partial\theta} + \frac{\partial\theta}{\partial z}\frac{\partial z}{\partial\theta} \\ &\quad = \left(\frac{\cos\phi\cos\theta}{R}\right) R\cos\theta\cos\phi \\ &\qquad + \left(\frac{\sin\phi\cos\theta}{R}\right) R\cos\theta\sin\phi + \frac{-\sin\theta}{R} - R\sin\theta = 1 \end{aligned}$$

Problem 5.03 Pauli Matrices and Coordinate Change

Coordinate transformation allows us to study physical quantities, e.g., gauge or its curvature on spherical, cylindrical, or other 2D curved surfaces. On the other hand, gauge fields appear oftentimes in the forms of

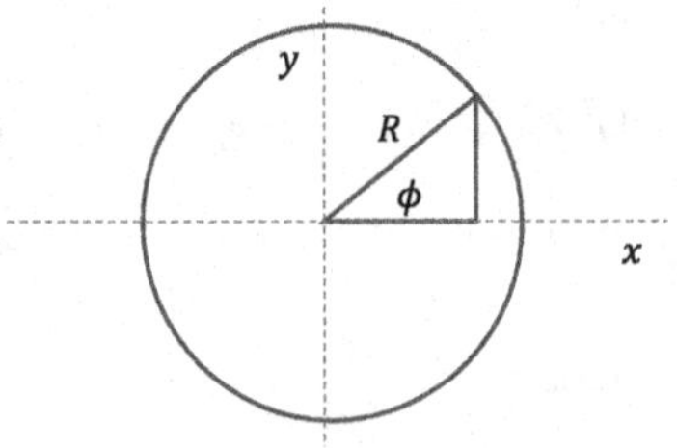

Fig. 1. A polar coordinate system, in which the Pauli matrices are defined and parameters ϕ, R are illustrated.

$SU(2)$, $U(1) \otimes U(1)$, or $SU(n)$. And it's also worth noting that 2D surfaces could be curved surfaces. It is, therefore, important to express the Pauli matrices in terms of surface metrics. Consider a simple 2D system in polar coordinates (see Figure 1). The Pauli matrix is a vector independent of coordinates and can be expressed as a set of orthogonal vectors

$$\boldsymbol{\sigma} = \sigma^{\phi}\, \boldsymbol{e_\phi} + \sigma^{R}\, \boldsymbol{e_R}$$

The covariant Pauli matrix oriented along ϕ is given by

$$\boldsymbol{\sigma} \cdot \boldsymbol{e_\phi} = \sigma_\phi = (\sigma^x\, \boldsymbol{e_x} + \sigma^y\, \boldsymbol{e_y}) \cdot \boldsymbol{e_\phi} \tag{A}$$

The contravariant counterpart Pauli matrix is given by

$$\boldsymbol{\sigma} \cdot \boldsymbol{e^\phi} = \sigma^\phi = (\sigma^x\, \boldsymbol{e_x} + \sigma^y\, \boldsymbol{e_y}) . \boldsymbol{e^\phi} \tag{B}$$

Alternatively, under coordinate transformations, one can write

$$\sigma_\phi = \partial^x_\phi \sigma^x + \partial^y_\phi \sigma^y \tag{C}$$

$$\sigma^\phi = \partial^\phi_x \sigma^x + \partial^\phi_y \sigma^y \tag{D}$$

Derive and show explicitly that the two following expressions are equivalent:

$$\sigma^\phi = \partial^\phi_x \sigma^x + \partial^\phi_y \sigma^y \text{ and } \sigma^\phi = g^{\phi\phi}\sigma_\phi + g^{\phi R}\sigma_R$$

Note: The metrics of the systems are $g_{\phi\phi} = \boldsymbol{e_\phi} \cdot \boldsymbol{e_\phi}$ *and* $g_{\phi R} = \boldsymbol{e_\phi} \cdot \boldsymbol{e_R}$. *Contravariant quantities are denoted by a raised index. Covariant quantities are denoted by a lower index. Metrics are used to raise or lower indices.*

Solution

Referring to Figure 1,

$$R\cos\phi = x, \quad R\sin\phi = y \rightarrow \frac{y}{x} = \tan\phi$$

What follows is

$$\frac{\partial\phi}{\partial y} = \frac{1}{x}\cos^2\phi = \frac{\cos\phi}{R}, \quad \frac{\partial\phi}{\partial x} = \frac{-y}{x^2}\cos^2\phi = -\frac{\sin\phi}{R} \tag{a}$$

and

$$\boldsymbol{R} = R\cos\phi\,\boldsymbol{e_x} + R\sin\phi\,\boldsymbol{e_y} \quad \rightarrow \ \boldsymbol{e_\phi} = \frac{\partial \boldsymbol{R}}{\partial\phi}$$

$$= -R\sin\phi\boldsymbol{e_x} + R\cos\phi\boldsymbol{e_y} \tag{b}$$

Substitute (b) into (A),

$$\boldsymbol{\sigma}\cdot\boldsymbol{e_\phi} \rightarrow \sigma_\phi = -R\sin\phi\,\sigma^x + R\cos\phi\,\sigma^y$$

Substitute (a) into (D) and $\sigma^\phi = \partial_x^\phi\sigma^x + \partial_y^\phi\sigma^y$ would lead to

$$\sigma^\phi = -\frac{\sin\phi}{R}\sigma^x + \frac{\cos\phi}{R}\sigma^y$$

Now, we turn to the metric of the system,

$$g^{\phi\phi} = \frac{1}{g_{\phi\phi}} = \frac{1}{\boldsymbol{e_\phi}.\boldsymbol{e_\phi}} = \frac{1}{R^2}; \quad g^{\phi R} = 0$$

Therefore, $\sigma^\phi = g^{\phi\phi}\sigma_\phi + g^{\phi R}\sigma_R$ would lead to

$$\sigma^\phi = \frac{1}{R^2}\sigma_\phi = \frac{1}{R^2}\left(-R\sin\phi\,\sigma^X + R\cos\phi\,\sigma^y\right)$$

$$= -\frac{\sin\phi}{R}\sigma^X + \frac{\cos\phi}{R}\sigma^y$$

Remarks and Reflections

Note that

$$(\sigma^x\,\boldsymbol{e_x} + \sigma^y\,\boldsymbol{e_y}) = \left(\sigma^\phi\,\boldsymbol{e_\phi} + \sigma^R\,\boldsymbol{e_R}\right)$$

It follows that

$$\sigma_\phi = (\sigma^x\,\boldsymbol{e_x} + \sigma^y\,\boldsymbol{e_y})\,.\boldsymbol{e_\phi} = \left(\sigma^\phi\,\boldsymbol{e_\phi} + \sigma^R\,\boldsymbol{e_R}\right)\cdot\boldsymbol{e_\phi}$$

$$= \sigma^\phi g_{\phi\phi} + \sigma^R g_{\phi R}$$

Berry-Pancharatnam Gauge

Problem 5.04 Frame Rotation Gauge Field

In modern spinor systems like graphene, spintronics, and topological surfaces, it is common to apply a rotational transformation to study the Berry–Pancharatnam phase of the systems. One simple example is to rotate the frame about an axis in the X–Y plane. Such a frame rotation would be mathematically described by

$$H' = U^\dagger H U$$

The gauge field is produced in the form of $U^\dagger(\partial_k U)$. Matrix U that rotates the spin is given as follows:

$$U = \begin{pmatrix} \cos\frac{\theta}{2} & \sin\frac{\theta}{2}e^{-i\phi} \\ -\sin\frac{\theta}{2}e^{i\phi} & \cos\frac{\theta}{2} \end{pmatrix}$$

Show that the matrix above would lead to the gauge fields of

$$\pm\frac{1}{2}(1-\cos\theta)\, i\frac{\partial\phi}{\partial \boldsymbol{k}}$$

Solution

$$U^\dagger(\partial_k U) = \begin{pmatrix} \cos\frac{\theta}{2} & -\sin\frac{\theta}{2}e^{-i\phi} \\ \sin\frac{\theta}{2}e^{i\phi} & \cos\frac{\theta}{2} \end{pmatrix} \partial_k \begin{pmatrix} \cos\frac{\theta}{2} & \sin\frac{\theta}{2}e^{-i\phi} \\ -\sin\frac{\theta}{2}e^{i\phi} & \cos\frac{\theta}{2} \end{pmatrix}$$

$$= \begin{pmatrix} \cos\frac{\theta}{2} & -\sin\frac{\theta}{2}e^{-i\phi} \\ \sin\frac{\theta}{2}e^{i\phi} & \cos\frac{\theta}{2} \end{pmatrix}$$

$$\times \begin{pmatrix} -\frac{1}{2}\sin\frac{\theta}{2}\partial_k\theta & \frac{1}{2}\cos\frac{\theta}{2}\partial_k\theta\, e^{-i\phi} - i\partial_k\phi\, e^{-i\phi}\sin\frac{\theta}{2} \\ -\frac{1}{2}\cos\frac{\theta}{2}\partial_k\theta\, e^{i\phi} - i\partial_k\phi\, e^{i\phi}\sin\frac{\theta}{2} & -\frac{1}{2}\sin\frac{\theta}{2}\partial_k\theta \end{pmatrix}$$

$$= \begin{pmatrix} +i\sin^2\frac{\theta}{2}\partial_k\phi & \frac{1}{2}\partial_k\theta\, e^{-i\phi} - i\sin\frac{\theta}{2}\cos\frac{\theta}{2}\partial_k\phi\, e^{-i\phi} \\ -\frac{1}{2}\partial_k\theta e^{i\phi} - i\sin\frac{\theta}{2}\cos\frac{\theta}{2}\partial_k\phi\, e^{i\phi} & -i\sin^2\frac{\theta}{2}\partial_k\phi \end{pmatrix}$$

Finally,

$$U^{\dagger}\left(\partial_{k}U\right)=\begin{pmatrix} +i\dfrac{(1-\cos\theta)}{2}\partial_{k}\phi & \dfrac{1}{2}\partial_{k}\theta\, e^{-i\phi}-\dfrac{i}{2}\sin\theta\partial_{k}\phi\, e^{-i\phi} \\ -\dfrac{1}{2}\partial_{k}\theta e^{i\phi}-\dfrac{i}{2}\sin\theta\partial_{k}\phi\, e^{i\phi} & -i\dfrac{(1-\cos\theta)}{2}\partial_{k}\phi \end{pmatrix}$$

But the above merely rotates the frame such that the Z axis coincides with the spin, which would only produce a pure gauge. To obtain the gauge field, an adiabatic theory is needed, in which the Hamiltonian (effective B field) evolves slowly and the spin tracks the field and is constantly aligned with it. In this way, only the diagonal components need to be considered as shown in the following:

$$U^{\dagger}\left(\partial_{k}U\right)=\begin{pmatrix} +i\dfrac{(1-\cos\theta)}{2}\partial_{k}\phi & 0 \\ 0 & -i\dfrac{(1-\cos\theta)}{2}\partial_{k}\phi \end{pmatrix}$$

The gauge fields are diagonal terms of $U^{\dagger}\left(\partial_{k}U\right)$:

$$\pm\frac{1}{2}\left(1-\cos\theta\right)i\frac{\partial\phi}{\partial\boldsymbol{k}}$$

These are the gauge fields that would lead to their non-trivial curvatures.

Problem 5.05 Path Integral and Gauge Field

The evolution of an eigenstate from initial to final state can be described by a path integral of the spatial propagators where $G\left(x_{n+1}t_{n+1},x_{0}t_{0}\right)$ is the propagator between times t_0 and t.

$$\psi\left(x_{n+1},t_{n+1}\right)=\int G\left(x_{n+1}t_{n+1},x_{0}t_{0}\right)\psi\left(x_{0},t_{0}\right)dx_{0}$$

Show that the wavefunction is characterized by the action of the system as shown in the following:

$$\psi\left(x_{n+1},t_{n+1}\right)=\int\left[\left(\frac{m}{2\pi i\hbar\Delta t}\right)^{\frac{n+1}{2}}e^{\frac{iS(t)}{\hbar}}dx_{n}\cdots dx_{1}\right]\psi\left(x_{0}t_{0}\right)dx_{0}$$

Show that if the states of the system are characterized by spinors which evolve with the changing b fields, a geometric phase aka the Berry–Pancharatnam arises in the action of the system.

Solution

The propagator in explicit spatial terms is written as follows:

$$G\left(x_{n+1}t_{n+1}, x_0t_0\right)$$
$$= \int \left\langle x_{n+1}\left|U\left(t_{n+1}t_n\right)\right|x_n\right\rangle\left\langle x_n\left|U\left(t_nt_{n-1}\right)\right|x_{n-1}\right\rangle$$
$$\ldots\left\langle x_1\left|U\left(tt_0\right)\right|x_0\right\rangle\ dx_1dx_2\ldots dx_n$$

where the component propagator between two spatial points is given by

$$\left\langle x_n\left|U\left(t_nt_{n-1}\right)\right|x_{n-1}\right\rangle = \sqrt{\frac{m}{2\pi i\hbar\Delta t}}\exp\left(\frac{im}{2\hbar}\left(\frac{x_n-x_{n-1}}{\Delta t}\right)^2\Delta t\right)$$
$$\times\exp\left(-\frac{iV}{\hbar}\Delta t\right)$$
$$\equiv\sqrt{\frac{m}{2\pi i\hbar\Delta t}}\gamma_{n-1}$$

Substituting this into wavefunction $\psi\left(x_{n+1}, t_{n+1}\right)$ yields

$$\psi\left(x_{n+1}, t_{n+1}\right) = \int\left(\frac{m}{2\pi i\hbar\Delta t}\right)^{\frac{n+1}{2}}\left[\gamma_n\cdots\gamma_0\, dx_n\cdots dx_1\right]\psi\left(x_0t_0\right)\, dx_0$$

One could determine that $\gamma_n\cdots\gamma_0 = e^{\frac{iS(t)}{\hbar}}$, where

$$S\left(t\right) = \frac{i}{\hbar}\int_0^T\frac{m}{2}\left(\frac{dx}{dt}\right)^2 - V\left(x_n\right)\, dt$$

is the action of the system that characterizes the propagation. Therefore,

$$\psi\left(x_{n+1}, t_{n+1}\right) = \int\left[\left(\frac{m}{2\pi i\hbar\Delta t}\right)^{\frac{n+1}{2}}e^{\frac{iS(t)}{\hbar}}dx_n\cdots dx_1\right]\,\psi\left(x_0t_0\right)dx_0$$

Take note that $\left(\frac{m}{2\pi i\hbar\Delta t}\right)^{\frac{n+1}{2}}$ has the dimension of $\left(\frac{1}{l}\right)^{n+1}$ which cancels the dimension of the volume element $dx_n\cdots dx_1dx_0$.

In a spinor system which evolves with the changing b fields, the infinitesimal propagator corresponding to $\sqrt{\frac{m}{2\pi i\hbar\Delta t}}\gamma_n = \left\langle x_{n+1}\left|U\left(t_{n+1}t_n\right)\right|x_n\right\rangle$

would be

$$C\gamma_n = \left\langle z_{n+1} \,|\, e^{-\frac{i}{\hbar}\delta t\,\mu\,\boldsymbol{b}.\boldsymbol{\sigma}} \,|\, z_n \right\rangle$$
$$= \langle z_{n+1} \mid z_n \rangle \, e^{-\frac{i}{\hbar}\delta t\,\mu\, b}$$

where n and $n+1$ correspond to intervals t and $t+\delta t$, respectively. One imagines the spinor evolves adiabatically as guided by a slow changing magnetic field $\boldsymbol{b} = b\boldsymbol{n}$ and $\boldsymbol{n}$ is the unit vector of the magnetic field. Use is then made of the following:

$$\langle z_{n+1} | z_n \rangle = 1 - z_{n+1}^{\dagger} \left(z_{n+1} - z_n \right)$$
$$\approx 1 - z_{n+1}^{\dagger} \partial_t z_n \delta t$$
$$\approx e^{-i\delta t\, z_n^{\dagger} \partial_t z_n}$$

Finally, one is led to

$$C\gamma_n = \exp\left(-\delta t\, z^{\dagger} \partial_t z\right) \cdot \exp\left(-\frac{i}{\hbar}\delta t\, \mu b\right)$$

And the action of the evolving spinor system is

$$S\,(t) = -\mu BT + i\hbar \int_0^T z_n^{\dagger}\,(t)\; \partial_t z_n(t)\; dt$$

The explicit expressions of the geometric phase as given in the second row of the following table depend on the explicit expressions of the eigenspinor as well as the parallel or anti-parallel alignment.

Table		$z_n^{\dagger}\,(t)\,\partial_t z_n(t)$
Parallel 1	$z_n^{(+)} = \begin{pmatrix} \cos\frac{\theta}{2} \\ e^{i\phi} \sin\frac{\theta}{2} \end{pmatrix}$	$\frac{1}{2}\,(1 - \cos\theta)\; i\frac{\partial\phi}{\partial t}$
Anti-parallel 1	$z_n^{(-)} = \begin{pmatrix} e^{-i\phi} \sin\frac{\theta}{2} \\ -\cos\frac{\theta}{2} \end{pmatrix}$	$\frac{-1}{2}\,(1 - \cos\theta)\; i\frac{\partial\phi}{\partial t}$
Parallel 2	$z_n^{(+)} = \begin{pmatrix} \cos\frac{\theta}{2} \\ e^{i\phi} \sin\frac{\theta}{2} \end{pmatrix}$	$\frac{1}{2}\,(1 - \cos\theta)\; i\frac{\partial\phi}{\partial t}$
Anti-parallel 2	$z_n^{(-)} = \begin{pmatrix} \sin\frac{\theta}{2} \\ -e^{i\phi} \cos\frac{\theta}{2} \end{pmatrix}$	$\frac{1}{2}\,(1 + \cos\theta)\; i\frac{\partial\phi}{\partial t}$

Remarks and Reflections

Note that the frame rotation approach produces gauge fields (upon the adiabatic theory) in their diagonal terms as shown in the following:

$$U^{\dagger}\left(\partial_k U\right) = \begin{pmatrix} +i\dfrac{(1-\cos\theta)}{2}\partial_k\phi & 0 \\ 0 & -i\dfrac{(1-\cos\theta)}{2}\partial_k\phi \end{pmatrix}$$

It thus follows that these gauge fields correspond to $z_n^{\dagger}(t)\,\partial_t\, z_n(t)$ as follows:

$$U^{\dagger}\left(\partial_k U\right) = \begin{pmatrix} z_n^{(+)\dagger}\partial_k\, z_n^{(+)} & 0 \\ 0 & z_n^{(-)\dagger}\partial_k\, z_n^{(-)} \end{pmatrix}$$

where

$$z_n^{(+)} = \begin{pmatrix} \cos\dfrac{\theta}{2} \\ e^{i\phi}\sin\dfrac{\theta}{2} \end{pmatrix}, \quad z_n^{(-)} = \begin{pmatrix} e^{-i\phi}\sin\dfrac{\theta}{2} \\ -\cos\dfrac{\theta}{2} \end{pmatrix}$$

Problem 5.06 Magnetic Monopole

Once the Berry–Pancharatnam phase in a system (e.g., graphene or topological materials) has been found, a curvature can be derived. In a system with $H = \boldsymbol{\sigma}\cdot\boldsymbol{b}$, answer the following questions:

(a) Show that the curvature in the along Z is

$$\Omega_B^z(\pm) = \pm\frac{1}{2}i\frac{b_z}{b^3}$$

(b) Likewise, derive $\Omega_B^x(+)$ and Ω_B^y.
(c) Show that the curvature in B space is a magnetic monopole (neglecting Dirac string).

$\pm$ indicates electron spin/pseudo-spin aligns parallel/anti-parallel to the magnetic field.

Solution

(a) The Hamiltonian of $H = \boldsymbol{\sigma}\cdot\boldsymbol{b}$ has eigenstates in arbitrary (θ, ϕ) as shown

$$|+\rangle = \begin{pmatrix} e^{-i\phi}\cos\dfrac{\theta}{2} \\ \sin\dfrac{\theta}{2} \end{pmatrix}, \quad |-\rangle = \begin{pmatrix} -e^{-i\phi}\sin\dfrac{\theta}{2} \\ \cos\dfrac{\theta}{2} \end{pmatrix}$$

The gauge fields in magnetic (B) space are given by

$$\boldsymbol{A}_{\boldsymbol{B}}^{+} = \langle + \,|\, \partial_{\boldsymbol{B}} \,|\, + \rangle = \frac{(1+\cos\theta)}{2}\begin{pmatrix} -i\dfrac{\partial \phi}{\partial b_x} \\ -i\dfrac{\partial \phi}{\partial b_y} \\ -i\dfrac{\partial \phi}{\partial b_z} \end{pmatrix},$$

$$\boldsymbol{A}_{\boldsymbol{B}}^{-} = \langle - \,|\, \partial_{\boldsymbol{B}} \,|\, - \rangle = \frac{(1-\cos\theta)}{2}\begin{pmatrix} -i\dfrac{\partial \phi}{\partial b_x} \\ -i\dfrac{\partial \phi}{\partial b_y} \\ -i\dfrac{\partial \phi}{\partial b_z} \end{pmatrix}$$

Referring to $b_x = b\sin\theta\cos\phi, b_y = b\sin\theta\sin\phi, b_z = b\cos\theta$, one can express (θ,ϕ,b) in terms of (b_x,b_y,b_z), e.g., $\tan\phi = \frac{b_y}{b_x}$ and $\left(\frac{\partial\phi}{\partial b_x}, \frac{\partial\phi}{\partial b_y}, \frac{\partial\phi}{\partial b_z}\right)$ can be found. Likewise, $b^2 = b_x^2 + b_y^2 + b_z^2$, and $\left(\frac{\partial b}{\partial b_x}, \frac{\partial b}{\partial b_y}, \frac{\partial b}{\partial b_z}\right)$ can be found. As for $\left(\frac{\partial\theta}{\partial b_x}, \frac{\partial\theta}{\partial b_y}, \frac{\partial\theta}{\partial b_z}\right)$, the relation $\tan^2\theta = \left(\frac{b_x}{b_z}\right)^2 + \left(\frac{b_y}{b_z}\right)^2$ is used.

With $\tan^2\theta = \left(\frac{b_x}{b_z}\right)^2 + \left(\frac{b_y}{b_z}\right)^2$, one has $2\tan\theta\sec^2\theta\left(\frac{\partial\theta}{\partial b_x}\right) = \frac{2b_x}{b_z^2}$, which leads to

$$\left(\frac{\partial\theta}{\partial b_x}\right) = \frac{\cos\theta\cos\phi}{b}, \left(\frac{\partial\theta}{\partial b_y}\right) = \frac{\cos\theta\sin\phi}{b}, \left(\frac{\partial\theta}{\partial b_z}\right) = -\frac{\cos^3\theta}{\sin\theta}\left(\frac{b_x^2+b_y^2}{b_z^3}\right)$$
$$= -\frac{\sin\theta}{b}$$

With $\tan\phi = \frac{b_y}{b_x}$, one has $\sec^2\phi\left(\frac{\partial\phi}{\partial b_x}\right) = -\frac{b_y}{b_x^2}$, which leads to

$$\left(\frac{\partial\phi}{\partial b_x}\right) = -\frac{b_y}{b_x^2}\cos^2\phi = \frac{\sin\phi}{-b\sin\theta}, \quad \left(\frac{\partial\phi}{\partial b_y}\right) = \frac{\cos\phi}{b\sin\theta}, \left(\frac{\partial\phi}{\partial b_z}\right) = 0$$

With $b^2 = b_x^2 + b_y^2 + b_z^2$, one has $\left(\frac{\partial b}{\partial b_x}\right) = \frac{b_x}{b}$, which leads to

$$\left(\frac{\partial b}{\partial b_x}\right) = \frac{b_x}{b} = \sin\theta\cos\phi, \quad \frac{\partial b}{\partial b_y} = \frac{b_y}{b} = \sin\theta\sin\phi, \quad \frac{\partial b}{\partial b_z} = \frac{b_z}{b} = \cos\theta$$

Garnering the relevant terms,

$$\frac{\partial \phi}{\partial b_y} = \frac{b_x}{b^2 \sin^2 \theta}; \quad \frac{\partial \phi}{\partial b_x} = \frac{-b_y}{b^2 \sin^2 \theta}$$

$$\frac{\partial \theta}{\partial b_y} = \frac{b_z b_y}{b^3 \sin \theta}; \quad \frac{\partial \theta}{\partial b_x} = \frac{b_z b_x}{b^3 \sin \theta}$$

The curvature is

$$\begin{aligned}
\Omega_B^z(+) &= \frac{\partial A_y^+}{\partial b_x} - \frac{\partial A_x^+}{\partial b_y} = \frac{1}{2} i \sin \theta \left(\frac{\partial \theta}{\partial b_x} \frac{\partial \phi}{\partial b_y} - \frac{\partial \theta}{\partial b_y} \frac{\partial \phi}{\partial b_x} \right) \\
&\quad - i \frac{(1 + \cos \theta)}{2} \left[\frac{\partial \phi}{\partial b_x}, \frac{\partial \phi}{\partial b_y} \right] \\
&= \frac{1}{2} i \sin \theta \left(\frac{b_z b_x^2}{b^5 \sin^3 \theta} + \frac{b_z b_y^2}{b^5 \sin^3 \theta} \right) - 0 \\
&= \left(\frac{1}{2} i \right) \left(+ \frac{b_z}{b^3} \right) \\
\Omega_B^z(-) &= \frac{\partial A_y^-}{\partial b_x} - \frac{\partial A_x^-}{\partial b_y} = \frac{1}{2} i \sin \theta \left(\frac{\partial \theta}{\partial b_y} \frac{\partial \phi}{\partial b_x} - \frac{\partial \theta}{\partial b_x} \frac{\partial \phi}{\partial b_y} \right) \\
&\quad - i \frac{(1 - \cos \theta)}{2} \left[\frac{\partial \phi}{\partial b_x}, \frac{\partial \phi}{\partial b_y} \right] \\
&= \frac{1}{2} i \sin \theta \left(- \frac{b_z b_y^2}{b^5 \sin^3 \theta} - \frac{b_z b_x^2}{b^5 \sin^3 \theta} \right) - 0 \\
&= \left(\frac{1}{2} i \right) \left(- \frac{b_z}{b^3} \right)
\end{aligned}$$

(b) Let's focus on the curvature for $\boldsymbol{A}_{\boldsymbol{B}}^+ = \langle + \,|\, \partial_{\boldsymbol{B}} \,|\, + \rangle$ and repeat the process for $\Omega_B^x(+)$, $\Omega_B^y(+)$. It follows that

$$\begin{aligned}
\Omega_B^x(+) &= \frac{\partial A_z^+}{\partial b_y} - \frac{\partial A_y^+}{\partial b_z} \\
&= \frac{1}{2} i \sin \theta \left(\frac{\partial \theta}{\partial b_y} \frac{\partial \phi}{\partial b_z} - \frac{\partial \theta}{\partial b_z} \frac{\partial \phi}{\partial b_y} \right) - i \frac{(1 + \cos \theta)}{2} \left[\frac{\partial \phi}{\partial b_y}, \frac{\partial \phi}{\partial b_z} \right] \\
&= \frac{1}{2} i \frac{\sin \theta \cos \phi}{b^2} + 0 = \left(\frac{1}{2} i \right) \left(+ \frac{b_x}{b^3} \right)
\end{aligned}$$

$$\Omega_B^y(+) = \frac{\partial A_x^+}{\partial b_z} - \frac{\partial A_z^+}{\partial b_x}$$
$$= \frac{1}{2} i \sin\theta \left(\frac{\partial \theta}{\partial b_z} \frac{\partial \phi}{\partial b_x} - \frac{\partial \theta}{\partial b_x} \frac{\partial \phi}{\partial b_z} \right) - i\frac{(1+\cos\theta)}{2} \left[\frac{\partial \phi}{\partial b_z}, \frac{\partial \phi}{\partial b_x} \right]$$
$$= \frac{1}{2} i \sin\theta \left(\frac{b_y}{b^3 \sin\theta} \right) + 0 = \left(\frac{1}{2} i \right) \left(+ \frac{b_y}{b^3} \right)$$

(c) Therefore, the curvature in B space for the entire system is

$$(\Omega_b)^2 = (\Omega_B^x)^2 + (\Omega_B^y)^2 + (\Omega_B^z)^2$$
$$\Omega_b = \left(\frac{1}{2} i \right) \frac{1}{b^2}$$
$$\boldsymbol{\Omega}_b = \frac{1}{2} i \frac{\boldsymbol{b}}{b^3}$$

Problem 5.07 Berry–Pancharatnam Phase in 2D Systems

Parameter Ω_{BP} is a geometric phase in the Berry–Pancharatnam (BP) context, where $\Omega_{BP} = \frac{e}{\hbar} \oint_C \boldsymbol{A}.d\boldsymbol{k}$ and $\boldsymbol{A}$ is the gauge field defined in the momentum space of a 2D material system. Show that in the Dirac-based electronic systems, e.g., the 2D graphene, 2D silicene, or Rashba, linear Dresselhaus semiconductor systems (see the following table), the geometric phase in the valence band is given by $n\pi$. Note that n is the number of times A is integrated over a closed surface contour. In the context of graphene, the gauge field is associated with valley K, with constant $\tau = 1$.

Table	Gauge potential	Ω
Linear Dresselhaus	$\boldsymbol{A} = \pm \frac{\hbar}{2ek^2}(k_y, -k_x, 0)$	$\mp n\pi$
Massless Graphene, Rashba	$\boldsymbol{A} = \pm \frac{\hbar}{2ek^2}(-k_y, k_x, 0)$	$\pm n\pi$
Massless Dirac (bilayer graphene)	$\boldsymbol{A} = \pm \frac{\hbar}{2ek^2}(-2k_y, 2k_x, 0)$	$\pm 2n\pi$

Note: The $\pm$ sign here refers to parallel or anti-parallel spin alignment in the valence band. In the event that only parallel spin alignment is considered, the gauge field takes on the + sign. The conduction band gauge field is not discussed here. On the other hand, gauge field sign changes due to the specific valley or Weyl node that is being considered.

Solution

To obtain the Berry–Pancharatnam (BP) phase, let us examine the gauge field of

$$\boldsymbol{A} = \pm\frac{\hbar}{2ek^2}\begin{pmatrix} pk_y \\ qk_x \\ 0 \end{pmatrix}$$

which is the general expression for the above-mentioned systems at $\theta{=}\pi/2$, i.e., in the plane of a real space (R) or a momentum space (K) 2D system. The BP phase can be derived with $\oint_C \boldsymbol{A}\left(\theta = \frac{\pi}{2}\right) .d\boldsymbol{k}$ where C represents a closed path in the plane surrounding the B space monopole at $(k_x, k_y) = 0$. One could equivalently evaluate

$$\oint_{SF}\left[\nabla\times\boldsymbol{A}\left(\theta{=}\frac{\pi}{2}\right)\right]\cdot d^2\boldsymbol{k}$$

on SF which represents the flat circular surface bound by C.

Method A: 2D Stokes' Theorem

The BP phase is given by $\Omega_{BP} = \frac{e}{\hbar}\oint_{SF}[\boldsymbol{\nabla}\times\boldsymbol{A_{2D}}]\cdot d\boldsymbol{S}$, which can be evaluated with Stokes' theorem as follows:

$$\begin{aligned}\Omega_{BP} &= \frac{e}{\hbar}\oint_{SF}[\boldsymbol{\nabla}\times\boldsymbol{A_{2D}}]\cdot d\boldsymbol{S} = \frac{e}{\hbar}\oint_{C}\boldsymbol{A_{2D}}\cdot d\boldsymbol{k} \\ &= \pm\frac{1}{2}\oint_C 1/(k^2)(-k_y, k_x)\cdot d\boldsymbol{k}\end{aligned}$$

By reparametrizing $\boldsymbol{k} = (k_x, k_y, 0) = (k\cos t, k\sin t, 0)$,

$$\begin{aligned}\Omega_{BP} &= \pm\frac{1}{2}\int_0^{2n\pi}\frac{1}{k^2}\left(-k_y, k_x\right)\cdot\frac{d\boldsymbol{k}}{dt}dt \\ &= \pm\frac{1}{2}\int_0^{2n\pi}\frac{1}{k^2}\left(-k\sin t, k\cos t\right)\cdot\frac{d\boldsymbol{k}}{dt}\,dt \\ &= \pm\frac{1}{2}\int_0^{2n\pi}\frac{1}{k^2}\left(-k\sin t, k\cos t\right)\cdot\left(-k\sin t, k\cos t\right)\,dt\end{aligned}$$

Finally,

$$\Omega_{BP} = \pm n\,\pi$$

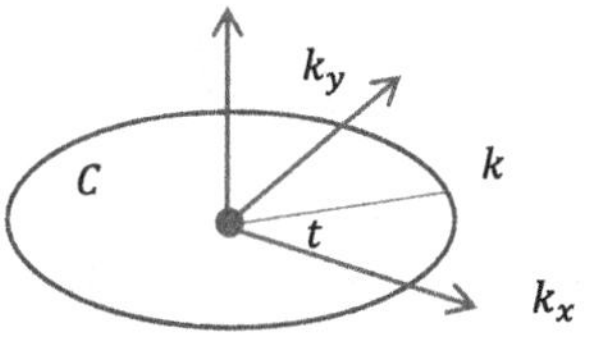

Fig. 2. Integration is carried out along the line contour C.

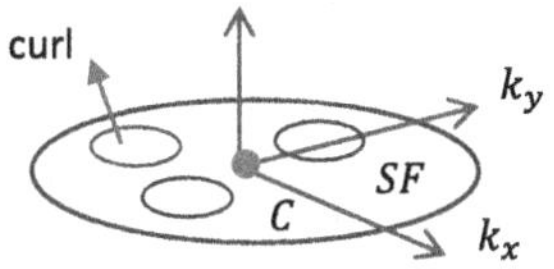

Fig. 3. Integration is carried out for the curl of the gauge field A across the surface SF bound by circle C.

Method B: 2D Direct approach

On the other hand, the BP phase can be obtained by evaluating the curvature of the gauge field directly as shown in Figure 3,

$$\Omega_{BP} = \frac{e}{\hbar}\oint_{SF}[\boldsymbol{\nabla}\times \boldsymbol{A_{2D}}]\cdot d\boldsymbol{S} = \pm\frac{1}{2}\oint_{SF}\left(\partial_{k_x}\left(\frac{k_x}{k^2}\right)+\partial_{k_y}\left(\frac{k_y}{k^2}\right)\right)\boldsymbol{e_z}\cdot d\boldsymbol{S}$$
$$= \pm\frac{1}{2}\oint_{SF}\boldsymbol{\nabla}\cdot\left(\frac{\boldsymbol{k}}{k^2}\right)dk_x\,dk_y$$

Define $R = k^2+\eta$, where $k^2 = k_x^2+k_y^2$, it follows that $\frac{\partial k}{\partial k_v} = \frac{k_v}{k}$. The BP phase would be

$$\Omega_{BP} = \pm\oint_{SF}\frac{1}{2}\lim_{\eta\to 0}\left(\partial_{k_x}\left(\frac{k_x}{R}\right)+\partial_{k_y}\left(\frac{k_y}{R}\right)\right)dk_x\,dk_y$$

$$\Omega_{BP} = \pm\oint_{SF}\frac{1}{2}\lim_{\eta\to 0}\left(\frac{R-2k_x^2+R-2k_y^2}{R^2}\right)dk_x\,dk_y$$

$$\Omega_{BP} = \pm\oint_{SF}\frac{1}{2}\lim_{\eta\to 0}\left(\frac{2\eta}{\left(k^2+\eta\right)^2}\right)dk_x\,dk_y$$

Moving on,

$$\Omega_{BP} = \pm\oiint\lim_{\eta\to 0}\left(\frac{\eta}{\left(k^2+\eta\right)^2}\right)k\,d\phi\,dk = \pm\lim_{\eta\to 0}\left(\pi-\frac{\pi\eta}{k^2+\eta}\right) = \pm\,\pi$$

Alternatively, recalling that $\lim_{\eta\to 0}\left(\frac{2\eta}{(k^2+\eta)^2}\right) = 2\pi\delta^2(k)$ for a 2D system,

$$\Omega_{BP} = \pm \oint_{SF} \frac{1}{2}\lim_{\eta\to 0}\left(\frac{2\eta}{(k^2+\eta)^2}\right) dk_x\, dk_y = \pm \oint_{SF} \pi\delta^2(k)\, dk_x\, dk_y = \pm\,\pi$$

Problem 5.08 Graphene Gauge Field

The Hamiltonian for a two-dimensional system, e.g., graphene or a Weyl-semi-metal is given by

$$H = \hbar v_F\left(\sigma_x \tau_z k_x + \sigma_y k_y\right) + \left(m_C + m_H \tau_z s_z\right)\sigma_z$$

where τ_z is the valley index and s_z the spin index of the system. The mass terms open a gap at the Dirac point and usher in the physics of topology. Graphene gauge field without a mass term was given in previous problems. Show that with mass terms, the gauge field for the 2D system is now given by

$$\boldsymbol{A} = \frac{\hbar}{2e}\left(1 - \frac{(m_C + m_H\tau_z s_z)}{\sqrt{k^2 + (m_C + m_H\tau_z s_z)^2}}\right)\left(-\frac{k_y}{k^2}, \frac{k_x}{k^2}\right)\tau_z$$

Solution

Gauge field in momentum space is given by

$$\boldsymbol{A} = \frac{\hbar}{2e}(1 - \cos\theta)\frac{\partial\phi}{\partial \boldsymbol{k}} \to \boldsymbol{A} = \frac{\hbar}{2e}\left(1 - \frac{n_z}{n}\right)\left(\frac{\partial\phi}{\partial k_x}, \frac{\partial\phi}{\partial k_y}, 0\right)$$

where (θ, ϕ) is defined by the effective magnetic field of the Hamiltonian as shown in Figure 4. For simplicity, $\hbar v_F$ is taken as 1. The Hamiltonian thus comprises effective magnetic fields as follows:

$$\begin{aligned} n_x &= n\,\sin\theta\cos\phi = \tau_z k_x,\ n_y = n\,\sin\theta\sin\phi = k_y,\ n_z = n\cos\theta \\ &= (m_C + m_H\tau_z s_z) \end{aligned}$$

The coordinate transformation is as follows:

$$(n_x, n_y, n_z) \rightleftharpoons (n, \theta, \phi)$$

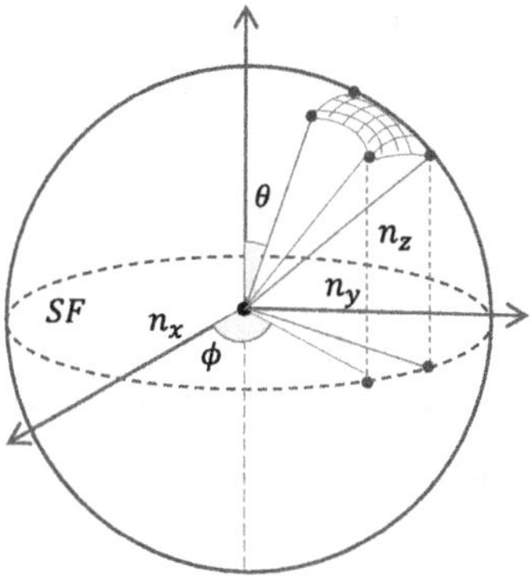

Fig. 4. A 3D illustration of the Bloch sphere with effective magnetic fields (n_x, n_y, n_z).

Taking note of $n_x = n\sin\theta\cos\phi$ and $n_y = n\sin\theta\sin\phi$, the following is derived:

$$\cos^2\phi = \left(\frac{n_x}{n}\right)^2 \frac{1}{\sin^2\theta} = \frac{n_x^2}{n_x^2 + n_y^2}$$

$$\sin^2\phi = \left(\frac{n_y}{n}\right)^2 \frac{1}{\sin^2\theta} = \frac{n_y^2}{n_x^2 + n_y^2}$$

Note that $\tan\phi = \frac{n_y}{n_x}$, and

$$\frac{\partial\phi}{\partial n_x} = -\frac{n_y}{n_x^2}\cos^2\phi = -\frac{n_y}{n_x^2 + n_y^2}$$

$$\frac{\partial\phi}{\partial n_y} = \frac{1}{n_x}\cos^2\phi = \frac{n_x}{n_x^2 + n_y^2}$$

Now, in momentum space,

$$\frac{\partial\phi}{\partial k_x} = \frac{\partial\phi}{\partial n_x}\frac{\partial n_x}{\partial k_x} + \frac{\partial\phi}{\partial n_y}\frac{\partial n_y}{\partial k_x}, \quad \frac{\partial\phi}{\partial k_y} = \frac{\partial\phi}{\partial n_x}\frac{\partial n_x}{\partial k_y} + \frac{\partial\phi}{\partial n_y}\frac{\partial n_y}{\partial k_y}$$

Therefore,

$$\frac{\partial\phi}{\partial k_x} = \left(\frac{1}{n_x^2 + n_y^2}\right)\left(n_x\frac{\partial n_y}{\partial k_x} - n_y\frac{\partial n_x}{\partial k_x}\right) = \frac{-k_y\tau_z}{k^2}$$

$$\frac{\partial\phi}{\partial k_y} = \left(\frac{1}{n_x^2 + n_y^2}\right)\left(n_x\frac{\partial n_y}{\partial k_y} - n_y\frac{\partial n_x}{\partial k_y}\right) = \frac{k_x\tau_z}{k^2}$$

In the above, note has been taken of $n_x = \tau_z k_x$, $n_y = k_y$. Finally, as $\boldsymbol{A} = \frac{\hbar}{2e}\left(1 - \frac{n_z}{n}\right)\left(\frac{\partial\phi}{\partial k_x}, \frac{\partial\phi}{\partial k_y}, 0\right)$, the gauge field in momentum space is

$$\boldsymbol{A} = \frac{\hbar}{2e}\left(1 - \frac{(m_C + m_H\tau_z s_z)}{\sqrt{k^2 + (m_C + m_H\tau_z s_z)^2}}\right)\left(-\frac{k_y}{k^2}, \frac{k_x}{k^2}\right)\tau_z$$

Problem 5.09 Graphene Berry–Pancharatnam Curvature

Graphene gauge field without a mass term was given in previous problems. It has also been shown that with the mass terms, the gauge field for the 2D systems, e.g., graphene, is given as

$$\boldsymbol{A} = \frac{\hbar}{2e}\left(1 - \frac{(m_C + m_H\tau_z s_z)}{\sqrt{k^2 + (m_C + m_H\tau_z s_z)^2}}\right)\left(-\frac{k_y}{k^2}, \frac{k_x}{k^2}\right)\tau_z$$

In the above, a gap has been opened at the Dirac point by the mass terms $(m_C + m_H\tau_z s_z)\,\sigma_z$. And the physics of topology is introduced via the curvature of the gauge field. Curvature has the physical significance of a magnetic field. One can therefore imagine that the mass terms introduce an effective magnetic field to the system. Show that the curvature in the Z axis in this context is given by

$$\Omega_z = \frac{\hbar}{2e}\left(\frac{\tau_z\,(m_C + m_H\tau_z s_z)}{\left(k^2 + (m_C + m_H\tau_z s_z)^2\right)^{\frac{3}{2}}}\right)$$

Note: Ω_{BP} denotes the BP phase in previous problems. In this problem, Ω_z denotes the BP curvature.

Solution

The curvature in the Z axis is

$$\Omega_z = \partial_{kx}A_y - \partial_{ky}A_x$$

Denote $(m_C + m_H\tau_z s_z)$ with Δ

$$\frac{\partial A_y}{\partial k_x} = \frac{\tau_z\hbar}{2e}\left[\frac{\Delta k_x^2}{k^2\left(k^2+\Delta^2\right)^{\frac{3}{2}}} + \left(1 - \frac{\Delta}{\left(k^2+\Delta^2\right)^{\frac{1}{2}}}\right)\frac{\left(k_y^2 - k_x^2\right)}{k^4}\right]$$

$$\frac{\partial A_x}{\partial k_y} = \frac{\tau_z\hbar}{2e}\left[\frac{-\Delta k_y^2}{k^2\left(k^2+\Delta^2\right)^{\frac{3}{2}}} + \left(1 - \frac{\Delta}{\left(k^2+\Delta^2\right)^{\frac{1}{2}}}\right)\frac{\left(k_y^2 - k_x^2\right)}{k^4}\right]$$

Note that in the above, use has been made of the following:

$$\frac{\partial k}{\partial k_x} = \frac{k_x}{k}; \quad \frac{\partial k}{\partial k_y} = \frac{k_y}{k}$$

$$\frac{\partial}{\partial k_x}\left(\frac{k_x}{k^2}\right) = \frac{k_y^2 - k_x^2}{k^4}; \quad \frac{\partial}{\partial k_y}\left(\frac{-k_y}{k^2}\right) = \frac{k_y^2 - k_x^2}{k^4}$$

Since we are dealing with a 2D real space system, it suffices to look into the Z axis curvature.

$$\Omega_z = \frac{\partial A_y}{\partial k_x} - \frac{\partial A_x}{\partial k_y} = \frac{\tau_z \hbar}{2e}\left[\frac{(m_C + m_H \tau_z s_z)}{\left(k^2 + (m_C + m_H \tau_z s_z)^2\right)^{\frac{3}{2}}}\right]$$

The Z BP curvature above is considered a magnetic field that produces Lorentz forces on particles (with both spin and charge degrees of freedom).

Remarks and Reflections

Note that in the above, a gap has been opened at the Dirac point by the mass terms $(m_C + m_H \tau_z s_z)\sigma_z$. The gauge field and its curvature have both been derived in the gapped system. And the physics of topology is introduced via the curvature of the gauge field. Without the mass terms, the gauge field is a forceless quantity everywhere except at the Dirac point. Therefore, curvature in graphene would be 0 everywhere surrounding the Dirac point, as shown earlier. With the mass, the system becomes forceful.

It's also worth noting that the mass term normally breaks inversion and time reversal symmetry. But it need not always do so. In the case of spin–orbit coupling, the mass term respects all symmetry.

Problem 5.10 Spin Hall Gauge in a 2D system 1

In a 2D Rashba spin–orbit system, the Hamiltonian is written in the sense of magnetic fields as follows:

$$H = \frac{p^2}{2m} + \alpha k\left(\sigma_x \frac{k_y}{k} - \sigma_y \frac{k_x}{k}\right) \rightarrow H = \frac{p^2}{2m} - \alpha k\, \boldsymbol{\sigma} \cdot \left(\boldsymbol{n} + \frac{\hbar}{\alpha k}\boldsymbol{n} \times \partial_t \boldsymbol{n}\right)$$

The spin Hall effect is better understood with the magnetic field that the spin aligns to, i.e., $\boldsymbol{n} = \left(-\frac{y}{r}, +\frac{x}{r}\right)$. But in this problem, we will write

$\boldsymbol{n'} = \left(\frac{y}{r}, -\frac{x}{r}\right)$, taking note that $\boldsymbol{n'}$=-$\boldsymbol{n}$. The definitions of the SHE gauge field are given by

$$C_x n'_x = \left(\frac{\partial n'_y}{\partial x}\frac{\partial n'_z}{\partial y} - \frac{\partial n'_z}{\partial x}\frac{\partial n'_y}{\partial y}\right) n'_x$$

$$C_y n'_y = \left(\frac{\partial n'_z}{\partial x}\frac{\partial n'_x}{\partial y} - \frac{\partial n'_x}{\partial x}\frac{\partial n'_z}{\partial y}\right) n'_y$$

$$C_z n'_z = \left(\frac{\partial n'_x}{\partial x}\frac{\partial n'_y}{\partial y} - \frac{\partial n'_y}{\partial x}\frac{\partial n'_x}{\partial y}\right) n'_z$$

(a) Show that $C_x n'_x = C_y n'_y = C_z n'_z = \frac{2\pi}{3}\delta(x)\,\delta(y)$
(b) As the SHE current density is given in the following, find C_z.

$$J^z_y = \int -\gamma g \frac{dk_x dk_y}{(2\pi)^2}\frac{e}{\hbar}\left(\frac{\partial \boldsymbol{n}}{\partial k_y} \times \frac{\partial \boldsymbol{n}}{\partial k_x}\right) E_x \cdot \boldsymbol{e_z} = \int -\gamma g \frac{dk_x dk_y}{(2\pi)^2}\frac{e}{\hbar}(-C_z)E_x$$

Note: $(x, y, z) \leftrightarrow (k_x, k_y, k_z)$ *and that* $r = x^2 + y^2 \leftrightarrow k$.

Solution

(a) Rewrite

$$\boldsymbol{n'} = \left(\frac{y}{r}, -\frac{x}{r}\right) \to \left(\frac{y}{R}, -\frac{x}{R}, \frac{v}{R}\right)$$

where $R^2 = \lim_{v\to 0} x^2 + y^2 + v^2 = \lim_{v\to 0} r^2 + v^2$. It follows that

$$\frac{\partial R}{\partial x} = \frac{x}{R}; \frac{\partial R}{\partial y} = \frac{y}{R}$$

which leads to

$$\frac{\partial n'_x}{\partial x} = -\frac{yx}{R^3}, \quad \frac{\partial n'_y}{\partial x} = -\left(\frac{1}{R} - \frac{x^2}{R^3}\right), \quad \frac{\partial n'_z}{\partial x} = -\frac{vx}{R^3}$$

and

$$\frac{\partial n'_x}{\partial y} = \left(\frac{1}{R} - \frac{y^2}{R^3}\right), \quad \frac{\partial n'_y}{\partial y} = \frac{yx}{R^3}, \frac{\partial n'_z}{\partial y} = -\frac{vy}{R^3}$$

Substitution leads to

$$C_x n'_x = \lim_{v\to 0}\frac{vy^2}{(r^2 + v^2)^{\frac{5}{2}}}, \; C_y n'_y = \lim_{v\to 0}\frac{vx^2}{(r^2 + v^2)^{\frac{5}{2}}}, \; C_z n'_z = \lim_{v\to 0}\frac{v^3}{(r^2 + v^2)^{\frac{5}{2}}}$$

Let us examine $C_x n_x$, which has the mathematical structure of a Dirac delta function (RHS) as shown in the following:

$$\lim_{v\to 0} \frac{vy^2}{\left(r^2+v^2\right)^{\frac{5}{2}}} \equiv A\delta^2\left(r\right)$$

Integrating both sides and recalling that $R^2 = r^2 + v^2$,

$$A = \int \lim_{v\to 0} \frac{vy^2\, dx\, dy}{\left(r^2+v^2\right)^{\frac{5}{2}}} = v \int \lim_{v\to 0} \frac{r^3 \sin^2\theta}{R^5} d\theta\, dr$$

$$A = v\,\pi \int \lim_{v\to 0} \frac{r^2}{R^4} dR = v\,\pi \left[-\frac{1}{R} + \frac{v^2}{3R^3}\right]_{R=v}^{R=\infty}$$

Therefore,

$$A = \frac{2\pi}{3}$$

Let us examine $C_y n_y$, which has the mathematical structure of a Dirac delta function (RHS) as shown in the following:

$$\lim_{v\to 0} \frac{v\,x^2}{\left(r^2+v^2\right)^{\frac{5}{2}}} \equiv A\delta^2\left(r\right)$$

Integrating both sides and recalling that $R^2 = r^2 + v^2$,

$$A = \int \lim_{v\to 0} \frac{v\,x^2\, dx\, dy}{\left(r^2+v^2\right)^{\frac{5}{2}}} = v \int \lim_{v\to 0} \frac{r^3 \cos^2\theta}{R^5} d\theta\, dr$$

$$A = v\,\pi \int \lim_{v\to 0} \frac{r^2}{R^4} dR = v\,\pi \left[-\frac{1}{R} + \frac{v^2}{3R^3}\right]_{R=v}^{R=\infty}$$

Therefore,

$$A = \frac{2\pi}{3}$$

Let us examine $C_z n_z$, which has the mathematical structure of a Dirac delta function (RHS) as shown in the following:

$$\lim_{v\to 0} \frac{v^3}{\left(r^2+v^2\right)^{\frac{5}{2}}} \equiv A\delta^2\left(r\right)$$

Integrating both sides and recalling that $R^2 = r^2 + v^2$,

$$A = \int \lim_{v\to 0} \frac{v^3\,dx\,dy}{(r^2+v^2)^{\frac{5}{2}}} = v \int \lim_{v\to 0} \frac{v^3}{R^5} d\theta\,dr$$

$$A = 2\,\pi \int \lim_{v\to 0} \frac{v^3}{R^4} dR = 2\,\pi \left[-\frac{v^3}{3R^3} \right]_{R=v}^{R=\infty}$$

Therefore,

$$A = \frac{2\pi}{3}$$

Finally,

$$C_x n'_x = \lim_{v\to 0} \frac{vy^2}{(r^2+v^2)^{\frac{5}{2}}} = \frac{2\pi}{3}\delta(x)\,\delta(y)$$

$$C_y n'_y = \lim_{v\to 0} \frac{vx^2}{(r^2+v^2)^{\frac{5}{2}}} = \frac{2\pi}{3}\delta(x)\,\delta(y)$$

$$C_z n'_z = \lim_{v\to 0} \frac{v^3}{(r^2+v^2)^{\frac{5}{2}}} = \frac{2\pi}{3}\delta(x)\,\delta(y)$$

(b) We will now look specifically into C_z.

$$C_z n'_z = \lim_{v\to 0} \frac{v^3}{(r^2+v^2)^{\frac{5}{2}}} = \lim_{v\to 0} \frac{v^2}{(r^2+v^2)^2} \left(\frac{v}{(r^2+v^2)^{\frac{1}{2}}} \right) \quad \rightarrow$$

$$C_z = \lim_{v\to 0} \frac{v^2}{(r^2+v^2)^2}$$

Compare and find D as follows:

$$C_z = \lim_{v\to 0} \frac{v^2}{(r^2+v^2)^2} \equiv D\,\delta(x)\,\delta(y)$$

It follows that

$$\int_{-\infty}^{+\infty} \int_{-\infty}^{+\infty} \lim_{v\to 0} \frac{v^2}{(r^2+v^2)^2} dx\,dy = D$$

And

$$D = \int_0^{2\pi} \int_0^{+\infty} \lim_{v\to 0} \frac{v^2}{(r^2+v^2)^2}\, r\,dr\,d\phi = 2\pi \left(0 + \frac{1}{2}(1) \right) = \pi$$

Therefore,

$$C_z = +\pi\delta(x)\,\delta(y)$$

Remarks and Reflections

If $\boldsymbol{n}$ instead of $\boldsymbol{n}'$ is used in the derivation above, the results would be

$$C_x n_x = \lim_{\delta\to 0}\frac{-\delta y^2}{\left(r^2+\delta^2\right)^{\frac{5}{2}}} = -\frac{2\pi}{3}\delta(x)\,\delta(y)$$

$$C_y n_y = \lim_{\delta\to 0}\frac{-\delta x^2}{\left(r^2+\delta^2\right)^{\frac{5}{2}}} = -\frac{2\pi}{3}\delta(x)\,\delta(y)$$

$$C_z n_z = \lim_{\delta\to 0}\frac{-\delta^3}{\left(r^2+\delta^2\right)^{\frac{5}{2}}} = -\frac{2\pi}{3}\delta(x)\,\delta(y)$$

Since C_x, C_y, C_z consist of an even number of $\boldsymbol{n}$ or $\boldsymbol{n}'$ components, they remain unchanged in sign for the opposite sign of n_a. However, $C_x n_x, C_y n_y, C_z n_z$ would take on opposite signs for the choice of $\boldsymbol{n}$ instead of $\boldsymbol{n}'$.

We will now look specifically into C_z. As C_z consists of an even number of $\boldsymbol{n}$ or $\boldsymbol{n}'$, its sign will not change with the choice of $\boldsymbol{n}$ or $\boldsymbol{n}'$, i.e.,

$$C_z = +\pi\,\delta(x)\,\delta(y)\;\; for\;\; \boldsymbol{n} or\; \boldsymbol{n}'$$

Problem 5.11 Spin Hall Gauge in a 2D System 11

The spin Hall curvature function is

$$\boldsymbol{\Omega} = \begin{pmatrix} 0 \\ 0 \\ 2\pi\delta(x)\,\delta(y) \end{pmatrix} \equiv \begin{pmatrix} \Omega^x \\ \Omega^y \\ \Omega^z \end{pmatrix} \equiv \begin{pmatrix} (\partial_y \boldsymbol{n}\times\partial_z \boldsymbol{n})\cdot\mathbf{n} \\ (\partial_z \boldsymbol{n}\times\partial_x \boldsymbol{n})\cdot\mathbf{n} \\ (\partial_x \boldsymbol{n}\times\partial_y \boldsymbol{n})\cdot\mathbf{n} \end{pmatrix}$$

Using the results above for $\boldsymbol{n}' = \left(\frac{y}{r}, -\frac{x}{r}\right)$, show that the spin Hall curvature is

$$\Omega^z = +2\pi\,\delta(x)\,\delta(y)$$

Solution

It has been shown that

$$C_x n_x' = \lim_{v\to 0}\frac{v y^2}{\left(r^2+v^2\right)^{\frac{5}{2}}} = \frac{2\pi}{3}\delta(x)\,\delta(y)$$

$$C_y n_y' = \lim_{v\to 0} \frac{vx^2}{(r^2+v^2)^{\frac{5}{2}}} = \frac{2\pi}{3}\delta(x)\,\delta(y)$$

$$C_z n_z' = \lim_{v\to 0} \frac{v^3}{(r^2+v^2)^{\frac{5}{2}}} = \frac{2\pi}{3}\delta(x)\,\delta(y)$$

Therefore,

$$\Omega^z = C_x n_x' + C_y n_y' + C_z n_z' = \frac{2\pi}{3}\delta(x)\,\delta(y) \times 3 = +2\pi\delta(x)\,\delta(y)$$

If one hadn't worked out the individual explicit expressions of $C_x n_x'$, $C_y n_y'$, $C_z n_z\prime$, one could, alternatively, start with

$$\Omega^z = C_x n_x' + C_y n_y' + C_z n_z'$$

$$= \lim_{v\to 0}\left[\frac{vy^2}{(r^2+v^2)^{\frac{5}{2}}} + \frac{vx^2}{(r^2+v^2)^{\frac{5}{2}}} + \frac{v^3}{(r^2+v^2)^{\frac{5}{2}}}\right]$$

$$\Omega^z = \lim_{v\to 0} \frac{v}{(r^2+v^2)^{\frac{3}{2}}}$$

The above has the mathematical structure of a Dirac delta function (RHS) as shown in the following:

$$\lim_{v\to 0} \frac{v}{(r^2+v^2)^{\frac{3}{2}}} \equiv A\delta(x)\,\delta(y)$$

Integrating both sides,

$$A = \int \lim_{v\to 0} \frac{v\,dx\,dy}{(r^2+v^2)^{\frac{3}{2}}}$$

$$= \int_0^{2\pi}\int_0^{\infty} \lim_{v\to 0} \frac{v\,r\,dr\,d\theta}{(r^2+v^2)^{\frac{3}{2}}} = 2\pi$$

Leading to

$$\Omega^z = +2\pi\,\delta(x)\,\delta(y)$$

One arrives at the same answer.

Remarks and Reflections

If it is $\boldsymbol{n}$ instead of $\boldsymbol{n}'$ that is used in the derivation, the results would be

$$\Omega^z = \lim_{\nu \to 0} \frac{-\nu}{(r^2+\nu^2)^{\frac{3}{2}}} \equiv A\,\delta(x)\,\delta(y)$$

$$A = \int \lim_{\nu \to 0} \frac{-\nu\; dxdy}{(r^2+\nu^2)^{\frac{3}{2}}} = \int_0^{2\pi}\int_0^{\infty} \lim_{\nu \to 0} \frac{-\nu\, r\, dr\, d\phi}{(r^2+\nu^2)^{\frac{3}{2}}} = -2\pi$$

Therefore,

$$\Omega^z = -2\pi\,\delta(x)\,\delta(y) \quad for\;\; \boldsymbol{n}$$

$$\Omega^z = +2\pi\,\delta(x)\,\delta(y) \quad for\;\; \boldsymbol{n}'$$

Problem 5.12 Spin Hall Gauge: Integral Calculus

Integrate the following function:

$$I = \iint f(x,y)\, dx\, dy$$

where

$$f(x,y) = \frac{y^2 - x^2}{(y^2+x^2)^2}$$

Solution

By partial fraction, function $f(x,y)$ can be written as follows:

$$\frac{y^2-x^2}{(y^2+x^2)^2} = \frac{1}{x^2+y^2} - \frac{2x^2}{(x^2+y^2)^2}$$

$$= \frac{d}{dx}\left(\frac{x}{x^2+y^2}\right)$$

Integration leads to

$$I_1 = \int_{-\infty}^{+\infty} \frac{y^2-x^2}{(y^2+x^2)^2}\, dy\, dx$$

$$= \int_{-\infty}^{+\infty} dy \left(\frac{x}{x^2+y^2}\right)_{-\infty}^{+\infty}$$

$$= \int_{-\infty}^{+\infty} dy\; 2 \lim_{x\to\infty}\left(\frac{x}{x^2+y^2}\right)$$

Let's take the x and y limits to an arbitrary a, instead of ∞,

$$I_1 = \lim_{y\to a}\int_{-y}^{+y} 2\lim_{x\to a}\left(\frac{x}{x^2+y^2}\right)dy$$
$$= 2\tan^{-1}\left(\frac{y}{a}\right)_{-a}^{+a} = \pi$$

As a matter of fact, careful examination shows that $f(x,y) = \frac{y^2-x^2}{(y^2+x^2)^2}$ can be a result of differentiation with respect to x or y. There are thus two ways to represent function $f(x,y)$ as follows:

$$f(x,y) = \frac{y^2-x^2}{(y^2+x^2)^2} = \frac{d}{dx}\left(\frac{x}{x^2+y^2}\right)$$
$$f(x,y) = \frac{y^2-x^2}{(y^2+x^2)^2} = \frac{d}{dy}\left(\frac{-y}{x^2+y^2}\right)$$

For comparison, integration of the above would yield, respectively,

$$I_1 = \int\int \frac{d}{dx}\left(\frac{x}{x^2+y^2}\right)dx\,dy = \lim_{y\to a}\int_{-y}^{+y} dy\, 2\lim_{x\to a}\left(\frac{x}{x^2+y^2}\right) = \pi$$
$$I_2 = \int\int \frac{d}{dy}\left(\frac{-y}{x^2+y^2}\right)dx\,dy = -\lim_{x\to a}\int_{-x}^{+x} dx\, 2\lim_{y\to a}\left(\frac{y}{x^2+y^2}\right) = -\pi$$

Since two different results are obtained, the above method is problematic. One now resorts to the spherical coordinates

$$\boldsymbol{I} = \int\int \frac{y^2-x^2}{(y^2+x^2)^2}dx\,dy = \int_0^{2\pi}\int_0^{\infty}\frac{\cos 2\theta}{r}dr\,d\theta$$
$$= \left(\frac{\sin 2\theta}{2}\right)_0^{2\pi}(lnr)_0^{\infty}$$
$$= 0\,(2\,\infty)_{slow} = 0$$

Remarks and Reflections

Fubini–Tonelli theorem has it that if integrals of $\int|f(x,y)|\,dx\,dy$, $\int|f(x,y)|\,dy\,dx$, and $\int f(x,y)\,d(x,y)$ are finite for all values of x, y, then the integrals satisfy the following:

$$\int f(x,y)\,dxdy = \int f(x,y)\,dy\,dx = \int f(x,y)\,d(x,y)$$

Non-Abelian Gauge

Problem 5.13 Non-Abelian Gauge Transformation 1

In a non-abelian gauge system, the transformation rule for the non-abelian gauge A_μ^a is such that

$$\left(\partial_\mu + \frac{ie}{\hbar}\left(\sigma^a A_\mu^a\right)'\right) U\psi = U\left[\left(\partial_\mu + \frac{ie}{\hbar}\sigma^a A_\mu^a\right)\psi\right]$$

where $A_\mu^{a'}$ is the transformed A_μ^a. Note that under $\psi \to U\psi$, the Pauli adjoint field transforms as $\bar{\psi} \to \bar{\psi}U^\dagger$. The transformation rule is thus deduced as follows:

$$(\sigma^a A_\mu^a)' = U\sigma^a A_\mu^a U^\dagger - i\frac{\hbar}{e}U(\partial_\mu U^\dagger)$$

and the term $-i\frac{\hbar}{e}U(\partial_\mu U^\dagger)$ is known as the emergent gauge. Derive the transformation formula

$$D_\mu\psi \to \left(\partial_\mu + \frac{ie}{\hbar}\left(\sigma^v A_\mu^v\right)'\right) U\psi = U\left[\left(\partial_\mu + \frac{ie}{\hbar}\sigma^v A_\mu^v\right)\psi\right]$$

where U is a unitary matrix, and

$$\frac{ie}{\hbar}\left(\sigma^v A_\mu^v\right)' = \frac{ie}{\hbar}U\sigma^v A_\mu^v U^\dagger + U(\partial_\mu U^\dagger)$$

Solution

One begins with

$$\left(\partial_\mu + \frac{ie}{\hbar}\sigma^v A_\mu^{v'}\right) U\psi = \left(\partial_\mu + \frac{ie}{\hbar}U\sigma^v A_\mu^v U^\dagger + U\left(\partial_\mu U^\dagger\right)\right) U\psi$$

$$= \partial_\mu\left(U\psi\right) + \frac{ie}{\hbar}U\sigma^v A_\mu^v\psi + U(\partial_\mu U^\dagger)U\psi$$

The above requires a separate effort to analyze the differential processes involving the unitary matrices. Take a digression and solve the following:

$$U\partial_\mu\left(U^\dagger U\right)\psi = U\left(\partial_\mu U^\dagger\right)U\psi + \partial_\mu\left(U\psi\right)$$

$$U\partial_\mu\psi = U\left(\partial_\mu U^\dagger\right)U\psi + \partial_\mu\left(U\psi\right)$$

A useful relation is found in the above. Substituting this into $\left(\partial_\mu + \frac{ie}{\hbar}\sigma^v A_\mu^{v'}\right) U\psi$, one is led to

$$\left(\partial_\mu + \frac{ie}{\hbar}\sigma^v A_\mu^{v'}\right) U\psi = U\partial_\mu\psi + \frac{ie}{\hbar}U\sigma^v A_\mu^v\psi$$

$$= U(\partial_\mu + \frac{ie}{\hbar}\sigma^v A_\mu^v)\psi$$

Remarks and Reflections

Sign convention used in non-abelian gauge is given in the following:

---

1. $E_\mu = -\frac{i\hbar}{e}U\left(\partial_\mu U^\dagger\right)$
2. $D_\mu = \left(\partial_\mu + i\frac{e}{\hbar}E_\mu\right) = -i\hbar\left(\partial_\mu + U\left(\partial_\mu U^\dagger\right)\right)$
   $= \left(-i\hbar\partial_\mu - \frac{i\hbar}{e}eU\left(\partial_\mu U^\dagger\right)\right)$
   $= (p_\mu + eE_\mu) = \left(\partial_\mu + U\left(\partial_\mu U^\dagger\right)\right)$
3. $[D_\mu\, , D_v] = i\frac{e}{\hbar}\left(\partial_\mu E_v - \partial_v E_\mu\right) + i\frac{e}{\hbar}i\frac{e}{\hbar}\left[E_\mu, E_v\right] = i\frac{e}{\hbar}F_{\mu v}$
   $U = e^{-iaF} = e^{-ie/\hbar \int E.dl}$

Problem 5.14 Non-Abelian Gauge Transformation 11

It has been shown that

$$\frac{ie}{\hbar}\left(\sigma^v A_\mu^v\right)' = \frac{ie}{\hbar}U\sigma^v A_\mu^v U^\dagger + U(\partial_\mu U^\dagger)$$

Now, in infinitesimal form, $U = \left(1 - \frac{1}{2}i\sigma^a\theta^a\right)$. Show that transformation leads to

$$\left(\sigma^a A_\mu^a\right)' = \sigma^a A_\mu^a + i\left[\sigma^b A_\mu^b, \frac{\sigma^c\theta^c}{2}\right] + \frac{\hbar}{e}\frac{\sigma^a}{2}\partial_\mu\theta^a$$

$$= \sigma^a A_\mu^a + \left(\boldsymbol{\theta} \times \boldsymbol{A_\mu}\right)^a \sigma^a + \frac{\hbar}{e}\frac{\sigma^a}{2}\partial_\mu\theta^a$$

Solution

Applying $U = \left(1 - \frac{1}{2} i\sigma^a\theta^a\right)$,

$$\left(\sigma^v A_\mu^v\right)' = U\sigma^v A_\mu^v U^\dagger + \frac{\hbar}{ie} U\left(\partial_\mu U^\dagger\right) = \left(1 - \frac{1}{2} i\sigma^b\theta^b\right)\left(\sigma^v A_\mu^v\right)$$
$$\times\left(1 + \frac{1}{2} i\sigma^c\theta^c\right) + \frac{\hbar}{ie}\left(\partial_\mu \frac{1}{2} i\sigma^c\theta^c\right)$$
$$= \sigma^v A_\mu^v + \sigma^v A_\mu^v \frac{1}{2} i\sigma^c\theta^c - \frac{1}{2} i\sigma^b\theta^b\sigma^v A_\mu^v + \frac{\hbar}{ie}\left(\partial_\mu \frac{1}{2} i\sigma^c\theta^c\right)$$

There are four terms on the RHS. Now, $v \to b$ for term 2 and term 3. But caution on term 3 that $b \to c$ is necessary so as to avoid term 3 to appear like $\frac{1}{2} i\sigma^b\theta^b\sigma^b A_\mu^b$. As a result,

$$\left(\sigma^v A_\mu^v\right)' = \sigma^v A_\mu^v + +\sigma^b A_\mu^b \frac{1}{2} i\sigma^c\theta^c - \frac{1}{2} i\sigma^c\theta^c\sigma^b A_\mu^b + \frac{\hbar}{e}\left(\partial_\mu \frac{1}{2}\sigma^c\theta^c\right)$$

As all four terms are independent, term 1 and term 4 can all be indexed by a for convenience. As a result,

$$\left(\sigma^v A_\mu^v\right)' = \sigma^a A_\mu^a + +\sigma^b A_\mu^b \frac{1}{2} i\sigma^c\theta^c - \frac{1}{2} i\sigma^c\theta^c\sigma^b A_\mu^b + \frac{\hbar}{e}\left(\partial_\mu \frac{1}{2}\sigma^a\theta^a\right)$$
$$= \sigma^a A_\mu^a + i\left[\sigma^b A_\mu^b, \frac{\sigma^c\theta^c}{2}\right] + \frac{\hbar}{e}\frac{\sigma^a}{2}\partial_\mu\theta^a$$

Expanding $\left[\sigma^b A_\mu^b, \frac{\sigma^c\theta^c}{2}\right]$, one observes that

$$\frac{1}{2}\left(\sigma^b A_\mu^b\right)\left(\sigma^c\theta^c\right) = \frac{1}{2} i\varepsilon_{abc}\,\sigma^a\, A_\mu^b\,\theta^c \tag{A}$$

$$\frac{1}{2}\left(\sigma^c\theta^c\right)\left(\sigma^b A_\mu^b\right) = -\frac{1}{2} i\varepsilon_{abc}\,\sigma^a\, A_\mu^b\,\theta^c \tag{B}$$

(A)–(B) leads to

$$i\left[\sigma^b A_\mu^b, \frac{\sigma^c\theta^c}{2}\right] = -\varepsilon_{abc}\sigma^a A_\mu^b\,\theta^c$$

Therefore,

$$\left(\sigma^a A_\mu^a\right)' = \sigma^a A_\mu^a + i\left[\sigma^b A_\mu^b, \frac{\sigma^c \theta^c}{2}\right] + \frac{\hbar}{e}\frac{\sigma^a}{2}\partial_\mu \theta^a$$

$$= \sigma^a A_\mu^a - \varepsilon_{abc}\sigma^a A_\mu^b \theta^c + \frac{\hbar}{e}\frac{\sigma^a}{2}\partial_\mu \theta^a$$

$$\left(\sigma^a A_\mu^a\right)' = \sigma^a A_\mu^a + (\boldsymbol{\theta} \times \boldsymbol{A}_\mu)^a \sigma^a + \frac{\hbar}{e}\frac{\sigma^a}{2}\partial_\mu \theta^a$$

Remarks and Reflections

Note that in the above, $U = \left(1 - \frac{1}{2} i\sigma^a \theta^a\right)$. In the event that $U = \left(1 - \frac{e}{\hbar} i\sigma^a \lambda^a\right)$, the following relationship is observed

$$\frac{1}{2}\theta^a = \frac{e}{\hbar}\lambda^a$$

Substituting this,

$$\left(\sigma^a A_\mu^a\right)' = \sigma^a A_\mu^a + \frac{2e}{\hbar}(\lambda \times \boldsymbol{A}_\mu)^a \sigma^a + \sigma^a \partial_\mu \lambda^a$$

Problem 5.15 Non-Abelian Gauge Curvature

Recently, the non-abelian, Yang–Mills-inspired physics expanded to many modern areas that straddle condensed matter and the quantum spin transport, e.g., 2D graphene and semiconductor nanostructures with strong spin–orbit coupling. The curvature of the non-abelian gauge is the origin of many new physics in modern fields.

Show that $F_{\mu\nu}^a \sigma^a = \partial_\mu A_\nu^a \sigma^a - \partial_\nu A_\mu^a \sigma^a + i\frac{e}{\hbar}[A_\mu^b \sigma^b, A_\nu^c \sigma^c]$ is consistent with

$$F_{\mu\nu}^a = \partial_\mu A_\nu^a - \partial_\nu A_\mu^a - \frac{2e}{\hbar}\epsilon_{abc} A_\mu^b A_\nu^c$$

With $F_{\mu v} = F_{\mu\nu}^a \sigma^a$ and $A_v = A_\nu^a \sigma^a$, rewrite the above as

$$F_{\mu v} = \partial_\mu A_v - \partial_\nu A_\mu + i\frac{e}{\hbar}[A_\mu, A_v]$$

With $\boldsymbol{F}_{\boldsymbol{\mu v}} = F_{\mu\nu}^a \boldsymbol{e_a}$ and $\boldsymbol{A_v} = A_\nu^a \boldsymbol{e_a}$, show that

$$\boldsymbol{F}_{\boldsymbol{\mu v}} = \partial_\mu \boldsymbol{A_v} - \partial_\nu \boldsymbol{A_\mu} - \frac{2e}{\hbar}(\boldsymbol{A_\mu} \times \boldsymbol{A_v})$$

Solution

$F^a_{\mu\nu}\sigma^a = \partial_\mu A^a_\nu\sigma^a - \partial_\nu A^a_\mu\sigma^a + i\frac{e}{\hbar}\left[A^b_\mu\sigma^b, A^c_\nu\sigma^c\right]$ leads to

$$\begin{aligned} F^a_{\mu\nu}\sigma^a &= \partial_\mu A^a_\nu\sigma^a - \partial_\nu A^a_\mu\sigma^a + i\frac{e}{\hbar}A^b_\mu A^c_\nu\left[\sigma^b, \sigma^c\right] \\ &= \partial_\mu A^a_\nu\sigma^a - \partial_\nu A^a_\mu\sigma^a + i\frac{e}{\hbar}A^b_\mu A^c_\nu\left(2i\sigma^a\epsilon_{abc}\right) \\ &= \left(\partial_\mu A^a_\nu - \partial_\nu A^a_\mu\right)\sigma^a - \frac{2e}{\hbar}\epsilon_{abc}A^b_\mu A^c_\nu\sigma^a \end{aligned}$$

Eliminating σ^a, the above shows consistency with $F^a_{\mu\nu} = \partial_\mu A^a_\nu - \partial_\nu A^a_\mu - \frac{2e}{\hbar}\epsilon_{abc}A^b_\mu A^c_\nu$.

Substituting $\mathcal{F}_{\mu v} = F^a_{\mu\nu}\sigma^a$ and $\mathcal{A}_v = A^a_\nu\sigma^a$ into $F^a_{\mu\nu}\sigma^a = \partial_\mu A^a_\nu\sigma^a - \partial_\nu A^a_\mu\sigma^a + i\frac{e}{\hbar}[A^b_\mu\sigma^b, A^c_\nu\sigma^c]$, and one would have

$$\mathcal{F}_{\mu v} = \partial_\mu\mathcal{A}_v - \partial_\nu\mathcal{A}_\mu + i\frac{e}{\hbar}[\mathcal{A}_\mu, \mathcal{A}_v]$$

Writing $\boldsymbol{F}_{\boldsymbol{\mu v}} = F^a_{\mu\nu}\,\boldsymbol{e}_{\boldsymbol{a}}$, one has

$$\boldsymbol{F}_{\boldsymbol{\mu v}} = \left(\partial_\mu A^a_\nu - \partial_\nu A^a_\mu\right)\boldsymbol{e}_{\boldsymbol{a}} - \frac{2e}{\hbar}\epsilon_{abc}A^b_\mu\,A^c_\nu\,\boldsymbol{e}_{\boldsymbol{a}}$$

As $\boldsymbol{A}_{\boldsymbol{v}} = A^a_\nu\boldsymbol{e}_{\boldsymbol{a}}$, expression $\boldsymbol{F}_{\boldsymbol{\mu v}} = \left(\partial_\mu A^a_\nu - \partial_\nu A^a_\mu\right)\boldsymbol{e}_{\boldsymbol{a}} - \frac{2e}{\hbar}\epsilon_{abc}\,A^b_\mu\,A^c_\nu\,\boldsymbol{e}_{\boldsymbol{a}}$ leads naturally to

$$\boldsymbol{F}_{\boldsymbol{\mu v}} = \partial_\mu\boldsymbol{A}_{\boldsymbol{v}} - \partial_\nu\boldsymbol{A}_{\boldsymbol{\mu}} - \frac{2e}{\hbar}\left(\boldsymbol{A}_{\boldsymbol{\mu}} \times \boldsymbol{A}_{\boldsymbol{v}}\right)$$

Topology and Gauge

Problem 5.16 Topological Models

(a) Explain why the quantity N as shown in the following is a quantized integer:

$$N = \frac{1}{2\pi i}\oint\frac{d\left(\log q\left(k\right)\right)}{dk}\,dk$$

Note that $q(k)$ is a complex function of k.

(b) Give examples of N exhibiting different integer values.

Solution

(a) As q cycles over a closed loop, it returns to its original value. But it may have acquired complex winding phases along the way, i.e.,

$q(k) \to q(k)\, e^{2\pi i n}$. In this case, $d(\log q(k)) = 2\pi i n$, and thus

$$N = \frac{1}{2\pi i} \oint \frac{dq}{q} = \frac{2\pi i n}{2\pi i} = n$$

(b) The simplest examples are $q(k) = e^{iNk}$. But any other expression that can be continuously deformed to them, i.e., is homotopically equivalent will work too.

Problem 5.17 Exact Solution to the Su–Schrieffer–Heeger (SSH) Model

The Su–Schrieffer–Heeger (SSH) model is one of the most common archetypal models for illustrating topological protection in 1D. In momentum space, its Hamiltonian is given by

$$H(k) = (t + \cos k)\, \sigma_x + \sin k\, \sigma_y \tag{A}$$

where σ_x, σ_y are Pauli matrices.

(a) By explicitly considering the SSH model in real space, show that when $|t| < 1$, there exists an eigenmode that is exponentially close to zero.
(b) Obtain the spatial profile of this eigenmode.
(c) Obtain the inverse of the SSH Hamiltonian in real space, one of the very lattice models which possess such an inverse.

Note: Momentum in the current and subsequent problems is dimensionless taking values between 0 and 2π.

Solution

(a) To convert an operator from momentum to real space with a total of N unit cells, we Fourier expand it and form the $(2N \times 2N)$ block Toeplitz matrix H with the (i, j) th block being the ith$-j$th Fourier coefficients of H. In this case, the 0th Fourier coefficient is $t\sigma_x$, the 1st Fourier coefficient is $\frac{\sigma_x + i\sigma_y}{2}$, and the -1th Fourier coefficient is $\frac{\sigma_x - i\sigma_y}{2}$. Putting them together, we obtain

$$H_{\mathrm{SSH}} = \begin{pmatrix} 0 & t & 0 & 0 & 0 & \cdots \\ t & 0 & 1 & 0 & 0 & \cdots \\ 0 & 1 & 0 & t & 0 & \cdots \\ 0 & 0 & t & 0 & 1 & \cdots \\ 0 & 0 & 0 & 1 & 0 & \cdots \\ \vdots & \vdots & \vdots & \vdots & \vdots & \ddots \end{pmatrix} \tag{B}$$

for the first few unit cells. In this model, we essentially only have alternating hoppings of 1 and t along the first upper and lower diagonals, which are physically the nearest neighbor couplings of adjacent sites. The lower right corner terminates with t hoppings.

To show that it contains an almost zero mode, we solve its eigenequation $Det\,[H_{SSH} - E\,I] = 0$. There are $2N$ solutions, $2N-2$ of which approximate Eq. (A) with certain k. To find the remaining two solutions with the smallest magnitudes of eigenenergy E, we neglect all higher powers of E in the eigenequation except for the linear term and obtain

$$E \approx \pm \frac{(-t)^N(1-t^2)}{1-t^{2[N/2]}}$$

which exponentially decays with N the system size. Evidently, it is close to 0 as long as $|t| < 1$.

(b) By substituting $E = 0$ in Eq. (B), we find a special eigensolution of the form

$$\psi_0 \propto (1, 0, -t, 0, t^2 \cdots)$$

This can be easily verified by direct substitution. Importantly, we see that while every even site is empty, every odd site has an amplitude that decays exponentially with the distance from the first unit cell. Since its eigenvalue is well separated from other eigensolutions, it is robust and is known as a robust boundary mode.

As an aside, the SSH model has the special property that the decay length $\xi = (\log|t|\,)^{-1}$ of its boundary mode coincides exactly with its imaginary gap, which is related to the imaginary part of k_x necessary for closing the gap:

$$\sqrt{1 + t^2 + 2t\cos k_x} = 0 \quad \rightarrow \quad e^{ik_x} = t \; or \; t^{-1}$$

By direct substitution, one can show that

$$H_{SSH}^{-1} = \frac{1}{-(-t)^n}$$

$$\times \begin{pmatrix} 1 & (-t)^{n-1} & -t & (-t)^{n-2} & (-t)^2 & (-t)^{n-3} & \cdots \\ (-t)^{n-1} & 0 & 0 & 0 & 0 & 0 & \cdots \\ -t & 0 & (-t)^2 & (-t)^{n-1} & (-t)^3 & (-t)^{n-2} & \cdots \\ (-t)^{n-2} & 0 & (-t)^{n-1} & 0 & 0 & 0 & \cdots \\ (-t)^2 & 0 & (-t)^3 & 0 & (-t)^4 & (-t)^{n-1} & \cdots \\ (-t)^{n-3} & 0 & (-t)^{n-2} & 0 & (-t)^{n-1} & 0 & \cdots \\ \vdots & \vdots & \vdots & \vdots & \vdots & \vdots & \cdots \end{pmatrix}$$

This expression for the inverse is important due to its significance as Green's function and represents one of the few topological lattice models with an exact Green's function expression.

Problem 5.18 Topological Characterization of the SSH Model

Here, we investigate the topological properties of the Su–Schrieffer–Heeger (SSH) model mentioned in the previous question. In momentum space, its Hamiltonian is given by

$$H(k) = (t + \cos k)\,\sigma_x + \sin k\,\sigma_y \tag{A}$$

where σ_x, σ_y are Pauli matrices. Show that the condition for robust boundary modes $|t| < 1$ originates from a non-trivial 1D topological winding. Also, show how this winding is related to the 2D Chern winding (2nd homotopy) on the Bloch sphere restricted by sub-lattice symmetry.

Solution

Equation (A) maps the 1D momentum, which lives in the circle S^1, to the space of two coefficients that cannot be simultaneously zero (when gapped), which also lives in a topologically S^1 space. More colloquially, it maps the 1D momentum onto the equation of the Bloch sphere, which is a circle.

Computing its normalized eigenvector $\phi = \phi(x)$, we can calculate its integer winding number

$$\begin{aligned} N_{1D} &= \frac{1}{2\pi}\oint \boldsymbol{A}\cdot dk \\ &= -\frac{i}{2\pi}\oint \phi^\dagger \nabla_\phi dk_x \\ &= \int_{k_x=0}^{k_x=2\pi} d\left[\tan^{-1}\left(\frac{\sin k_x}{t+\cos k_x}\right)\right] \\ &= \theta(1-t) \end{aligned} \tag{B}$$

The second step of Eq. (B) relies on the absence of a σ_z term in $H_{SSH}(k)$, which is enforced by the sub-lattice symmetry of the circuit (every node looks the same up to a left-right reflection). This winding number N_{1D} is related to the Chern number in the following sense. Consider a 2D extension to H_{SSH} (Eq. (A)). The contribution to N_{2D} from a momentum space region R is

$$\begin{aligned} N_{2D} &= \frac{1}{2\pi}\int_R \nabla\times \boldsymbol{A} d^2k \\ &= \frac{1}{2\pi}\oint_{\partial R} \boldsymbol{A}\cdot dk \\ &= \frac{1}{4\pi}\int_R \boldsymbol{d}\cdot(\partial_x \boldsymbol{d}\times\partial_y \boldsymbol{d})d^2k \end{aligned}$$

where $\nabla\times\boldsymbol{A}$ is the Berry flux. Now suppose that there is sub-lattice symmetry, i.e., that there is no σ_3 term so that $\boldsymbol{d}\perp\boldsymbol{e_x}$. Then, the second line reduces to a line integral along the equator of the Bloch sphere, which is mathematically S^1. As such, we can define a 1D topological invariant of the winding of the mapping from the 1D torus ∂R to the equator $S^1 \to S^1$. This invariant needs the protection of sub-lattice symmetry; upon breaking it by adding a small σ_3 term, for instance, the $\boldsymbol{d}$ vector will not be confined to the equator, and a second homotopy invariant instead of a first homotopy invariant is required.

Problem 5.19 Topological Polarization of the 2D Chern Insulator

One of the simplest models for a 2D Chern insulator is given by $H(\boldsymbol{k}) = d(\boldsymbol{k})\cdot\sigma$, where $d(\boldsymbol{k}) = (\sin k_x, \sin k_y, m+\cos k_x+\cos k_y)$. Compute its topological polarization. Through an approximation or otherwise, show that it evolves by the quantized amount of one unit as an inserted flux cycles over one period, when $0 < m < 2$.

Solution

The topological polarization for a given band $\phi(k)$ shows how much the center-of-mass of its x-localized Wannier function will move in the y-direction as k_y is cycled through a period (x and y labels are freely interchangeable). Through the Laughlin gauge argument, this state pumping through flux insertion should coincide with the Chern number. By definition, the topological polarization is given by

$$P(k_y) = \frac{1}{2\pi}\int_0^{2\pi i} \phi^\dagger \partial_{k_x} \phi dk_x$$

For a 2-component model, a few simplifications can be made. We first write the 2-component state as $\phi = \frac{a+bi}{N}$, where a and b are real vectors and N is the normalization factor. As $|\phi|^2 = a^2 + b^2 = 1$ is a constant, the real parts of $\phi^\dagger \partial_{k_x} \phi$ must disappear. Hence,

$$\begin{aligned}
\phi^\dagger \partial_{k_x} \phi &= i((b/N)\partial_{k_x}(a/N) - (a/N)\partial_{k_x}(b/N)) \\
&= \frac{i}{N^2}(b\partial_{k_x} a - a\partial_{k_x} b) \\
&= \frac{i}{N^2}(-\sin k_y \partial_{k_x} \sin k_x) \\
&= \frac{-i \sin k_y \cos k_x}{2\lambda((m + \cos k_x + \cos k_y) + \lambda)}
\end{aligned}$$

where $\lambda^2 = (m + \cos k_x + \cos k_y)^2 + \sin^2 k_x + \sin^2 k_y$. Thus, the exact integral expression for the polarization is

$$P(k_y) = -\frac{1}{2\pi}\int_0^{2\pi} \frac{\sin k_y \cos k_x}{\lambda(\lambda + m + \cos k_x + \cos k_y)} dk_x \tag{A}$$

This polarization function jumps at $k_y = \pi$. To analytically study the jump in the regime $0 < m < 2$ shown above, we Taylor expand the integrand about $k_y = \pi$ so that $Y = k_y - \pi$ and $\cos k_y \approx -1 + \frac{Y^2}{2}$. This integrates to a closed-form albeit complicated formula for the polarization which is extremely accurate for $|Y| = |k_y - \pi| < 1$. To make things more tractable, we shall also do an approximation on the integrated variable $X = x + \pi$. But here we have to be very careful as the integral (A) has different limits for different values of m. For the current simplest case of $0 < m < 2$, where the only singularity in the BZ $[0, 2\pi]^2$ occurs at $(\pi\pi)$. We can thus expand about $X = x - \pi = 0$ so that $\cos k_x \approx -1 + \frac{X^2}{2}$ and

$$\begin{aligned}
\lambda &= \sqrt{(m-2)^2 + (m-1)(X^2+Y^2) + \left(\frac{X^2+Y^2}{2}\right)^2} \\
&\approx (2-m)\sqrt{1 + \frac{m-1}{(m-2)^2}(X^2+Y^2)} \\
&\approx (2-m)\left(1 + \frac{m-1}{2(m-2)^2}(X^2+Y^2)\right)
\end{aligned}$$

and

$$\begin{aligned}
\lambda + (m + \cos x + \cos y) &= \lambda + m - 2 + \frac{1}{2}\left(X^2+Y^2\right) \\
&\approx (2-m)\left(1 + \frac{m-1}{2\left(m-2\right)^2}\left(X^2+Y^2\right)\right) \\
&\quad + m - 2 + \frac{1}{2}\left(X^2+Y^2\right) \\
&\approx \frac{1}{2(2-m)}\left(X^2+Y^2\right)
\end{aligned}$$

All in all, the polarization integral (A) can be approximated to the lowest non-trivial order in $X^2 + Y^2$ by

$$P(Y) = -\int_{-\pi}^{\pi} \frac{Y dX}{\pi(X^2+Y^2)} = -\frac{1}{\pi} tan^{-1}(\pi/Y)$$

or, in an analytically continued form,

$$Y = -\pi cot(\pi P(Y)) \text{ where } Y = k_y - \pi$$

This is an exceedingly simple expression obtained by expanding λ up to the first order in $X^2 + Y^2$; an analogous expansion up to the second order yields a much more accurate curve also expressible in closed form. In any case, it is obvious that $P(Y)$ jumps by a unit as k_y is cycled.

Problem 5.20 Berry Curvature of the Dirac Model

We have dealt with problems of Berry–Pancharatnam curvature for graphene with and without the mass terms. Here, we will relook at the Berry curvature in general models like the Dirac and the d-wave.

(a) Find the Berry curvature of the Dirac model with d-vector given by

$$d = (\sin k_x, \sin k_y, 2 + m - \cos k_x - \cos k_y)$$

(b) Similarly, find the Berry curvature of the d-wave model with

$$d = (\sin k_x - \sin k_y, \sin k_x + \sin k_y, 2\cos k_x \cos k_y - M)$$

having a winding number of 2 around $(0, 0)$.

Solution

(a) For the Dirac model with d-vector

$$d = (\sin k_x, \sin k_y, 2 + m - \cos k_x - \cos k_y)$$

we have

$$\partial_{k_x}\mathbf{d} = (\cos k_x, 0, \sin k_x)$$
$$\partial_{k_y}\mathbf{d} = (0, \cos k_y, \sin k_y)$$

The Berry curvature is hence given by

$$\begin{aligned} F_{xy} &= \frac{\mathbf{d}\,(\partial_{k_x}\mathbf{d} \times \partial_{k_y}\mathbf{d})}{|\mathbf{d}|^3} \\ &= \frac{-\cos k_x - \cos k_y + (2+m)\cos k_x \cos k_y}{(\sin^2 k_x + \sin^2 k_y + (2 - m - \cos k_x - \cos k_y)^2)^{3/2}} \\ &\approx \frac{-m}{(m^2 + k_x^2 + k_y^2)^{3/2}} \end{aligned}$$

We manifestly see a gapped Dirac cone at $\boldsymbol{k} = (0, 0)$, which contributes approximately half of a Chern number. It can be shown that the UV regularization of the model also integrates to give the remaining approximate half of the Chern number, yielding a Chern number of one in total.

(b) Analogous to the previous part, we have for

$$d = (\sin k_x - \sin k_y,\ \sin k_x + \sin k_y,\ 2\cos k_x \cos k_y - M)$$

$$F_{xy} = \frac{\mathbf{d} \cdot (\partial_{k_x}\mathbf{d} \times \partial_{k_y}\mathbf{d})}{|\mathbf{d}|^3}$$

$$= \frac{4\cos^2 k_y \sin^2 k_x + 4\cos^2 k_x \sin^2 k_y + 4\cos^2 k_x \cos^2 k_y}{(2(\sin^2 k_x + \sin^2 k_y) + (M - 2\cos k_x \cos k_y)^2)^{3/2}}$$

Appendix 5A. Spherical Coordinates in Different Bases

Holonomic Basis (Standard)	**Magnitude of the Basis Vector**
$dl = \frac{\partial l}{\partial \theta} d\theta + \frac{\partial l}{\partial \phi} d\phi + \frac{\partial l}{\partial r} dr$	$\lvert \boldsymbol{e_\theta} \rvert = R$
$dl = \boldsymbol{e_\theta}\, d\theta + \boldsymbol{e_\phi}\, d\phi + \boldsymbol{e_R}\, dr$	$\lvert \boldsymbol{e_\phi} \rvert = R\sin\theta$
$dl^2 = R^2 d\theta^2 + R^2 \sin^2\theta d\phi^2 + dR^2$	$\lvert \boldsymbol{e_R} \rvert = 1$
Non-holonomic Basis	**Magnitude of the Basis Vector**
$dl = \frac{\partial l}{\partial l_\theta} dl_\theta + \frac{\partial l}{\partial l_\phi} dl_\phi + \frac{\partial l}{\partial l_r} dl_r$	$\lvert \boldsymbol{e'_\theta} \rvert = 1$
$dl = \boldsymbol{e'_\theta}\, dl_\theta + \boldsymbol{e'_\phi}\, dl_\phi + \boldsymbol{e'_R}\, dl_r$	$\lvert \boldsymbol{e'_\phi} \rvert = 1$
$\boldsymbol{dl} = \boldsymbol{e'_\theta}\,(R\, d\theta) + \boldsymbol{e'_\phi}\,(R\sin\theta\, d\phi) + \boldsymbol{e'_R}\, dR$	$\lvert \boldsymbol{e'_R} \rvert = 1$
$dl^2 = (Rd\theta)^2 + (R\sin\theta\, d\phi)^2 + dR^2$	*Note*: The non-holonomic basis is commonly used in standard undergrad text. But they are almost always written without the prime signs.
From the above, the basis vectors for the holonomic basis can be found for their relationship to the Cartesian basis vectors as follows:	
$\boldsymbol{e_\theta} = \frac{\partial l}{\partial \theta} = R\cos\theta\cos\phi\, \boldsymbol{e_x} + R\cos\theta\sin\phi\, \boldsymbol{e_y} - R\sin\theta\, \boldsymbol{e_z}$	
$\boldsymbol{e_\phi} = \frac{\partial l}{\partial \phi} = -R\sin\theta\sin\phi\, \boldsymbol{e_x} + R\sin\theta\cos\phi\, \boldsymbol{e_y}$	
$\boldsymbol{e_R} = \frac{\partial l}{\partial R} = \sin\theta\cos\phi\, \boldsymbol{e_x} + \sin\theta\sin\phi\, \boldsymbol{e_y} + \cos\theta\, \boldsymbol{e_z}$	

Chapter 6

Advanced Condensed Matter Physics

This chapter comprises numerous sub-topics in condensed matter physics. It consists of conventional topics like the band structure, as well as selective problems of heat capacity and magnetic susceptibility. Towards the end, the modern and advanced topics of entanglement entropy are presented in numerous different contexts. These are advanced sub-topics and problems in this chapter could also be used for postgraduate courses in quantum mechanics and condensed matter physics.

Crystal Lattice and Bandstructure

Problem 6.01	Crystal Unit Cell
Problem 6.02	Rotational Symmetry
Problem 6.03	Wannier Functions
Problem 6.04	Heat Capacity of a 3D Electron Gas
Problem 6.05	Magnetic Susceptibility
Problem 6.06	Quantum Harmonic Oscillator
Problem 6.07	Heat Capacity: D-Dimensional Lattice of Phonons
Problem 6.08	Quantum Heisenberg Chain
Problem 6.09	Bandstructure of a 2D Lattice

Entanglement Entropy

Problem 6.10	Entanglement Entropy of the Free Fermions
Problem 6.11	Entanglement Entropy of a Critical System
Problem 6.12	Entanglement Entropy and Conformal Mapping

Crystal Lattice and Band structure

Problem 6.01 Crystal Unit Cell

Consider a lattice spanned by lattice vectors $\boldsymbol{a_1}, \boldsymbol{a_2}, \boldsymbol{a_3}$. Obtain the expression for the volume of a crystal unit cell and show that it remains unchanged under basis transforms.

Solution

Its unit cell is a parallelepiped spanned by vectors $\boldsymbol{a_1}, \boldsymbol{a_2}, \boldsymbol{a_3}$. Its volume V is given by the product of its base area

$$|\boldsymbol{a_2} \times \boldsymbol{a_3}|$$

and height

$$\boldsymbol{a_1}.\boldsymbol{n} = \frac{\boldsymbol{a_2} \times \boldsymbol{a_3}}{|\boldsymbol{a_2} \times \boldsymbol{a_3}|}$$

the latter being the normal unit vector to the base. Multiplying, we obtain

$$V = \boldsymbol{a_1}(\boldsymbol{a_2} \times \boldsymbol{a_3})$$

The above expression is manifestly basis independent when written in tensor notation:

$$V = \epsilon_{ijk} a_i b_j c_k$$

A basis transformation $a_i \to U_{ij} a_j$ and analogously for vectors b, c just rotates the anti-symmetric tensor which, by virtue of being a tensor, remains invariant.

Problem 6.02 Condensed Matter: Rotational Symmetry

Show that 2-, 3-, 4-, and 6-fold rotational symmetries are the only possible discrete rotation symmetries of a 2D lattice.

Solution

Suppose that a 2D lattice has n-fold rotation symmetry. Let's work in complex coordinates. If 1 is a basis vector, so are ω, ω^2, etc., where $\omega = e^{\frac{2\pi i}{n}}$.

Since there can only be two independent basis vectors in the 2D plane, ω^2 must be expressible as an integer superposition of ω and 1, i.e.,

$$\omega^2 = A + B\omega$$

where A and B are integers. Taking the imaginary part of this equation, we obtain the necessary condition

$$\cos\frac{2\pi}{n} = \frac{B}{2}$$

It is trivial to see that the LHS is not equal to half of any integer when $n = 5$ or $n > 6$.

Problem 6.03 Wannier Functions

In each band n, orthogonalities of Wannier functions are superpositions of plane wave eigenstates $|kn\rangle$ such that the phase $\theta(k)$ in front of each eigenstate labeled by k is chosen to localize the resultant superposition. Explicitly, Wannier functions can be written as

$$W_n(R) = \sum\nolimits_k e^{i\theta(k)} e^{-i\boldsymbol{k}\cdot\boldsymbol{R}} \, |kn\rangle$$

Show that $\langle R'\, n' | \, Rn\rangle = \delta_{R'R}\delta_{n'n}$, i.e., the Wannier functions are all orthogonal.

Solution

Since plane waves are orthogonal, i.e., $\langle k' | \, k\rangle = \delta_{k'k}$ and so are different bands, i.e., $\langle n' | \, n\rangle = \delta_{n'n}$,

$$\begin{aligned}
\langle R'\, n' | \, Rn\rangle &= \sum_{k,k'} e^{i(\theta(k)-\theta(k'))} e^{i\left(\boldsymbol{k}'\cdot\boldsymbol{R}' - \boldsymbol{k}\cdot\boldsymbol{R}\right)} \langle k'\, n' | \, kn\rangle \\
&= \sum_{k,k'} e^{i(\theta(k)-\theta(k'))} e^{i\left(\boldsymbol{k}'\cdot\boldsymbol{R}' - \boldsymbol{k}\cdot\boldsymbol{R}\right)} \delta_{k'k}\delta_{n'n} \\
&= \sum_{k} e^{i\boldsymbol{k}\cdot(\boldsymbol{R}' - \boldsymbol{R})} \delta_{n'n} = \delta_{R'R}\delta_{n'n}
\end{aligned}$$

Problem 6.04 Heat Capacity of a 3D Electron Gas

Find the heat capacity of a 3D electron gas via the Sommerfeld expansion. It will be helpful to know that the electronic density of states is $g(\epsilon) = \frac{m\sqrt{2m}}{\pi^2 \hbar^3}\epsilon^{\frac{1}{2}}$, where m is the electronic mass.

Solution

The Sommerfeld expansion is a systematic approximation approach for computing the temperature dependence of quantities involving electrons filled according to the Fermi–Dirac distribution $f(\epsilon)$. In principle, such a quantity can be written as $I = \int^{\infty} d\epsilon f(\epsilon)$, where $H(\epsilon)$ is the marginal value of the quantity at energy ϵ. However, the Fermi–Dirac distribution is almost a step function at chemical potential μ at room temperature. Hence, finite temperature corrections are small and can be expressed as higher-order corrections. Explicitly, the Sommerfeld expansion approximates an integral weighted by $f(\epsilon)$ by

$$
\begin{aligned}
I &= \int^{\infty} d\epsilon f(\epsilon) H(\epsilon) \\
&\approx \int^{\mu} d\epsilon H(\epsilon) + \frac{\pi^2}{6}(kT)^2 H'(\mu) + \frac{7\pi^4}{360}(kT)^4 H'''(\mu)
\end{aligned}
$$

In this problem, we will first need to find the internal energy $U(T)$ as a function of temperature T. Since $I = \int^{\infty} \epsilon g(\epsilon) f(\epsilon) d\epsilon$, we can let $H(\epsilon) = \epsilon g(\epsilon)$ and $I = U(T)$. This gives the heat capacity

$$
\begin{aligned}
c &= \frac{dU}{dT} \\
&= \frac{d}{dT}\left(\frac{\pi^2}{6}(kT)^2 H'(\mu) + \frac{7\pi^4}{360}(kT)^4 H'''(\mu)\right) + 0 \\
&= A\left(\frac{(\pi k)^2}{2} T\sqrt{\mu} - \frac{7(k\pi T)^3 k\pi}{240\mu^{3/2}}\right)
\end{aligned}
$$

where A is a constant defined via

$$
H(\epsilon) = \frac{m\sqrt{2m}}{\pi^2\hbar^3}\epsilon^{3/2} = A\epsilon^{3/2}
$$

Problem 6.05 Magnetic Susceptibility

Using the Sommerfeld expansion, calculate the magnetic susceptibility of free electrons in a magnetic field B. The energy associated with pointing along/opposite the direction of the magnetic field is $\mp g\mu_B B$.

Solution

The magnetization is given by $M = \frac{g\mu_B}{2}(n_+ - n_-)$, where $n_\pm$ is the number of electron spins with magnetization energy $g^\mp \mu_B B$. Since this magnetization energy is very small, we can approximate

$$\begin{aligned} n_\pm &= \frac{1}{2}\int d\epsilon g(\epsilon) f(\epsilon \mp g\mu_B B) \\ &\approx \frac{1}{2}\int d\epsilon g(\epsilon)\left[f(\epsilon) \mp g\mu_B B\frac{df(\epsilon)}{d\epsilon}\right] \end{aligned}$$

Hence, the magnetization is given by

$$\begin{aligned} M &= \frac{g\mu_B}{2}(n_+ - n_-) \approx -\frac{(g\mu_B)^2}{4}B\int d\epsilon\, g(\epsilon)\,\frac{df(\epsilon)}{d\epsilon} \\ &= \frac{(g\mu_B)^2}{4}B\int d\epsilon\, f(\epsilon)\,\frac{dg(\epsilon)}{d\epsilon} \end{aligned}$$

where we have integrated by parts in the last step. Applying Eq. (6) with

$$H(\epsilon) = \frac{dg}{d\epsilon} = \frac{m\sqrt{2m}}{2\pi^2\hbar^3}\epsilon^{-\frac{1}{2}} = \frac{A}{2}\epsilon^{-\frac{1}{2}}$$

with A defined in the previous problem, we obtain the (para)magnetic susceptibility as

$$\begin{aligned} \chi &= \frac{\partial M}{\partial B} \\ &\approx \frac{(g\mu B)^2}{4}\left(g(\mu) - \frac{\pi^2}{6}(kT)^2\frac{A}{4\mu^{3/2}}\right) \end{aligned}$$

To leading order, the susceptibility depends only on the density of states $g(\mu)$ at μ. However, it decreases quadratically with temperature at subleading order.

Problem 6.06 Quantum Harmonic Oscillator

The Hamiltonian of a quantum harmonic oscillator chain is given by

$$H = \sum_{k\neq 0}\omega_k\left(a_k^\dagger a_k + \frac{1}{2}\right)$$

with eigenenergies $\omega_k = \left(n_k + \frac{1}{2}\right)$. Find the partition function of this system at arbitrary non-zero temperature and obtain its heat capacity.

Solution

At temperature $= (k_B\beta)^{-1}$, the partition function is given by

$$\begin{aligned}
Z &= Tr(e^{-\beta H}) \\
&= \sum_{n_1,\dots,n_{N-1}} exp\left(-\beta\left(E_0 n_0^2 + \sum_{j\neq 0}\omega_j\left(n_j+\frac{1}{2}\right)\right)\right) \\
&= \sum_{n_0} e^{-\beta E_0 n_0^2}\prod_{j\neq 0}\sum_{n_j} e^{-\beta\omega_j(n_j+\frac{1}{2})} \\
&\approx \int dn_0 e^{-\beta E_0 n_0^2}\prod_{j\neq 0}\frac{e^{-\beta\omega_j/2}}{1-e^{-\beta\omega_j}} \\
&\propto \frac{1}{\sqrt{\beta}}\prod_{j\neq 0}\frac{e^{-\beta\omega_j/2}}{1-e^{-\beta\omega_j}}
\end{aligned}$$

where E_0 is an unimportant zero energy level. With a partition function of this form, the heat capacity for N particles is given by energy level offset

$$\begin{aligned}
C &= \frac{1}{N}\frac{\partial(\langle E\rangle)}{\partial T} \\
&= -\frac{\beta^2}{N}\frac{\partial(\langle E\rangle)}{\partial\beta} \\
&= \frac{\beta^2}{N}\frac{\partial^2 lnZ}{\partial\beta^2} \\
&= \frac{1}{N}\left(\frac{1}{2}+\sum_{j\neq 0}\left(\frac{\beta\omega_j/2}{\sinh(\beta\omega_j/2)}\right)^2\right) \\
&\approx \frac{1}{N}\sum_{j\neq 0}\left(\frac{\beta\omega_j/2}{\sinh(\beta\omega_j/2)}\right)^2
\end{aligned}$$

In the large N limit, the term $\frac{1}{2}$ arising from the $j=0$ mode is negligibly small compared to the summation term which contains $N-1$ terms.

Problem 6.07 Heat Capacity: D-Dimensional Lattice of Phonons

Building on the results of the previous problem, compute the heat capacity of a d-dimensional lattice of phonons in the low-temperature limit. Note the following:

(a) We shall assume the linear dispersion relation $\omega_k = v\,|k|$, which is valid in the small k regime.
(b) The Hamiltonian decouples such that there are now d identical independent copies of modes that obey $\omega_k = v\,|k|$, and $\ln Z = \ln Tr\,(e^{-\beta H})$ is thus multiplied by a factor d.

Solution

First, we must count the states correctly. In d-dimensions, the momentum summation becomes a multidimensional integral, and care has to be taken to sum it correctly. In particular, the number of allowed points in k-space is the same as the number of particles N. If we use the Debye approximation such that the sum is over a spherical region in k-space, we obtain the upper limit of momentum integral k_D, where

$$N = \frac{V}{(2\pi)^d}\int_{s^d(k_D)} dk = \frac{V}{(2\pi)^d}[Vol(S^d(1))]k_D^d$$

where V is the real space volume of the system and $S^d(r)$ denotes a d-dimensional sphere of radius r in momentum space. Also, note that $\int_{S^d(k_D)} dk = Area\left(S^d(1)\right)\int_0^{k_D} k^{d-1}dk$ so that $Area\left(S^d(1)\right) = d \times Vol(S^d(1))$. With this straightened out, the heat capacity takes the form

$$\begin{aligned}
C &= \frac{Vd}{(2\pi)^d N} Area(S^d(1))\int_0^{k_D} k^{d-1}dk\left(\frac{\beta v k/2}{\sinh(\beta v k/2)}\right)^2 \\
&= \frac{d^2}{(vk_D\beta)^d}\int_0^{vk_D\beta} dx\frac{x^{d+1}e^x}{(e^x-1)^2} \\
&\approx \frac{d^2}{(vk_D\beta)^d}\int_0^{\infty} dx\frac{x^{d+1}e^x}{(e^x-1)^2} \\
&= d^2(d+1)\zeta(d+1)\left(\frac{T}{T_D}\right)^d \\
&\propto T^d
\end{aligned}$$

where $T_D = vk_D$ is the Debye temperature. In setting the upper limit of the integral on the third line to be ∞, we have assumed the low-temperature limit.

This ubiquitous power law result for the heat capacity does not hold in the high-temperature limit, since a small upper limit of the integral causes the heat capacity to saturate to a constant.

Problem 6.08 Quantum Heisenberg Chain

Consider a periodic quantum Heisenberg chain given by

$$H = H_{xx,yy} + H_{zz} = J \sum_i (\sigma_i^x \sigma_{i+1}^x + \sigma_i^y \sigma_{i+1}^y) \\ + J_z \sum_i \sigma_i^z \sigma_i^z + 1$$

Use the Jordan–Wigner transformation to bring it into a fermionic chain form.

Solution

The Jordan–Wigner transformation is defined as

$$\begin{aligned} \sigma_i^z &= 2n_i - 1 \\ \sigma_i^x &= (c_i^\dagger + c_i) \prod_{j<i} (2n_j - 1) \\ \sigma_i^y &= i(-c_i^\dagger + c_i) \prod_{j<i} (2n_j - 1) \end{aligned}$$

The non-local products in σ_i^x and σ_j^y are necessary to keep track of fermionic statistics, which is non-local in origin too. With that, we have

$$\begin{aligned} H_{xx,yy} &= J \sum_i (\sigma_i^x \sigma_{i+1}^x + \sigma_i^y \sigma_{i+1}^y) \\ &= J \sum_{i<N} ((c_i^\dagger + c_i)(c_{i+1}^\dagger + c_{i+1}) - (c_i^\dagger - c_i)(c_{i+1}^\dagger - c_{i+1}))(2n_i - 1) \\ &\quad \times \prod_{j<i} (2n_j - 1)^2 \\ &= 2J \sum_{i<N} (c_i^\dagger c_{i+1} + c_i c_{i+1}^\dagger)(2n_i - 1) \\ &= 2J \sum_{i<N} (c_i^\dagger c_{i+1} + c_i c_{i+1}^\dagger)(2c_i^\dagger c_i - 1) \\ &= -2J \sum_{i<N} (c_i^\dagger c_{i+1} + c_{i+1}^\dagger c_i) \end{aligned}$$

This is a free fermion Hamiltonian as it can be diagonalized via a Bogoliubov transformation. The product term is trivially equal to unity because

$(2n_j - 1)^2 = (2n - 1)^2 = 1$. With the ZZ interaction term, the Hamiltonian acquires an additional quartic interaction term:

$$\begin{aligned} H_{zz} &= J_z \sum_{i<N} (2n_i - 1)(2n_{i+1} - 1) \\ &= J_z \sum_{i<N} (4n_i n_{i+1} - 2(n_i + n_{i+1}) + 1) \\ &= -4J_z \sum_{i<N} c_i^\dagger c_i + \frac{1}{2} \sum_{i<N} 8J_z c_i^\dagger c_i c_{i+1}^\dagger c_{i+1} + const. \end{aligned}$$

It is interesting that starting from a chain with only two-spin interactions, we obtain a fermionic chain with not just onsite and nearest-neighbor couplings but also a 4-operator (2-body) interaction term that is repulsive (attractive) if $J_z > 0 (J_z < 0)$.

Problem 6.09 Band structure of a 2D Lattice

This problem provides a pedagogical guide for analyzing the band structure of a non-trivial 2D crystal lattice. Consider the CuO lattice, which comprises Cu ions at the corners of a 2D square lattice and O ions at the centers of both the horizontal and vertical separations between the Cu ions.

For simplicity, assume that the hopping amplitude between the Cu and O ions is $+t$ if it is downward or rightwards, and $-t$ if it is upwards or leftwards. Also, let the hopping between two O ions be t' if it is towards the top-right/bottom-left, and $-t'$ if it is towards the bottom-right/top-left. Also, assuming on-site energy $_C$ and $_O$ for Cu and O ions, respectively.

(a) Write down the momentum–space Hamiltonian for this CuO lattice.
(b) Assuming for the moment that $t' = 0$, solve for the band eigenenergies and analyze its eigenvector structure.
(c) Solve for the eigenenergies for the generic $t' \neq 0$ case.

Solution

(a) We first review how to in general write down the momentum–space Hamiltonian $H(k)$ corresponding to a given real-space hopping Hamiltonian. This is done via a lattice Fourier transform described as follows. We start from

$$H = \sum_{ij} t_{ij}\, c_i^\dagger c_j$$

where t_{ij} is the hopping (or overlap integral) in the real-space (Wannier) basis. We switch to the momentum basis via $c_i^\dagger = \sum_k e^{i\boldsymbol{k}.\boldsymbol{r}_i} c_k^\dagger$ so that

$$
\begin{aligned}
H &= \sum_{ij} t_{ij} c_i^\dagger c_j \\
&\propto \sum_{ij} \sum_{\boldsymbol{k},\boldsymbol{k}'} t_{ij} e^{i\boldsymbol{k}\cdot \boldsymbol{r}_i} e^{-i\boldsymbol{k}'\cdot \boldsymbol{r}_j} c_{\boldsymbol{k}}^\dagger c_{\boldsymbol{k}'} \\
&= \sum_{\boldsymbol{k},\boldsymbol{k}'} \sum_{i,\Delta r} t_{ij} e^{i(\boldsymbol{k}-\boldsymbol{k}')\cdot \boldsymbol{r}_i} e^{-i\boldsymbol{k}'\cdot \Delta \boldsymbol{r}} c_{\boldsymbol{k}}^\dagger c_{\boldsymbol{k}'} \\
&\propto \sum_{\boldsymbol{k},\boldsymbol{k}'} \sum_{\Delta r} t_{\Delta r} \delta_{\boldsymbol{k},\boldsymbol{k}'} e^{-i\boldsymbol{k}'\cdot \Delta \boldsymbol{r}} c_{\boldsymbol{k}}^\dagger c_{\boldsymbol{k}'} \\
&= \sum_{\boldsymbol{k}} \left[\sum_{\Delta r} t_{\Delta r} e^{-i\boldsymbol{k}\cdot \Delta \boldsymbol{r}} \right] c_{\boldsymbol{k}}^\dagger c_{\boldsymbol{k}} \\
&= \sum_{\boldsymbol{k}} H(\boldsymbol{k}) c_{\boldsymbol{k}}^\dagger c_{\boldsymbol{k}} \qquad \text{(A)}
\end{aligned}
$$

where $\Delta \boldsymbol{r} = \boldsymbol{r}_j - \boldsymbol{r}_i$ only depends on the displacement between sites r_i and r_j since the lattice is translationally invariant. In other words, $H(k)$ is just the Fourier transform of the various hopping contributions. For the CuO lattice, in particular, we call the Cu site A and the O sites left of and above the Cu atom sites B and C, respectively. The hopping from site A to B site on its left is t and that to B site on its right is $-t$. Setting the spacing between a Cu and an O ion to unity, we have

$$
\begin{aligned}
H_{AB} &= t(e^{ik_x} - e^{-ik_x}) = 2it \sin k_x \\
H_{AC} &= t(-e^{ik_y} + -e^{-ik_y}) = -2it \sin k_y \qquad \text{(B)}
\end{aligned}
$$

For the hoppings between the O ions, we have

$$
\begin{aligned}
H_{BC} &= t'(e^{i(k_x+k_y)} - e^{i(k_x-k_y)} + e^{i(-k_x-k_y)} - e^{i(-k_x+k_y)}) \\
&= 2t'(\cos(k_x + k_y) - \cos(k_x - k_y)) \\
&= 4t' \sin k_x \sin k_y \qquad \text{(C)}
\end{aligned}
$$

There are no hoppings between different sites of the same sub-lattice, so there are no exponential factors in H_{AA}, H_{BB}, and H_{CC}. All in all, we have

$$H(k) = \begin{bmatrix} \epsilon_C & 2it\sin k_x & -2it\sin k_y \\ -2it\sin k_x & \epsilon_O & 4t'\sin k_x \sin k_y \\ 2it\sin k_y & 4t'\sin k_x\sin k_y & \varepsilon_O \end{bmatrix}$$

$$= \begin{bmatrix} \epsilon_C & H_{AB} & H_{AC} \\ H_{AB}^* & \epsilon_O & H_{BC} \\ H_{AC}^* & H_{BC}^* & \epsilon_O \end{bmatrix}$$

(b) When $t' = 0$, the characteristic equation $\det(H - \epsilon I) = 0)$ is factorizable into $(\epsilon - \epsilon_O)$ and a quadratic expression. Hence, we easily obtain the eigenenegies

$$\epsilon = \epsilon_0$$

and

$$\epsilon = \frac{\epsilon_O + \epsilon_C}{2} \pm \sqrt{\left(\frac{\epsilon_C - \epsilon_O}{2}\right)^2 + |\alpha|^2}$$

where $|\alpha|^2 = |H_{AB}|^2 + |H_{AC}|^2$ is the combined Cu-O hoppings in both directions. There is one flat band $\epsilon = \epsilon_0$ that depends on neither ϵ_C nor k, and its corresponding eigenstate is

$$|\psi_1\rangle = \frac{2t}{|\alpha|}\left(\sin k_y\,|B\rangle + \sin k_x\,|C\rangle\right)$$

It lives entirely in the O sub-space and hence feels none of the effects of ϵ_C. As an eigenvector, $|\psi_1\rangle$ is orthogonal to $|\psi_1\rangle = |A\rangle$ and

$$|\psi_3\rangle = \frac{2t}{|\alpha|}\left(\sin k_x\,|B\rangle - \sin k_y\,|C\rangle\right)$$

That $|\psi_{1,2,3}\rangle$ form an orthonormal basis is easy to see in spherical coordinates: $|\psi_2\rangle$ points along one polar axis while the other two are in the equatorial plane at a right angle from each other. In real space, this basis corresponds to taking some complex linear combination of the neighboring O sites of the original unit cell. In the $|\psi_{1,2,3}\rangle$ basis, the Hamiltonian takes the block-diagonal form

$$H'(k) = \begin{bmatrix} \epsilon_O & 0 & 0 \\ 0 & \epsilon_O & i|\alpha| \\ 0 & -i|\alpha| & \epsilon_C \end{bmatrix}$$

We can see why there are two band eigenenergies that are manifestly symmetric in ϵ_C and ϵ_0. The system is equivalent to one involving a decoupled O site and a staggered lattice where the O and C sites are of equal standing. In particular, we see that it becomes gapless when $\epsilon_C = \epsilon_0$.

(c) For the case when $t' \neq 0$, the characteristic eigenenergy equation is now cubic:

$$\epsilon^3 - \epsilon^2(\epsilon_C + 2\epsilon_O) + \epsilon(\epsilon_O^2 + 2\epsilon_O\epsilon_C - |\alpha|^2 - H_{BC}^2)$$
$$+ (H_{BC}^2\epsilon_C + |\alpha|^2\epsilon_O - 2H_{AB}H_{BC}H_{CA}) = 0$$

This equation can in fact be solved analytically. First, we let $\epsilon \to \epsilon + \frac{\epsilon_C + 2\epsilon_0}{3}$ so that the quadratic term disappears. The resultant equation is of the form $\epsilon^3 + P\epsilon + Q = 0$, and it always has solutions

$$\epsilon = X + Y, \quad \epsilon = \omega X + \omega^2 Y \quad \textit{and} \quad \epsilon = \omega^2 X + \omega Y \quad \textit{where} \quad \omega^3 = 1$$

and

$$X = \sqrt[3]{-\frac{Q}{2} + \sqrt{(P/3)^3 + (Q/2)^2}}$$
$$Y = \sqrt[3]{-\frac{Q}{2} - \sqrt{(P/3)^3 + (Q/2)^2}}$$

The eigenenergies are still real even though they may not appear so here.

Entanglement Entropy

Problem 6.10 Entanglement Entropy of the Free Fermions

Starting from the definition $S = -Tr[\rho \log \rho]$ for the entanglement entropy S of free fermions, where ρ is the reduced density matrix, derive the formula

$$S = -Tr[C \log C + (I - C) \log (I - C)]$$

which expresses S in terms of the single-particle correlator C.

Solution

By definition,

$$\rho = \frac{e^{-H_E}}{Z}$$

where H_E is the entanglement Hamiltonian and Z enforces the normalization $Tr[\rho] = 1$. Z can also be interpreted as a partition function with respect to the entanglement "temperature" $\beta_E = 1$. Since free fermions obey Wick's theorem and only have occupancy numbers of either 0 or 1, H_E is related to the single-particle correlator C *via*

$$C = \frac{1}{1 + e^{H_E}}$$

Writing the *single-particle* eigenvalues of H_E as ϵ_l, we have, by performing a trace over many-body states,

$$S = -Tr\rho \log \rho = Tr\rho\,(H_E + \log Z)$$

It follows that

$$\begin{aligned} S &= Z^{-1} Tr H_E e^{-H_E} + (Tr\rho)\log Z \\ &= \prod_l \left(1 + e^{-\epsilon_l}\right)^{-1} \left[\sum_l \epsilon_l e^{-\epsilon_l} \prod_{j \neq l} \left(1 + e^{-\epsilon_j}\right)\right] \\ &\quad + \log \prod_l \left(1 + e^{-\epsilon_l}\right) \\ &= \sum_l \left[\frac{\epsilon_l}{1 + e^{\epsilon_l}} + \log\left(1 + e^{-\epsilon_l}\right)\right] \end{aligned}$$

This intermediate formula expresses the entanglement entropy S in terms of the single-particle entanglement eigenenergies ϵ_l. Continuing the derivation,

$$\begin{aligned} S &= \sum_l \left[\frac{\epsilon_l}{1 + e^{\epsilon_l}} + \log(1 + e^{-\epsilon_l})\right] \\ &= \sum_l \left[\frac{\epsilon_l}{1 + e^{\epsilon_l}} + \log(1 + e^{-\epsilon_l}) - \epsilon_l\right] \\ &= \sum_l \left[\left(\frac{1}{1 + e^{\epsilon_l}} + \frac{1}{1 + e^{-\epsilon_l}}\right)\log(1 + e^{\epsilon_1}) + \epsilon_1 \left(\frac{1}{1 + e^{\epsilon_l}} - 1\right)\right] \\ &= \sum_l \left[\frac{1}{1 + e^{\epsilon_l}}\log(1 + e^{\epsilon_l}) + \frac{1}{1 + e^{-\epsilon_l}}\log(1 + e^{-\epsilon_l})\right] \\ &= -Tr[C \log C + (\mathbb{I} - C)\log(\mathbb{I} - C)] \end{aligned}$$

Note that the trace on the last line is performed only over single-particle states.

Problem 6.11 Entanglement Entropy of a Critical System

Derive the expression $S_A = \frac{c}{3}\log l + Y$ for the entanglement entropy of a critical system when a sub-region A of length l is embedded in infinite 1D space. Y is a non-universal constant and c is the central charge of the system.

Solution

We shall derive the entanglement entropy via the "replica" trick and proceed with the help of conformal field theory. Given a (1+1)-dim system spatially partitioned into regions A and B, we have

$$S_A = -Tr\left(\rho_A \log \rho_A\right) = -\lim_{n\to 1} \partial_n \, Tr\rho_A^n$$

with ρ_A the reduced density matrix and

$$Tr\,\rho_A^n = e^{(1-n)\,S_R(n)}$$

where $S_R(n)$ is the Renyi entropy. $Tr\,\rho_A^n$ can be expanded as a path integral over real space x and imaginary time τ. The n copies of ρ_A correspond to n complex sheets. When we perform the partial trace over A, we are equivalently joining up the sheets such that the fields on the jth and $(j+1)$th sheets obey $\phi_j(x, \tau = \beta^-) = \phi_{j+1}(x, \tau = 0^+)$ for $x \in A$ only. This looks topologically like n sheets successively connected to each other via branch cuts $(x, \tau = 0)$ with $x \in A$. Call this n-sheeted Riemann surface R_n. The above construction can be summarized by writing

$$Tr\,\rho_A^n|_{R_n} = \langle \Phi(x_1)\bar{\Phi}(x_2)\rangle_C \tag{A}$$

where C is the complex plane and $\langle \Phi(x_1)\bar{\Phi}(x_2)\rangle_C$ is taken over one of the n sheets with ϕ, ϕ being branch-point twist fields that implement cyclic permutations

$$\Phi : \phi_i \to \phi_{i+1} \; and \; \bar{\Phi} : \phi_i \to \phi_{i-1}$$

All the $i's$ are taken mod n. Equation (A) can be evaluated by computing the stress–energy tensor T on R_n in two different ways, which when combined will give us the desired result on entanglement entropy. One way is a path integral approach with twist fields, like in the above, while another way is by direct conformal transformation from the complex plane. With the path integral approach, we have by definition

$$T|_{R_n} = \frac{\langle\Phi(x_1)\bar{\Phi}(x_2)T_j\rangle_C}{\langle\Phi(x_1)\bar{\Phi}(x_2)\rangle_C} = \frac{\langle\Phi(x_1)\bar{\Phi}(x_2)\rangle_C T_j\rangle_c}{Tr\ \rho_A^n|_{R_n}} \tag{B}$$

where j represents one of the n copies of the complex sheets. To evaluate the stress–energy tensor T by conformal transformation, one considers $T(z)$, the holomorphic part of the T with $z \in C$. We know that $T(z) = 0$ due to global conformal invariance. Now, let's map C to the $n-$sheeted surface parametrized by $\omega = x + i\tau \in R_n$ via $z = \left(\frac{\omega - x_1}{\omega - x_2}\right)^{\frac{1}{n}}$. In other words, we implement branch cuts from 0 to ∞ on the Riemann surface z^n, and these branch cuts correspond to our original region A from $\omega = x_1$ to $\omega = x_2$. Through standard manipulations with the Schwarzian derivative

$$\begin{aligned}
T(w)|_{R_n} &= T(z)\left(\frac{dz}{dw}\right)^2 + \frac{c}{12}\{z, w\} \\
&= \frac{c}{12}\{z, w\} \\
&= \frac{c}{12}\frac{z'''z' - \frac{3}{2}z''^2}{z'^2} \\
&= \frac{c(n - 1/n)}{24n}\frac{(x_1 - x_2)^2}{(w - x_1)^2(w - x_2)^2}
\end{aligned}$$

where c is the central charge. We next compare Eq. (B) with the conformal Ward identity on the stress–energy tensor:

$$\begin{aligned}
&\langle\Phi(x_1)\bar{\Phi}(x_2)T_j(\omega)\rangle_C \\
&= n\left(\frac{\partial_{x_1}}{\omega - x_1} + \frac{h_\Phi}{(\omega - x_1)^2} + \frac{\partial_{x_2}}{\omega - x_2} + \frac{h_{\bar{\Phi}}}{(\omega - x_2)^2}\right)\langle\Phi(x_1)\bar{\Phi}(x_2)\rangle_C \\
&= n\left(\frac{\partial_{x_1}}{\omega - x_1} + \frac{h_\Phi}{(\omega - x_1)^2} + \frac{\partial_{x_2}}{\omega - x_2} + \frac{h_{\bar{\Phi}}}{(\omega - x_2)^2}\right)|x_1 - x_2|^{-2n\Delta}
\end{aligned} \tag{C}$$

where $\Delta = h_\Phi = h_{\bar{\Phi}}$ is the scaling dimension. The factor of n arises from summing over n identical copies of T. Comparing the two equations (B) and (C), we obtain

$$\Delta = \frac{c}{12n}\left(n - \frac{1}{n}\right)$$

which implies that

$$\begin{aligned} Tr\,\rho_A^n|_{R_n} &= \langle \Phi(x_1)\bar{\Phi}(x_2)\rangle_C \\ &\propto |x_2 - x_1|^{-2n\Delta} \sim l^{-\frac{c}{6}\left(n-\frac{1}{n}\right)} \end{aligned}$$

From this,

$$S_A = -\lim_{n\to 1}\partial_n\, Tr\rho_A^n = \frac{c}{3}\log l + Y$$

where Y is a non-universal constant. This result holds for a sub-region A with length l embedded in the (infinite) real line.

Problem 6.12 Entanglement Entropy and Conformal Mapping

From the results of the previous problem, derive the critical entanglement entropy formula when the sub-region of length l is instead situated in a *finite* system of length L.

Solution

When boundary conditions are imposed on the critical system, its entanglement entropy formula can be derived by conformally mapping the results from the original geometry to that of the new geometry. Under a generic map $\omega \to v$, $Tr\,\rho_A^n = \langle \Phi(x_1)\bar{\Phi}(x_2)\rangle_C$ is transformed into

$$\langle \Phi(v_1,\bar{v}_1)\bar{\Phi}(v_2,\bar{v}_2)\rangle = |\omega_1'(v_1)\,\omega_2'(v_2)|^{n\Delta}\,\langle \Phi(\omega_1,\bar{\omega}_1)\bar{\Phi}(\omega_2,\bar{\omega}_2)\rangle$$

since the twist fields Φ and $\bar{\Phi}$ have scaling dimension $n\Delta$. For this problem, we consider the case where the system has a finite periodicity of L, which requires that we perform the mapping $v = \frac{L}{2\pi i}\log\omega$ so that each sheet in the w plane becomes an infinite cylinder with circumference L. This leads to the transformation

$$\begin{aligned}
Tr\rho_A^n &\to \left|\omega_1'(v_1)\,\omega_2'(v_2)\right|^{n\Delta}\,\langle\Phi(\omega_1,\bar{\omega}_1)\bar{\Phi}(\omega_2,\bar{\omega_2})\rangle \\
&= \left(\frac{2\pi}{L}\right)^{2n\Delta}\langle\Phi(\omega_1,\bar{\omega}_1)\bar{\Phi}(\omega_2,\bar{\omega}_2)\rangle \\
&= \left(\frac{2\pi}{L}\right)^{2n\Delta}\left|\omega_1-\omega_2\right|^{-2n\Delta} = \left(\frac{2\pi}{L}\right)^{2n\Delta}\left|e^{\frac{2\pi i v_1}{L}} - e^{\frac{2\pi i v_2}{L}}\right|^{-2n\Delta} \\
&= \left(\frac{2\pi}{L}\right)^{2n\Delta}\left|2\sin\left(\frac{\pi\,(v_2-v_1)}{L}\right)\right|^{-2n\Delta} = \left(\frac{L}{\pi}\sin\left(\frac{\pi l}{L}\right)\right)^{-2n\Delta}
\end{aligned}$$

yielding the expression

$$S_A = \frac{c}{3}\log\left(\frac{L}{\pi}\sin\frac{\pi l}{L}\right) + Y'$$

for a sub-region A of length l in a finite system with periodicity L.

Chapter 7

General Physics

This is a cross-disciplinary chapter that covers miscellaneous topics from general to mathematical physics. Problems here are not necessarily pure quantum by nature, but many deal with issues that have inspired the development of quantum mechanics and quantum fields. For example, the concept of minimization leads to the Euler–Lagrange equations that also form the foundation of quantum field theory. Questions are also raised in problems to stimulate thought about physical issues based on quantum physics, e.g., visualization of the electrons.

Minimization Physics

Problem 7.01 Minimum Pricniples in Physics
Problem 7.02 Shortest Time Trajectory
Problem 7.03 The Hamilton Principle

Electron Physics and Visualizations

Problem 7.04 What is the Size and Shape of an Electron?
Problem 7.05 What is Spin?
Problem 7.06 Single Electronic Motion: Master Equation
Problem 7.07 Semi-classical Electronic Transport

Mathematical Methods

Problem 7.08 Integral Method
Problem 7.09 Divergence Theorem
Problem 7.10 Barycentric Coordinates

Minimization Physics

Problem 7.01 Minimum Principles in Physics

In general, nature tends to minimize certain quantities when a physical process takes place. Therefore, it is always interesting to identify the quantities that nature would minimize and work out the consequences that minimization may lead to. The various physical processes are given in the left-most column of Table 1. Elaborate on the minimized quantities and the mathematical expressions and the physical outcome of the respective processes by filling in the blanks of Table 1.

Table 1. A consolidation of numerous physical processes and their minimization.

Physical Processes	Minimized Quantities and Mathematical Expressions	Physical Outcome
Light ray reflecting off of a mirror		Law of reflection
Fermat's reformulation of the light ray traveling physics in a medium		Law of reflection, Snell's law on refraction
Brachistochrone		Cycloid trajectory of the path
Geodesic on the curved surface of the sphere, e.g., earth		Great circle trajectory
Hamilton's principle		Equations of motion

Solution

Blanks in Table 1 are filled as follows:

Physical Processes	Minimized Quantities and Mathematical Expressions	Physical Outcome
Light ray reflecting off of a mirror	Light chooses the shortest possible path	Law of reflection
Fermat's reformulation of the light ray traveling physics in a medium	Light chooses the path that takes the shortest time	Law of reflection, Snell's law on refraction

(*Continued*)

Physical Processes	Minimized Quantities and Mathematical Expressions	Physical Outcome
Brachistochrone	Travel time sliding down a curved path under gravity $$t = \int_{0,0}^{x_2,y_2} \frac{ds}{v} = \int \left(\frac{1+(y')^2}{2gx}\right)^{\frac{1}{2}} dx$$	Cycloid trajectory of the path: The cycloid trajectory is identified as the path that gives the shortest time for a particle sliding down it under gravity
Geodesic on the curved surface of the sphere, e.g., earth	The path connecting two points (a, b) on the curved surface $$s = \int_a^b ds = R\int_a^b \left(\sin^2\theta + \left(\frac{d\theta}{d\phi}\right)^2\right)^{\frac{1}{2}} d\phi$$	Great circle trajectory: Great circle trajectory (e.g., equator) is identified as the shortest path between two points on a curved spherical surface
Hamilton's principle	The action of the system, which is defined as the integral of the Lagrangian $$S = \int T - V\, dt$$	Equations of motion: Equations of motion for different systems are found depending on the Lagrangian of the systems

Problem 7.02 Shortest Time Trajectory

In the Brachistochrone problem, a particle moves from origin $(0, 0)$ to point (x_2, y_2) along paths that could possibly be contained in the $x - y$ plane as shown in Figure 1. To find the path along which the particle takes the shortest time to reach the bottom (x_2, y_2), one needs to first establish the particle's kinetic and potential energies which are a constant zero as the particle starts from rest and is measured from $(0, 0)$ as its origin.

The particle's kinetic and potential energies are zero at the origin. As this is a conservative system, the total energy should remain a constant zero at any point along the path. Therefore, the speed of the particle at any one point is given by

$$\frac{1}{2}mv^2 - mgx = 0 \rightarrow v = \sqrt{2gx}$$

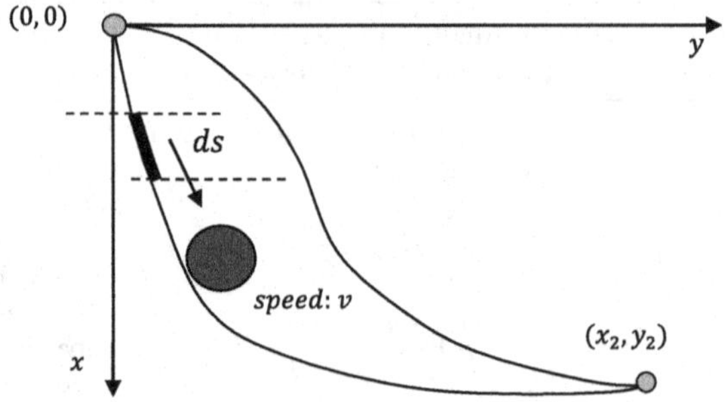

Fig. 1. Schematic illustrations of the different trajectories through which a massive particle moves from origin to destination under the pull of a downward force (gravity).

(a) As the distance traveled within a short period of time is $ds = (dx^2 + dy^2)^{\frac{1}{2}}$, show that the time spent traveling from $(0, 0)$ to (x_2, y_2) is given by

$$t = \int \left(\frac{1 + (y')^2}{2gx}\right)^{\frac{1}{2}} dx, \quad \text{where } y' = \frac{dy}{dx}$$

(b) Write down Euler's equation and explain how it can be used to find the minimum transit time.

(c) Write down the x and y expressions that form the cycloid equation for the shortest-time trajectory. Show (see Figure 2) by analyzing the geometry of the circle that at point $P(x, y)$ one can deduce the x and y expressions that correspond exactly to the ones derived.

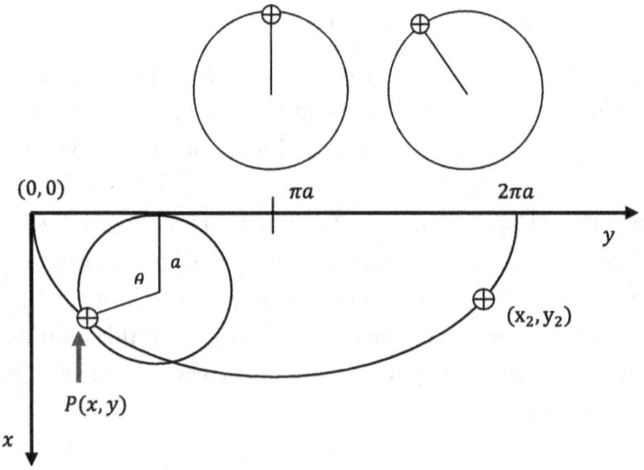

Fig. 2. The cycloid trajectory through which the massive particle moves is the shortest time trajectory for the particle to reach the bottom.

Solution

(a) Distance traveled within the short period of time is $ds = (dx^2 + dy^2)^{\frac{1}{2}}$. Since the particle speed is $v = \sqrt{2gx}$, the time taken over this short distance is

$$t = \int_{0,0}^{x_2,y_2} \frac{ds}{v} = \int_{0,0}^{x_2,y_2} \left(\frac{dx^2 + dy^2}{2gx}\right)^{\frac{1}{2}} = \int \left(\frac{1 + (y')^2}{2gx}\right)^{\frac{1}{2}} dx$$

(b) Euler's equation of

$$\frac{\partial f}{\partial y} - \frac{d}{dx}\left(\frac{\partial f}{\partial y'}\right) = 0$$

is a necessary condition for $J = \int_{x1}^{x_2} f\{y(x), y'(x); x\}dx$ to take on an extremum value. Time of transit is intended to be minimum, thus one could figure from

$$t = \int \left(\frac{1 + (y')^2}{2gx}\right)^{\frac{1}{2}} dx$$

that

$$f = \left(\frac{1 + (y')^2}{x}\right)^{\frac{1}{2}}$$

Since f is a function of y' and x, it is clear that $\frac{\partial f}{\partial y} = 0$. As a result, Euler's equation becomes $\frac{d}{dx}\left(\frac{\partial f}{\partial y'}\right) = 0$. This translates to

$$\frac{\partial f}{\partial y'} = \text{CONSTANT}$$

Constant is chosen to be $1/\sqrt{2a}$ for convenience:

$$\frac{\partial f}{\partial y'} = \frac{1}{\sqrt{2a}}$$

Squaring both sides,

$$\frac{y'^2}{x(1 + y'^2)} = \frac{1}{2a} \to (y')^2 = \frac{x}{2a - x}$$

The above leads to integral

$$y = \int \frac{x}{(2ax - x^2)^{\frac{1}{2}}} dx$$

With the change of variable: $x = a(1 - \cos\theta)$, $dx = a\sin\theta\, d\theta$, the integral would now be

$$y = \int \frac{\sin\theta(1-\cos\theta)}{(2(1-\cos\theta)-(1-\cos\theta)^2)^{\frac{1}{2}}}\, d\theta$$

Finally,

$$y = \int (1-\cos\theta)\, d\theta$$

(c) As the circle rolls along axis Y, the distance R marked by the red bar is equal to the arc of the circle spanned by θ. Therefore, referring to the dotted lines in Figure 3, one can deduce as follows:

along axis X:

$$x = a + a\cos(180 - \theta) = a - a\cos\theta$$

along axis Y:

$$y = R - a\sin(180 - \theta) = R - a\sin\theta$$

Finally,

$$y = a\,\theta - a\sin\theta$$

See Figure 3 for illustration.

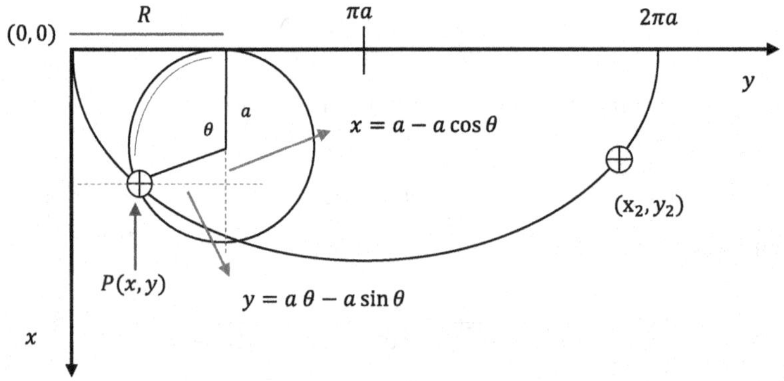

Fig. 3. The cycloid trajectory through which the massive particle moves is the shortest time trajectory for the particle to reach the bottom.

Problem 7.03 The Hamilton Principle

Hamilton's principle states that of all the possible paths a dynamical system might take moving from one point to another within a specified time interval and subject to the constraints that prevail, the actual path chosen would be that which minimizes a special time integral for a quantity known as the Lagrangian — the difference between the kinetic and the potential energy.

(a) The physical outcome of Hamilton's principle is the Euler–Lagrange equation which can be used to derive the equation of motion for a mechanical system. Show that Hamilton's principle is used to derive the equation of motion for a one-dimensional harmonic oscillator as shown in Figure 4.

$$m\frac{d^2x}{dt^2} + kx = 0$$

(b) The physical outcome of Hamilton's principle is the Euler–Lagrange equation which can be used to derive the equation of motion for a mechanical system. Show that Hamilton's principle is used to derive the equation of motion for a pendulum as shown in Figure 5.

$$\frac{d^2\theta}{dt^2} + \frac{g}{l}\sin\theta = 0$$

(c) The projectile inscribes a projectile trajectory in a 2D (x, y) system under the downward pull $(-y)$ of the gravity as shown in Figure 6. Use the Euler–Lagrange method to derive the equations of motion for the projectile in the x and y dimensions, where g is the gravity and u

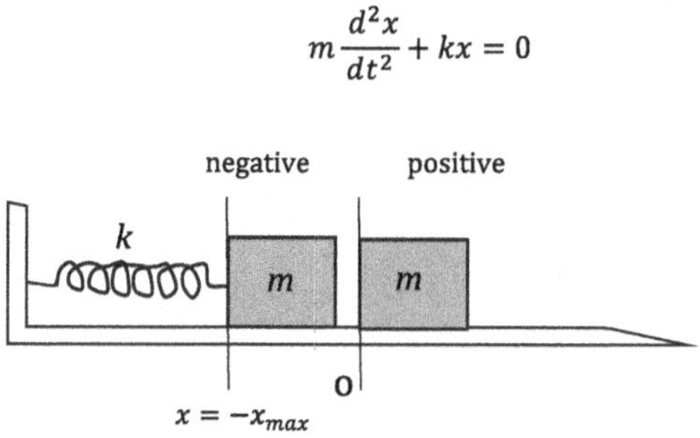

Fig. 4. A spring–mass system that performs the simple harmonic oscillation.

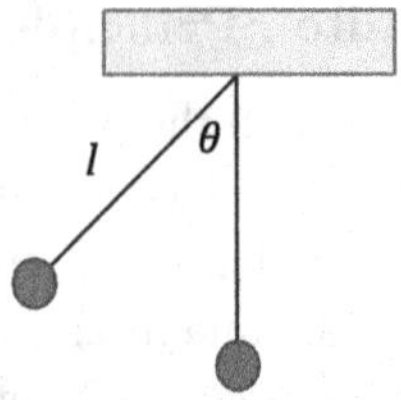

Fig. 5. A string–mass pendulum performs the simple harmonic oscillation.

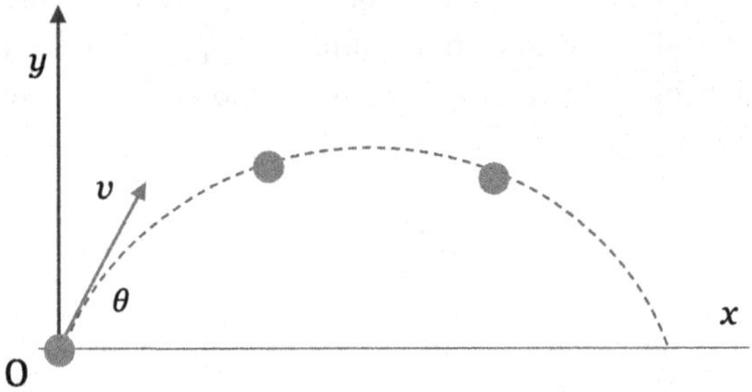

Fig. 6. The projectile inscribes a projectile trajectory in a 2D (x, y) system under the downward pull $(-y)$ of the gravity.

and v are, respectively, the initial and the final velocity as follows:

$$x: \; v_x = u_x$$

$$y: \; v_y = u_y - gt$$

Solution

(a) A one-dimensional harmonic oscillator is characterized by the kinetic and potential energies as follows:

$$T = \frac{1}{2}m\dot{x}^2 \quad and \quad U = \frac{1}{2}kx^2$$

The Lagrangian of the system is

$$L = T - U = \frac{1}{2}m\dot{x}^2 - \frac{1}{2}kx^2$$

Therefore,

$$\frac{\partial L}{\partial x} = -kx; \quad \frac{\partial L}{\partial \dot{x}} = m\dot{x} \rightarrow \frac{d}{dt}\left(\frac{\partial L}{\partial \dot{x}}\right) = m\ddot{x}$$

Using the Euler–Lagrange equation,

$$\frac{\partial L}{\partial x} - \frac{d}{dt}\left(\frac{\partial L}{\partial \dot{x}}\right) = 0 \rightarrow m\frac{d^2x}{dt^2} + kx = 0$$

(b) A pendulum has the Lagrangian as follows:

$$L = \frac{1}{2}ml^2\dot{\theta} - mgl(1 - \cos\theta)$$

$$\frac{\partial L}{\partial \theta} = -mgl\sin\theta; \quad \frac{\partial L}{\partial \dot{\theta}} = ml^2\dot{\theta} \rightarrow \frac{d}{dt}\left(\frac{\partial L}{\partial \dot{\theta}}\right) = ml^2\ddot{\theta}$$

Using the Euler–Lagrange equation,

$$\frac{\partial L}{\partial \theta} - \frac{d}{dt}\left(\frac{\partial L}{\partial \dot{\theta}}\right) = 0 \rightarrow \frac{d^2\theta}{dt^2} + \frac{g}{l}\sin\theta = 0$$

(c) The kinetic and potential energies of the projectile are

$$T = \frac{1}{2}m\acute{x}^2 + \frac{1}{2}m\acute{y}^2$$

$$U = mgy, \quad U = 0 \; at \; y = 0$$

The Lagrangian is

$$L = T - U = \frac{1}{2}m\acute{x}^2 + \frac{1}{2}m\acute{y}^2 - mgy$$

Equation of motion along X:

$$\frac{\partial L}{\partial x} - \frac{d}{dt}\left(\frac{\partial L}{\partial \dot{x}}\right) = 0 \rightarrow \frac{d}{dt}(m\dot{x}) = 0 \rightarrow m\ddot{x} = 0$$

In integral form,

$$\int \frac{d}{dt}\left(\frac{dx}{dt}\right) dt = 0$$

$$\int_{u_x}^{v_x} d\left(\frac{dx}{dt}\right) = 0 \rightarrow v_x - u_x = 0$$

The velocity along X is a constant.

Equation of motion along Y:

$$\frac{\partial L}{\partial y} - \frac{d}{dt}\left(\frac{\partial L}{\partial \dot{y}}\right) = 0 \rightarrow -mg - \frac{d}{dt}(m\dot{y}) = 0 \rightarrow m\ddot{y} = -g$$

In integral form,

$$\int \frac{d}{dt}\left(\frac{dy}{dt}\right) dt = \int -gdt$$

$$\int_{u_y}^{v_y} d\left(\frac{dy}{dt}\right) = \int_0^t -g \to v_y = u_y - gt$$

Electron Physics and Visualizations

Problem 7.04 What is the Size and Shape of an Electron?

Electron, as is any particle in quantum physics, is commonly pictured by physicists as some wave that permeates space, based on the theory of the wave–particle duality.

(a) Therefore, some physicists like to imagine that the electron's volume extends infinitely. What do you think?
(b) It is very convenient to imagine that electrons are round like little balls. While this picture is handy and practical, it is a very inaccurate representation. What do you think?

Note: What follows in the Solution section are really not solutions per se. They are mostly arguments and reasoning based on knowledge of quantum physics.

Solution

(a) The electron wave that permeates space does not represent an electron stretching in space. The wave of an electron is related to the probability of locating an electron, but it is actually not the electron in the sense of filling up space. In fact, whenever an electron is found at some point in space, it is always found in its entirety. It has never been found with half of its mass at one point and the other half in a different location nor has one been found with mass and spin at two different locations in space.

What the electron should be seen in a space volume should be the volume over which its fundamental *properties* of mass, charge, and spin

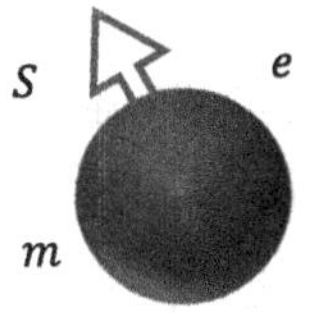

Fig. 7. An imagined picture of an electron with mass, spin, and charge.

spread (Figure 7). The volume is defined by the boundaries encircling these properties. But unfortunately, there is no theoretical evidence that points to an electron spreading its *properties* in space. The wavefunction that leads one to find an electron gives no indication either that the electron spreads its mass or charge before it is found. Wavefunction merely gives the spread of a particle's chance of being found in space.

From this reasoning, one could form a simple picture that electron is a collection of mass, charge, and spin at one space point. These properties do not spread in space and they are bound at one point by a zero-dimension boundary.

(b) Since electron is pictured as a point particle with mass, charge, and spin centered at that point, it has a zero dimension in its size. Since it has no dimension, shape cannot be defined. It may be tempting to liken the electron's wavefunction spread to defining volume and possibly shape. For example, one may argue that since the wavefunction of an electron permeates the entire space — and wavefunction represents the probability of locating the electron — electron should be considered to permeate the entire space. Therefore, electron should have the volume of the entire space.

The difficulty with this argument lies in the wavefunction, which without doubt permeates the entire space. But the wavefunction is not like a fundamental property of the electron. The volume of a particle should be defined by the space that contains the distribution of stuff like mass, spin, and charge. Therefore, the simple answer is that electron has no shape. The more elaborate answer is that electron has never been known to distribute its mass, charge, or spin in space. The short answer is that electrons or any other particles are not round like balls.

Remarks and Reflections

Points to ponder: There are indeed effective particles with fractional charge or spin–charge in separation. Does that mean such particle has a large volume if the distribution of property in space is anything to go by?

Points to ponder: What about exotic particles with fractional charge and spin–charge in separation? Does that mean such particle has a large volume if the distribution of property in space is anything to go by?

Problem 7.05 What is Spin?

(a) Does the electron really need to spin like a spinning top?
(b) How does an electron spin react to the magnetic field?

Note: What is presented in the Solution section is not really a solution to a problem per se. They are mostly arguments and reasoning based on our understanding of quantum physics.

Solution

(a) Momentum P is a quantity that is intrinsic to a particle. It can be related or unrelated to a particle's motion. For example, a photon with a constant P can slow down in a medium without much of a problem. In our big classical world, P is directly proportional to velocity; that's probably how P got the name that sounds so motion-like.

The same goes for angular momentum L, which measures angular velocity in a classical, massive sense like a spinning top or our dear planet earth. But in the case of a small particle, L can just exist on its own without having to measure any angular velocity. That L, which measures not the angular velocity, is called spin S to distinguish it from the L that measures angular velocity. In short, P or L can exist even in particles without motion. As for their multiples, integer, or half, I cannot think of any good explanation yet, although we can say a lot about the physical consequences of being such multiple. In my opinion, the electron doesn't need to spin in the mechanical sense of the word in order to carry the physical property of spin.

(b) Spin of a charged massive particle reacts to the electromagnetic field in a way known as the Zeeman interaction. Thus, one should suspect that "spin" when existing in simultaneity with "charge" and "mass" rotates (in a crude language) the "charge" and the "mass" to produce a charge dynamic. This explains why electron spin promises an intrinsic magnetic moment. Indeed, out of simplicity, one can imagine the electron as spinning to generate a magnetic moment. But as reasoned above, it does not need to spin to carry the physical property that mimics mechanical spinning.

Problem 7.06 Single Electronic Motion: Master Equation

Figure 8 shows a singly or doubly occupied single electronic device comprising a structure of FM/QD/FM, where FM stands for ferromagnetic leads and QD stands for the quantum dot central. The left and right leads comprise a pair of non-collinear magnetization with an angular deviation denoted by θ. Energy terms U_e and U_h are the Coulomb potential due to the presence of electron and hole, respectively, in the quantum dot.

The master equation here comprises a set of six equations that describe the rate of change of the particle density (electron or hole) in the quantum dot as shown in the following. At equilibrium, particle density is a constant translating to a zero rate of change.

$$\frac{dn_\sigma}{dt} = \Gamma^{in}_{n\sigma}\left[1 - n_\sigma - n_{\bar{\sigma}} - h_\sigma - h_{\bar{\sigma}} + n_{\uparrow\downarrow} + h_{\uparrow\downarrow}\right] + \tilde{\Gamma}^{in}_{n\sigma}\left[n_{\bar{\sigma}} - n_{\uparrow\downarrow}\right]$$
$$- \Gamma^{out}_{n\sigma}\left[n_\sigma - n_{\uparrow\downarrow}\right] - \tilde{\Gamma}^{out}_{n\sigma}\, n_{\uparrow\downarrow} - \Gamma^{f}_{n\sigma}\left[n_\sigma - n_{\bar{\sigma}}\right] \tag{A}$$

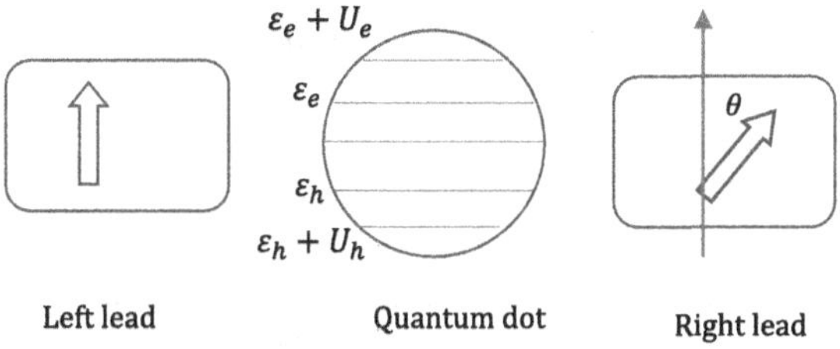

Fig. 8. A quantum dot device with Coulomb blockade tracks electron motion to the single-electron level.

As n denotes either the electron or hole and σ denotes up or down, Eq. (A) is a compact expression of four independent equations.

$$\frac{dn_{\uparrow\downarrow}}{dt} = \tilde{\Gamma}^{in}_{n\uparrow}\left[n_{\downarrow} - n_{\uparrow\downarrow}\right] + \tilde{\Gamma}^{in}_{n\downarrow}\left[n_{\uparrow} - n_{\uparrow\downarrow}\right] - \left[\tilde{\Gamma}^{out}_{n\uparrow} + \tilde{\Gamma}^{out}_{n\downarrow}\right] n_{\uparrow\downarrow} \qquad \text{(B)}$$

As n denotes either the electron or hole, Eq. (B) is a compact expression of two independent equations.

(a) Explain the physical meaning of the tunneling rates as shown in the following table.

	Tunneling Rate	Physical Meaning
1.	$\Gamma^{in}_{n\sigma}$	
2.	$\tilde{\Gamma}^{in}_{n\sigma}$	
3.	$\Gamma^{out}_{n\sigma}$	
4.	$\tilde{\Gamma}^{out}_{n\sigma}$	
5.	$\Gamma^{f}_{n\sigma}$	

(b) Explain the physical meanings of all the states of occupation in the quantum dot as shown in the following table.

	Particle Occupation States	Physical Meaning
1.	$[1 - n_\sigma - n_{\bar{\sigma}} - h_\sigma - h_{\bar{\sigma}} + n_{\uparrow\downarrow} + h_{\uparrow\downarrow}]$	
2.	$[n_{\bar{\sigma}} - n_{\uparrow\downarrow}]$	
3.	$[n_\sigma - n_{\uparrow\downarrow}]$	
4.	$n_{\uparrow\downarrow}$	
5.	n_σ	

(c) Derive the expression for the current that passes through the quantum dot.

(d) Write down the expression for the total current and explain your answer.

(e) Expressions for the individual current are given in the following table. Explain the physical meanings of the current terms.

	Current	Physical Meaning
1.	$\Gamma^{out}_{e\sigma R}\,(n_\sigma - n_{\uparrow\downarrow})$	
2.	$\tilde{\Gamma}^{out}_{e\sigma R}\, n_{\uparrow\downarrow}$	
3.	$\tilde{\Gamma}^{in}_{h\sigma R}\,(h_{\bar{\sigma}} - h_{\uparrow\downarrow})$	
4.	$+\Gamma^{in}_{h\sigma R}(1 - n_\sigma - n_{\bar{\sigma}} - h_\sigma - h_{\bar{\sigma}}$ $+n_{\uparrow\downarrow} + h_{\uparrow\downarrow})$	
5.	$\Gamma^{out}_{h\sigma R}\,(h_\sigma - h_{\uparrow\downarrow})$	
6.	$\tilde{\Gamma}^{out}_{h\sigma R}\, h_{\uparrow\downarrow}$	
7.	$\tilde{\Gamma}^{in}_{n\sigma R}\,(n_{\bar{\sigma}} - n_{\uparrow\downarrow})$	
8.	$\Gamma^{in}_{e\sigma R}(1 - n_\sigma - n_{\bar{\sigma}} - h_\sigma - h_{\bar{\sigma}}$ $+n_{\uparrow\downarrow} + h_{\uparrow\downarrow})$	

Solution

(a) Physical meaning of the different tunneling rates through quantum dot:

	Tunneling Rate	Physical Meaning
1.	$\Gamma^{in}_{n\sigma}$	Tunneling of a spin particle into an empty QD
2.	$\tilde{\Gamma}^{in}_{n\sigma}$	Tunneling of a spin particle into a singly occupied QD
3.	$\Gamma^{out}_{n\sigma}$	Tunneling of a spin particle out of a singly occupied QD
4.	$\tilde{\Gamma}^{out}_{n\sigma}$	Tunneling of a spin particle out of a doubly occupied QD
5.	$\Gamma^{f}_{n\sigma}$	Spin flipping rate of a spin sigma particle into the opposite spin state

(b) The particle occupation states are explained as follows:

	Particle Occupation States	Physical Meaning
1.	$[1 - n_\sigma - n_{\bar\sigma} - h_\sigma - h_{\bar\sigma} + n_{\uparrow\downarrow} + h_{\uparrow\downarrow}]$	The quantum dot is empty
2.	$[n_{\bar\sigma} - n_{\uparrow\downarrow}]$	The quantum dot is occupied by one particle of spin bar sigma only
3.	$[n_\sigma - n_{\uparrow\downarrow}]$	The quantum dot is occupied by one particle of spin sigma only
4.	$n_{\uparrow\downarrow}$	The quantum dot is doubly occupied by particle of two opposite spin
5.	n_σ	The quantum dot is occupied by either one particle of spin sigma or two particles of opposite spin

(c) Current through the quantum dot can be obtained by reviewing electrons and holes tunneling in and out of the dot across one specific junction — in this case, between the right electrode and the quantum dot — as shown in the following:

$$\vec{I} = \frac{e}{h}\sum_\sigma \Gamma^{out}_{e\sigma R}\,(n_\sigma - n_{\uparrow\downarrow}) + \tilde{\Gamma}^{out}_{e\sigma R}\,n_{\uparrow\downarrow}$$

$$+\tilde{\Gamma}^{in}_{h\sigma R}\,(h_{\bar\sigma} - h_{\uparrow\downarrow}) + \Gamma^{in}_{h\sigma R}(1 - n_\sigma - n_{\bar\sigma} - h_\sigma - h_{\bar\sigma} + n_{\uparrow\downarrow} + h_{\uparrow\downarrow})$$

$$\overleftarrow{I} = \left(-\frac{e}{h}\right)\sum_\sigma \Gamma^{out}_{h\sigma R}\,(h_\sigma - h_{\uparrow\downarrow}) + \tilde{\Gamma}^{out}_{h\sigma R}\,h_{\uparrow\downarrow} + \tilde{\Gamma}^{in}_{n\sigma R}\,(n_{\bar\sigma} - n_{\uparrow\downarrow})$$

$$+\Gamma^{in}_{e\sigma R}(1 - n_\sigma - n_{\bar\sigma} - h_\sigma - h_{\bar\sigma} + n_{\uparrow\downarrow} + h_{\uparrow\downarrow})$$

(d) The total current is

$$I = \vec{I} + \overleftarrow{I}$$

It can be explained as follows:

$\vec{I}$ — Electron flow from dot to the right electrode
$\overleftarrow{I}$ — Electron flow from the right electrode to dot

(e) The individual current associated with each occupation state is explained as follows:

Current	Physical Meaning
1. $\Gamma^{out}_{e\sigma R}\,(n_\sigma - n_{\uparrow\downarrow})$	Electron current of spin flows from QD to R when QD is singly occupied by an electron of spin
2. $\tilde{\Gamma}^{out}_{e\sigma R}\, n_{\uparrow\downarrow}$	Electron current of spin flows from QD to R when QD is doubly occupied by two electrons of opposite spin
3. $\tilde{\Gamma}^{in}_{h\sigma R}\,(h_{\bar{\sigma}} - h_{\uparrow\downarrow})$	Hole current of spin flows into QD from R when QD is singly occupied by a hole of opposite spin
4. $+\Gamma^{in}_{h\sigma R}(1 - n_\sigma - n_{\bar{\sigma}} - h_\sigma - h_{\bar{\sigma}} + n_{\uparrow\downarrow} + h_{\uparrow\downarrow})$	Hole current of spin flows into QD from R when QD is empty
5. $\Gamma^{out}_{h\sigma R}\,(h_\sigma - h_{\uparrow\downarrow})$	Hole current of spin flows from QD to R when QD is singly occupied by a hole of spin
6. $\tilde{\Gamma}^{out}_{h\sigma R}\, h_{\uparrow\downarrow}$	Hole current of spin flows from QD to R when QD is doubly occupied by two holes of opposite spin
7. $\tilde{\Gamma}^{in}_{n\sigma R}\,(n_{\bar{\sigma}} - n_{\uparrow\downarrow})$	Electron current of spin flows into QD from R when QD is singly occupied by an electron of opposite spin
8. $\Gamma^{in}_{e\sigma R}(1 - n_\sigma - n_{\bar{\sigma}} - h_\sigma - h_{\bar{\sigma}} + n_{\uparrow\downarrow} + h_{\uparrow\downarrow})$	Electron current of spin flows into QD from R when QD is empty

Remarks and Reflections

It was reasoned in (c) above that current through the quantum dot can be obtained by reviewing electrons and holes tunneling in and out of the dot across one specific junction — in this case, between the right electrode and the quantum dot. By contrast, the electron density rate of change of the quantum dot was studied taking into account all electrodes.

This can be understood from the continuity equation of

$$\frac{\partial n}{\partial t} + \nabla . j = 0$$

In the case of a two-electrode (right and left) device, density rate of change is approximately

$$\frac{\partial n}{\partial t} = \frac{j_R - j_L}{\delta l}$$

Note that to obtain the steady-state electron or hole density, $\frac{\partial n}{\partial t} = 0$. This would result in

$$j_R = j_L$$

Therefore, focusing our attention on one junction alone would be enough to calculate the current through the quantum dot.

Problem 7.07 Semi-classical Electronic Transport

Under the framework of Boltzmann transport in bulk electron system, the expression of current is

$$\mathcal{J} = -en\boldsymbol{v} = -\frac{e}{V}\sum_{\lambda k} \langle G \,|\, \hat{n}_{k\lambda} \,|\, G \rangle \, \boldsymbol{v}$$

where n is the particle density (per unit volume) and $\langle G \,|\, \hat{n}_{k\lambda} \,|\, G \rangle$ is the number of particles for one quantum state of k, λ.

(a) Show that the current can be expressed in terms of the equilibrium and the linear response parts

$$\mathcal{J} = \frac{-e^2\tau}{\hbar 4\pi^3} \int \frac{dS_F}{v_F} \, \boldsymbol{v_F}(\boldsymbol{E}.\boldsymbol{v_F}) + \int \boldsymbol{v} \frac{\theta(E_F - E)}{\tau} dk \, dS$$

Linear response | Equilibrium

(b) Show that the linear response part of the current is

$$\mathcal{J}_x = \frac{-e^2 l_F}{\hbar 3\pi^2} E_x(k_F^2)$$

(c) Show that the equilibrium part of the current is 0.

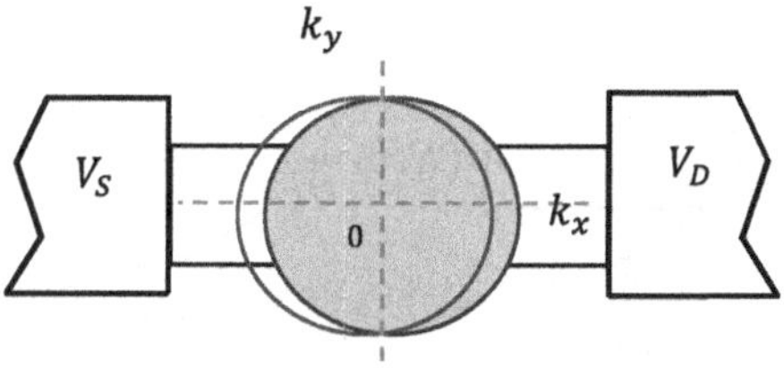

Fig. 9. Schematic illustration of the Fermi circle shift in the central region of a nano-electronic device.

Solution

In the presence of the electric field (applied from source to drain as shown in Figure 9), and assuming electronic charge is homogenously distributed, the Fermi distribution is perturbed as follows:

$$-\frac{\Delta f}{\tau} = 0 + 0 - e\boldsymbol{E}.\boldsymbol{v}\frac{\partial f_0}{\partial E}$$

(a) From the quantum expression of $\mathcal{J} = -\frac{e}{V}\sum_{\lambda k}\langle G\,|\,\hat{n}_{k\lambda}\,|\,G\rangle\,\boldsymbol{v}$, total electron current, accounting for both spins can be deduced, i.e.,

$$\mathcal{J} = -\frac{e(2)V}{V}\int\frac{d^3k}{(2\pi)^3}\,f\,v$$

where two electrons take up a volume of $\left(\frac{2\pi}{\mathrm{L}}\right)^3$ in k-space. Substituting f into the above leads to

$$\mathcal{J} = \frac{-e}{4\pi^3}\int d^3k\,f\,v = \frac{-e^2\tau}{4\pi^3}\int d^3k\,v\left(\boldsymbol{E}.\boldsymbol{v}\frac{\partial f_0}{\partial E} + \frac{f_0}{\tau}\right)$$

Now, use is made of the following:

$$\frac{-\partial f_0}{\partial E} = \frac{kT/4}{\cosh^2\left[\frac{kT}{2}(E-E_F)\right]} \to \lim_{T\to 0}\frac{-\partial f_0}{\partial E} = \delta(E_F - E)$$

Integration of the above in Cartesian coordinates of $\int dk_x dk_y dk_z$ can be carried out in the spherical coordinates of $\int k^2\,\sin\theta\,dk\,d\theta\,d\varnothing$, which yields

$$\mathcal{J} = \frac{-e^2\tau}{4\pi^3}\int dk\, dS\, v\left((\boldsymbol{E}.\boldsymbol{v})\frac{\partial f_0}{\partial E} + \frac{f_0}{\tau}\right)$$
$$= \frac{-e^2\tau}{\hbar 4\pi^3}\int \frac{dE}{v}\, dS\, v\left((\boldsymbol{E}.\boldsymbol{v})\,\delta(E-E_F) + \frac{\theta(E_F - E)}{\tau}\right)$$

Where $dS = k^2\, sin\theta\, d\theta\, d\phi$. Note that $E = \frac{(\hbar k)^2}{2m} \to dE = \frac{\hbar^2 k}{m} dk = \hbar v\, dk$. Recalling that

$$\int f(x)\,\delta(x-a) = f(a)$$

one arrives at

$$\boldsymbol{\mathcal{J}} = \frac{-e^2\tau}{\hbar 4\pi^3}\int \frac{dS_F}{v_F}\,\boldsymbol{v_F}(\boldsymbol{E}.\boldsymbol{v_F}) + \int \boldsymbol{v}\frac{\theta(E_F - E)}{\tau} dk\, dS$$

Linear response　　Equilibrium

Note also that $NdE = \frac{V}{4\pi^3}\int dk\, dS_F$, where N is the number of states per unit energy. The current expression of the above consists of two parts; the second part vanishes as we will show later.

(b) The first part can be written as

$$\mathcal{J}_i = \frac{-e^2\tau}{\hbar 4\pi^3}\int \frac{dS_F}{v_F} v_i v_j E_j = \sigma_{ij} E_j$$

Taking a digression, the Einstein summation convention is used, i.e., $v_j E_j = \sum_j^d v_j E_j$, where d is the dimension of the system. Using the fact that $v_x^2 = v_F^2 sin^2\theta cos^2\phi\; v_y^2 = v_F^2 sin^2\theta\, sin^2\phi\; v_z^2 = cos^2\theta$ show that the conductivity should not depend on the label of the reference axis, i.e., $\sigma^{xx} = \sigma^{yy} = \sigma^{zz}$ and the current expressions are

$$\mathcal{J}_x = \sigma^{xx} E_x, \quad \mathcal{J}_y = \sigma^{yy} E_y, \quad \mathcal{J}_z = \sigma^{zz} E_z$$

One could then show that the current in the longitudinal direction, or the direction along which voltage is applied, is

$$\mathcal{J}_x = \frac{-e^2\tau}{\hbar 4\pi^3}\int \frac{dS_F}{v_F}\, v_x v_x\, E_x$$

As $v_x^2 = v_F^2 sin^2\theta\, cos^2\phi$, one has

$$\mathcal{J}_x = \frac{-e^2\tau}{\hbar 4\pi^3} E_x \iint_{0,0}^{\pi,\,2\pi} v_F\, sin^3\theta\, cos^2\phi\, E_x\, k_F^2\, d\theta d\phi$$
$$= \frac{-e^2\tau v_F}{\hbar 4\pi^3} E_x \left(\frac{4}{3}\pi k_F^2\right) = \frac{-e^2\, l_F}{\hbar 3\pi^2} E_x(k_F^2)$$

(c) Getting back to the second part, where $\boldsymbol{\mathcal{J}} = \frac{-e^2\tau}{\hbar 4\pi^3} \int \boldsymbol{v} \frac{\theta(E_F - E)}{\tau} dk\, dS$, one has

$$\mathcal{J}_x = \frac{-e^2}{\hbar 4\pi^3} \int v_x\, \theta(E_F - E) dkdS$$
$$= \frac{-e^2}{\hbar 4\pi^3} \iint_{0,0}^{\pi,\,2\pi} k^2 v_F sin^2\theta\, cos\phi\; \theta(E_F - E)\, d\theta\, d\phi dk$$

Finally,

$$\mathcal{J}_x = \frac{-e^2}{\hbar 4\pi^3} \int_{-\infty}^{\infty} \int_0^{\pi} \int_0^{2\pi} k^2 v_F\, sin^2\theta cos\phi\; \theta(E_F - E)\, d\theta\; d\phi dk = 0$$

Remarks and Reflections

Taking a digression, one notes that in the Fermi gas, one defines the *de Broglie wavelength* (λ) as $\lambda = \frac{2\pi}{k}$, where $p(k)$ is the typical electron momentum (wavevector). For Fermi gas, the characteristic momentum is just the Fermi momentum. For the case of a single filled band in 2DEG where n_s is the sheet density, one can write $\lambda = \frac{2\pi}{k_F} = \sqrt{\frac{2\pi}{n_s}}$. For the Boltzmann gas, $p \approx \sqrt{2mkT}$, thus the Fermi wavelength is $\lambda = \frac{2\pi\hbar}{\sqrt{2mkT}}$. The non-vanishing part of the current expression can also be written in a more compact form, i.e.,

$$\mathcal{J}_x = \frac{-e^2\tau v_F}{\hbar 3\pi^2}(k_F^2)\, E_x = \frac{-e^2\tau}{m}\frac{k_F^3}{3\pi^2} E_x = \frac{-ne^2\tau}{m} E_x$$

It can be shown that $N = \sum_{k\lambda} \langle G\,|\, \hat{n}_{k\lambda}\,|\, G\rangle = \sum_{k\lambda} \theta(k_F - k)$, which simply leads to $N = \frac{V}{(2\pi)^3}\int_0^{k_F} d^3k \sum_\lambda \theta(k_F - k)$ and the highly useful and handy expression of the Fermi wavevector in terms of electron density $k_F^3 = 3\pi^2 n$.

The above has been carried out in the relaxation time approximation, i.e., all microscopic effect is lumped into τ.

Mathematical Methods

Problem 7.08 Integral Method

Show that

$$\int_0^1 x^{-x}dx = 1 - \frac{1}{2^2} + \frac{1}{3^3} - \frac{1}{4^4} + \cdots$$

Solution

At first sight, this looks like a problem involving some kind of discrete approximation to an integral, i.e., trapezium rule. However, that approach is likely doomed.

Instead, we can expand the integrand as a series:

$$\begin{aligned}\int_0^1 x^{-x}dx &= \int_0^1 e^{-x\log x}dx \\ &= \sum_{j=0} \int_0^1 \frac{(-x\log x)^j}{j!}dx \\ &= -\sum_{j=0} \int_0^\infty y^j e^{-(j+1)y}dy \\ &= \sum_{j=0} \frac{(-1)^j}{(j+1)^{j+1}}\end{aligned}$$

where in the last step, we have used the standard integral representation of the gamma function $\Gamma(j+1) = j!$

Problem 7.09 Divergence Theorem

Stokes' and the divergence theorems are commonly used in physics that involves geometry, e.g., electrodynamics and geometrical physics. We will take a look at a simple problem that uses the divergence theorem. The surface integral for A is given by

$$\oint_S \boldsymbol{A.ds}$$

where $\boldsymbol{A} = 2x\boldsymbol{e}_x + y\boldsymbol{e}_y - 2\boldsymbol{e}_z$ and $\boldsymbol{S}$ is the closed surface of a cylinder defined by $C^2 = x^2 + y^2$. The height of the cylinder is D. Find the volume integral, i.e., over the volume by surface $\boldsymbol{S}$ as follows:

$$\int_V (\nabla . \boldsymbol{A})\, dv$$

Show that the surface and the volume integral is related as follows:

$$\oint_S \boldsymbol{A} . \boldsymbol{ds} = \int_V (\nabla . \boldsymbol{A})\, dv$$

Solution

Lets' consider the volume integral on the RHS,

$$\int_V (\nabla . \boldsymbol{A})\; dv = 3 \int_V \; dv$$

Since the cylinder has a uniform surface throughout the height of D, the integral for one quadrant of the circle is

$$I = 3D \int_S dx\, dy = \int_0^C \int_0^{\sqrt{C^2 - y^2}} \; dxdy$$

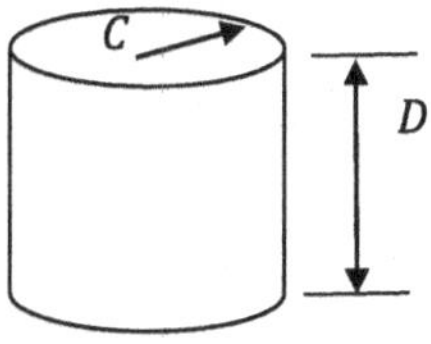

Fig. 10. A cylindrical structure with radius C and height D.

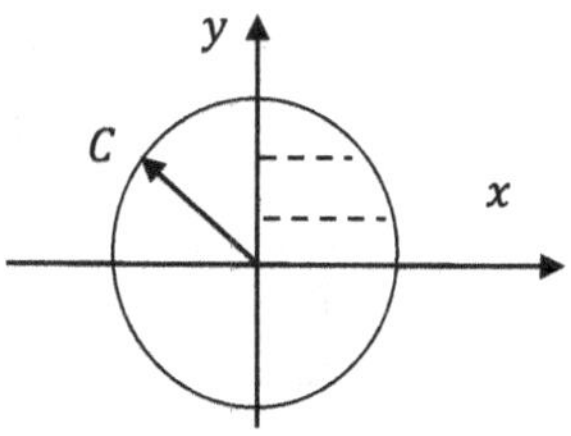

Fig. 11. Top view taken of the cylindrical structure of Figure 10.

$$I = 3D \int_S dx\, dy = 3D \int_0^C \int_0^{\sqrt{C^2-y^2}} dxdy$$
$$= 3D \int_0^C \sqrt{C^2 - y^2}\, dy$$

Using the substitution method, where $y = C \sin\theta$,

$$I = 3D \int_0^C C^2 \cos^2\theta\, d\theta = 3DC^2 \left[\frac{\sin 2\theta}{4} + \frac{\theta}{2}\right]_0^{2\pi} = \frac{3DC^2\pi}{4}$$

Summing over all four quadrants, the results are

$$I = 3DC^2\pi$$

Lets' consider the surface integral on the LHS

This integral will be carried out separately for the cylinder surface and the top and bottom surfaces (Figure 12).

The infinitesimal surface on the cylinder is

$$\boldsymbol{dS} = \boldsymbol{dl} \times \boldsymbol{dz}$$
$$\int \boldsymbol{A.dS} = \int \boldsymbol{A.}\ ((dx\, \boldsymbol{e_x} + dy\, \boldsymbol{e_y}) \times \boldsymbol{dz})$$
$$= \int \boldsymbol{A.} \left(\left(-\frac{y}{x}\, dy\, \boldsymbol{e_x} + dy\, \boldsymbol{e_y}\right) \times \boldsymbol{dz}\right)$$

The top view shows that $x^2 + y^2 = C^2 \rightarrow dx = -\frac{y}{x}\, dx$. Thus,

$$\int \boldsymbol{A.dS} = \int \left(2x + \frac{y^2}{x}\right) dy\, dz = D \int \left(2x + \frac{y^2}{x}\right) dy$$

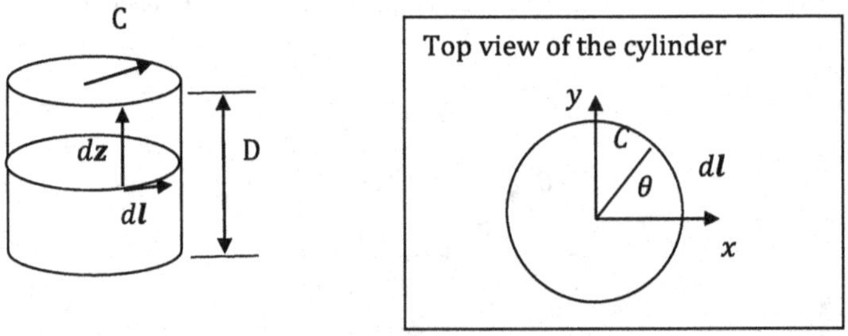

Fig. 12. Integration to be performed on the cylindrical surfaces with respect to the parameters.

The top view shows that $y = C\sin\theta$

$$\int_{Cylinder} \boldsymbol{A.dS} = 4D.\int_0^{\frac{\pi}{2}} \frac{2C^2 - y^2}{\sqrt{C^2 - y^2}}\,dy$$

$$= 4D.\int_0^{\frac{\pi}{2}} \frac{C^2 + C^2\cos^2\theta}{C\cos\theta}\,C\cos\theta\,dy$$

$$= 2DC^2\pi + 4DC^2\left(\frac{\pi}{4}\right) = 3DC^2\pi$$

Top and bottom surfaces:

Because $\boldsymbol{A.dS}$ is opposite on the top and bottom surfaces, i.e.,

$$\int_{Top} \boldsymbol{A.dS} + \int_{Bottom} \boldsymbol{A.dS} = 0$$

Finally,

$$I = \int_{Cylinder} \boldsymbol{A.dS} + \int_{Top} \boldsymbol{A.dS} + \int_{Bottom} \boldsymbol{A.dS} = 3DC^2\pi$$

Problem 7.10 Barycentric Coordinates

Barycentric coordinates symmetrically label N degrees of freedom when in fact only $N-1$ of them are linearly independent. They are useful, for instance, in symmetrically representing the center-of-mass coordinates of N particles, when their center-of-mass degree of freedom is already fixed. In the simple $N = 3$ case, they label the coordinates on a planar triangle. Derive the change of basis formulae between ordinary Cartesian coordinates and barycentric coordinates.

Solution

We first describe how to construct a manifestly $\mathrm{S_N}$-symmetric embedding of the tuple $(\bar{n}_1, \ldots, \bar{n}_N)$ onto the $(N-1)$-dimensional simplex in $\mathcal{R}^{N-1}$. We first express a state vector x as

$$\mathrm{x} = \kappa \sum_{k=1}^{N} \bar{n}_k \beta_k \tag{A}$$

where $x\epsilon\mathcal{R}^{N-1}$ and $\{\beta_k\}$, $k = 1, \ldots, N$ forms a linearly dependent basis set, also in $\mathcal{R}^{N-1}$, normalized so that

$$\beta_j \cdot \beta_k = \frac{N}{N-1}\delta_{jk} - \frac{1}{N-1} \tag{B}$$

Geometrically, the $\beta_k s$ define the vertices of a simplex and are angled $\cos^{-1}\left(-\frac{1}{N-1}\right)$ from one another. A basis consistent with the above requirements can be realized with

$$\begin{aligned}
\beta_1 &= (1, 0, \ldots, 0)\\
\beta_2 &= (C_1, S_1, \ldots, 0)\\
\beta_3 &= (C_1, S_1C_2, S_1S_2, \ldots, 0)\\
\beta_4 &= (C_1, S_1C_2, S_1S_2C_3, S_1S_2S_3, \ldots, 0)\\
&\vdots\\
\beta_{N-1} &= (C_1, S_1C_2, S_1S_2C_3, \ldots, S_1 \cdots S_{N-2})\\
\beta_N &= (C_1, S_1C_2, S_1S_2C_3, \ldots, -S_1 \cdots S_{N-2})
\end{aligned} \tag{C}$$

with $S_k^2 + C_k^2 = 1$, $1 \leq k \leq N-2$. Upon enforcing the scalar product constraint in Eq. (B), we require that

$$S_{k+1} = 1 - (1 + 1/N)/\left(\prod_{j=1}^{k} S_k\right)^2$$

which also implies that $C_k^2 + S_k^2\, C_{k+1} = C_k$. These requirements fortunately possess a simple solution:

$$C_k = -\frac{1}{N-k} \tag{D}$$

which implies that the kth projected component of the relative angles between the position vectors of the vertices approaches π as k and N increase. We proceed by substituting Eq. (D) into the explicit expressions of vertex positions β_k and expressing the latter in terms of the spherical coordinates. From Eq. (A), we can easily verify that

$$\bar{n}_k = \frac{N-1}{N_\kappa}\mathrm{x} \cdot \beta_k \tag{E}$$

and that

$$W^2 = |\mathrm{x}|^2 = \frac{N\kappa^2}{N-1}\sum_{k}^{N}\left(n_k - \frac{n}{N}\right)^2 = \frac{N\kappa^2}{N-1}\sum_{k}^{N}\bar{n}_k^2$$

We next express the state vector in Cartesian spherical coordinates, so as to explicitly connect the Cartesian and the barycentric coordinate representations. x takes the form

$$\begin{aligned}
x_1 &= W\cos\varphi_1\\
x_2 &= W\sin\varphi_1\cos\varphi_2\\
&\vdots\\
x_{N-2} &= W\cos\varphi_k\prod_{k=1}^{N-3}\sin\varphi_k\\
x_{N-1} &= W\prod_{k=1}^{N-2}\sin\varphi_k
\end{aligned}\tag{F}$$

Substituting the explicit expressions from Eqs. (C) and (F) into (E), we finally obtain

$$\begin{aligned}
\bar{n}_1 &= \frac{N-1}{N_\kappa}W\cos\varphi_1\\
\bar{n}_2 &= \frac{W}{N_\kappa}\left(-\cos\varphi_1 + \sqrt{N(N-2)}\sin\varphi_1\cos\varphi_2\right)
\end{aligned}$$

and

$$\begin{aligned}
\bar{n}_k = \frac{W}{N_\kappa}\Bigg[&-\cos\varphi_1 - \sqrt{\frac{N}{N-2}}\sin\varphi_1\cos\varphi_2\\
&-\sum_{j=2}^{k-2}\sqrt{\frac{N!}{(N-j-2)!}}\frac{(\prod_{i=1}^{j}\sin\varphi_i)\cos\varphi_{j+1}}{(N-j)(N-j-1)}\\
&+\sqrt{\frac{N!}{(N-k-1)!}}\frac{\left(\prod_{i=1}^{k-1}\sin\varphi_i\right)\cos\varphi_k}{N-k+1}\Bigg]
\end{aligned}\tag{G}$$

as well as

$$\bar{n}_{N-1} = \frac{W}{N\kappa}\left[-\cos\varphi_1 - \sqrt{\frac{N}{N-2}}\sin\varphi_1\cos\varphi_2\right.$$
$$-\sum_{j=2}^{N-3}\sqrt{\frac{N!}{(N-j-2)!}\frac{\left(\prod_{i=1}^{j}\sin\varphi_i\right)\cos\varphi_{j+1}}{(N-j)(N-j-1)}}$$
$$\left.+\frac{\sqrt{N!}}{2}\left(\prod_{i=1}^{N-2}\sin\varphi_i\right)\right]$$

$$\bar{n}_{N} = \frac{W}{N\kappa}\left[-\cos\varphi_1 - \sqrt{\frac{N}{N-2}}\sin\varphi_1\cos\varphi_2\right.$$
$$-\sum_{j=2}^{N-3}\sqrt{\frac{N!}{(N-j-2)!}\frac{\left(\prod_{i=1}^{j}\sin\varphi_i\right)\cos\varphi_{j+1}}{(N-j)(N-j-1)}}$$
$$\left.-\frac{\sqrt{N!}}{2}\left(\prod_{i=1}^{N-2}\sin\varphi_i\right)\right]$$

These are the explicit expressions for transcribing the $\bar{n}_j$, $1 \le j \le N$, indices directly into spherical coordinates. In (G), k ranges from 3 to $N-2$.

Chapter 8

Quantum Computation and Information

Chapter 8 is dedicated to the field of Quantum Computation and Information (QCI). The required quantum mechanical foundation of spin could be found in Chapter 2. This chapter begins with the introduction of the density matrix and its statistical and entanglement connotation. On the application side, the quantum gates of NOT, Hadamard, and CNOT are introduced. The concepts of quantum parallelism, entanglement, and no-cloning are introduced prior to the design of quantum circuits. The latter part of this chapter would introduce more advanced circuit designs like the teleportation and the implementation of algorithm. On the whole, this chapter can be used to teach quantum mechanics while showcasing its applications in QCI at the same time.

Quantum States of Qubits

Problem 8.01 Mixed State: Statistical Ensemble
Problem 8.02 Density Matrix and Expectation 1
Problem 8.03 Density Matrix and Expectation 11
Problem 8.04 Density Matrix of a 2-Qubit System
Problem 8.05 Quantum Measurement 1
Problem 8.06 Quantum Measurement 11
Problem 8.07 Projection Operator

Quantum Gates and Circuits

Problem 8.08 Quantum Logic Gates: NOT Gate
Problem 8.09 Quantum Logic Gates: Hadamard Gate
Problem 8.10 CNOT Gate
Problem 8.11 Toffoli Gate
Problem 8.12 Parallelism: Concept

Problem 8.13 Parallelism: Quantum Circuit and Bell States
Problem 8.14 No-Cloning Principle
Problem 8.15 Quantum Circuit: No-Cloning
Problem 8.16 Quantum Circuit: Swapping
Problem 8.17 Mathematical Methods: Boolean Algebra
Problem 8.18 Quantum Communication: Teleportation
Problem 8.19 Deutsch Algorithm

Quantum States of Qubits

Problem 8.01 Mixed State: Statistical Ensemble

In statistical quantum mechanics, the density operator is very important. Also known as the density matrix, the density operator is formulated to describe a system that cannot be described by a pure state. Such system is also known as the mixed state, which comprises an ensemble of pure states. A typical example of a mixed state is the canonical thermal distribution characterized by density operator $\rho = \frac{e^{-\beta H}}{\text{Trace}\,[e^{-\beta H}]}$. Assuming $\text{Trace}\,[e^{-\beta H}] = 1$ and E_v the energy of the eigenstate, show that

(a) the density operator can be written as

$$\rho = e^{-\beta H} = \sum_v |v\rangle\langle v| e^{-\beta E_v}$$

(b) the partition function is written as follows:

$$Z = \text{Trace}\,[e^{-\beta H}] = \sum_v e^{-\beta E_v}$$

Note: Density matrix and density operator are used in an interchangeable manner.

Solution

(a) In terms of the projection operators of the individual eigenstates, $\rho = e^{-\beta H}$ can be written as follows:

$$\rho = e^{-\beta H} \sum_v |v\rangle\langle v|$$

which leads upon H acting on the eigenstates to

$$\rho = \sum_v |v\rangle\langle v| e^{-\beta E_v}$$

(b) The partition function is

$$Z = \text{Trace}\left[e^{-\beta H}\right] = \sum_v \langle v|e^{-\beta H}|v\rangle$$
$$= \sum_n \sum_v \langle v|n\rangle\langle n|e^{-\beta E_n}|v\rangle = \sum_v e^{-\beta E_v}$$

where use is made of $\langle v|n\rangle = \delta_{vn}$ as a Kronecker delta.

Remarks and Reflection

Now, we will write down the matrix explicitly,

$$Z = \sum_v \langle v|e^{-\beta H}|v\rangle$$

Consider Z as the sum of some matrix components in the following manner:

$$Z = \text{Trace}\begin{bmatrix} \langle 1|e^{-\beta H}|1\rangle & \langle 1|e^{-\beta H}|2\rangle & \langle 1|e^{-\beta H}|3\rangle \\ \langle 2|e^{-\beta H}|1\rangle & \langle 2|e^{-\beta H}|2\rangle & \langle 2|e^{-\beta H}|3\rangle \\ \langle 3|e^{-\beta H}|1\rangle & \langle 3|e^{-\beta H}|2\rangle & \ldots\ldots \end{bmatrix}$$
$$= \langle 1|e^{-\beta H}|1\rangle + \langle 2|e^{-\beta H}|2\rangle + \cdots$$

Therefore,

$$Z = \text{Trace}\left[e^{-\beta H}\right]$$

Problem 8.02 Density Matrix and Expectation 1

The density matrix is an alternative formulation for a system consisting of an ensemble of pure states. It can therefore be used to describe the expectation of an observable in that ensemble state. Show that the expectation value of the observable A given in matrix form will be

$$\langle A\rangle = \text{Trace}\left[\rho A\right]$$

where ρ is the density operator.

Note: $\rho = e^{-\beta H} = \sum_v |v\rangle\langle v|e^{-\beta E_v}$

Solution

One can write down Trace $[\rho A]$ as follows:

$$\text{Trace}\,[\rho A] = \text{Trace}\left[\sum_v |v\rangle\langle v|e^{-\beta E_v} A\right]$$

Following the definition of Trace$[R]$, as shown in the following:

$$\text{Trace}\,[R] = \text{Trace}\begin{bmatrix} \langle 1|R|1\rangle & \langle 1|R|2\rangle & \langle 1|R|3\rangle \\ \langle 2|R|1\rangle & \langle 2|R|2\rangle & \langle 2|R|3\rangle \\ \langle 3|R|1\rangle & \langle 3|R|2\rangle & \ldots. \end{bmatrix}$$
$$= \langle 1|R|1\rangle + \langle 2|R|2\rangle + \cdots$$

one could deduce that the expectation of A is given by

$$\text{Trace}\,[\rho A] = \sum_a \langle a| \sum_v |v\rangle\langle v|e^{-\beta E_v} A|a\rangle$$
$$= \sum_a \delta_{av} \sum_v \langle v|e^{-\beta E_v} A|a\rangle$$

Taking note that $\langle a|v\rangle = \delta_{av}$ is the Kronecker delta,

$$\text{Trace}\,[\rho A] = \sum_v e^{-\beta E_v} \langle v|A|v\rangle$$

In a general format, the expectation of an operator A is given by

$$\langle A\rangle = \text{Trace}\,[\rho A] = \sum_v P_v \langle v|A|v\rangle$$

where P_v is the probability of existence of a state denoted by the subscript in the entire ensemble.

Remarks and Reflections

Similarly, note that for $\rho = \sum_v |v\rangle\langle v|e^{-\beta E_v}$, one has

$$\rho_{ab} = \langle a| \sum_v |v\rangle\langle v|e^{-\beta E_v}|b\rangle = \delta_{ab} e^{-\beta E_a}$$

Note that in the above, v is the running index, while a is a fixed number. Therefore,

$$\sum_a \rho_{aa} = \text{Trace}\,[\rho]$$

$$= \sum_a \langle a| \sum_v |v\rangle\langle v| e^{-\beta E_v} |a\rangle$$

$$= \sum_v e^{-\beta E_v} = Z$$

The normalized expectation of an observable in the ensemble is thus given by

$$\langle A\rangle = \frac{\text{Trace}\,[\rho A]}{\text{Trace}\,[\rho]}$$

Problem 8.03 Density Matrix and Expectation 11

The expectation value of the observable A is given in terms of its density matrix ρ as follows:

$$\text{Trace}\,[\rho A]$$

Show that the above can also be written using the Einstein convention for tensor summation as

$$\text{Trace}\,[\rho A] = \rho_{ab} A_{ba}$$

where ρ_{ab} is a general representation of matrix components $[\rho]$. The same is true for $[A]$.

Solution

Noting that

$$\rho = \sum_v |v\rangle\langle v| e^{-\beta E_v}$$

one can write down $\text{Trace}\,[\rho A]$ as follows:

$$\text{Trace}\,[\rho A] = \text{Trace}\left[\sum_v |v\rangle\langle v| e^{-\beta E_v} A\right]$$

Following the definition of Trace $[R]$, as shown in the following:

$$\text{Trace}\,[R] = \text{Trace}\begin{bmatrix} \langle 1|R|1\rangle & \langle 1|R|2\rangle & \langle 1|R|3\rangle \\ \langle 2|R|1\rangle & \langle 2|R|2\rangle & \langle 2|R|3\rangle \\ \langle 3|R|1\rangle & \langle 3|R|2\rangle & \dots \end{bmatrix}$$

$$= \langle 1|R|1\rangle + \langle 2|R|2\rangle + \cdots$$

one could deduce that the expectation of A is given by

$$\text{Trace}\,[\rho A] = \sum_{a} \langle a| \sum_{v} |v\rangle\langle v| e^{-\beta E_v} A|a\rangle$$

Using the resolution of identity,

$$\text{Trace}[\rho A] = \sum_{a} \langle a| \sum_{vb} |v\rangle\langle v| e^{-\beta_v} |b\rangle\langle b|A|a\rangle.$$

Now,

$$\langle a| \sum_{v} |v\rangle\langle v| e^{-\beta_v} |b\rangle = \rho_{ab}$$

$$\langle b|A|a\rangle = A_{ba}$$

Therefore,

$$\text{Trace}\,[\rho A] = \sum_{ab} \rho_{ab} A_{ba}$$

For convenience, the Einstein notation is used where double indices imply summation, so the brief form of the above is

$$\text{Trace}\,[\rho A] = \rho_{ab} A_{ba}$$

Remarks and Reflections

It also follows from the above that

$$\text{Trace}\,[\rho A] = \sum_{abv} \langle a \mid v\rangle\langle v| e^{-\beta E_v} |b\rangle\langle b|A|a\rangle$$

$$= \sum_{abv} \delta_{av} \langle v| e^{-\beta E_v} |b\rangle\langle b|A|a\rangle$$

$$= \sum_{bv} \langle v|e^{-\beta E_v}|b\rangle\langle b|A|v\rangle$$

$$= \sum_{v} e^{-\beta E_v} \langle v|A|v\rangle$$

which is the expectation value of A.

Problem 8.04 Density Matrix of a 2-Qubit System

In a quantum system that comprises two sub-systems, A and B, the density matrix proves to be the most suitable formulation for its description. The density matrix for the greater system is denoted by ρ^{AB}. The sub-systems would have dimensions N_a and N_b. Sub-system A is spanned by orthonormal basis vectors $|a\rangle$ and sub-system B by orthonormal basis vectors $|b\rangle$. The density operators in sub-spaces $A(B)$ of the 2-qubit system are given by

$$\rho^A = Tr_B[\rho^{AB}] = \sum_{b}^{N_b} (I_A \otimes \langle b|)\rho^{AB}(I_A \otimes |b\rangle)$$

$$\rho^B = Tr_A[\rho^{AB}] = \sum_{a}^{N_A} (\langle a| \otimes I_B)\rho^{AB}(|a\rangle \otimes I_B)$$

known as the partial trace. Show that if the 2-qubit system is separable, i.e., the greater state is a product state (see Remarks and Reflections) instead of an entangled state,

$$\rho^{AB} = \rho_A \otimes \rho_B$$

The partial trace of B results in

$$\rho^A = \rho_A$$

Solution

Noting that

$$\rho^A = Tr_B[\rho^{AB}] = \sum_{b}^{N_b} (I_A \otimes \langle b|)\rho^{AB}(I_A \otimes |b\rangle)$$

Applying this to sub-system A, one has

$$\rho^A = \sum_b^{N_b} (I_A \otimes \langle b|)\rho^{AB}(I_A \otimes |b\rangle) = \sum_b^{N_b} (I_A \otimes \langle b|)(\rho_A \otimes \rho_B)(I_A \otimes |b\rangle)$$

Use is made of the following:

$$([A] \otimes [B]) \circ ([C] \otimes [D]) = ([A] \circ [C]) \otimes [B] \circ [D])$$

where $\otimes$ is the direct product multiplication and $\circ$ denotes the ordinary matrix multiplication. Therefore,

$$\begin{aligned}\rho^A &= \sum_b^{N_b} (I_A \otimes \langle b|) \circ (\rho_A \otimes \rho_B) \circ (I_A \otimes |b\rangle) \\ &= \sum_b^{N_b} (I_A \otimes \langle b|) \circ ((\rho_A \circ I_A) \otimes (\rho_B \circ |b\rangle)) \\ &= \sum_b^{N_b} (I_A \otimes \langle b|) \circ (\rho_A \otimes (\rho_B \circ |b\rangle))\end{aligned}$$

Finally,

$\rho^A = \sum_b^{N_b} \rho_A \otimes \langle b| \circ \rho_B \circ |b\rangle = \rho_A \otimes \sum_b^{N_b} \langle b|\rho_B|b\rangle$	Note that $\circ$ is simply matrix multiplication

Now, ρ_B is the density operator of original system B. Since

$$\sum_b^{N_b} \langle b|\rho_B|b\rangle = 1 \rightarrow \rho^A = \rho_A$$

As a result,

$$\rho^A = \rho_A \sum_b^{N_b} \langle b|\rho_B|b\rangle = \rho_A$$

The same would apply to sub-system B, resulting in

$$\rho^B = \rho_B$$

Remarks and Reflections

It was mentioned that when the greater state is a product state of the individual system, the greater system is considered unentangled, i.e.,

$$\rho^{AB} = \rho_A \otimes \rho_B \to \textit{system unentangled}$$

For a more vivid illustration, we consider the pure state of a 2-qubit system (qubit A and qubit B). The pure state of the greater system is

$$|\psi\rangle = a|00\rangle + b|01\rangle + c|10\rangle + d|11\rangle$$

If the above can be factorized, e.g., $|\psi\rangle = |\psi_A\rangle \otimes |\psi_B\rangle$, its density matrix would be $\rho = \rho_A \otimes \rho_B$ and the system is unentangled.

$$|\psi_{AB}\rangle = |\psi_A\rangle \otimes |\psi_B\rangle \to \rho^{AB} = \rho_A \otimes \rho_B \to \textit{system unentangled}$$

Proof is as follows:

$$|\psi_{AB}\rangle = |\psi_A\rangle \otimes |\psi_B\rangle$$

Density matrix is

$$\begin{aligned}\rho^{AB} &= |\psi_{AB}\rangle\langle\psi_{AB}| = (|\psi_A\rangle \otimes |\psi_B\rangle) \circ (\langle\psi_A| \otimes \langle\psi_A|)\\ &= (|\psi_A\rangle \circ \langle\psi_A|) \otimes (|\psi_B\rangle \circ \langle\psi_B|) = \rho_A \otimes \rho_B\end{aligned}$$

The system is "unentangled". The converse is true for an entangled system, in which case

$$\rho^{AB} \neq \rho_A \otimes \rho_B$$

In perspectives,

$$\begin{aligned}\rho^A &= Tr_B[|\psi_{AB}\rangle\langle\psi_{AB}|] = \rho_A \ (\textit{entangled})\\ \rho^A &= Tr_B[|\psi_{AB}\rangle\langle\psi_{AB}|] \neq \rho_A \ (\textit{not entangled})\end{aligned}$$

Problem 8.05 Quantum Measurement 1

An arbitrary 2D in-plane spin state is characterized in a quantum manner by eigenvectors:

$$|\eta_+\rangle = \frac{1}{\sqrt{2}}\begin{pmatrix} e^{-\frac{i\phi}{2}} \\ e^{\frac{i\phi}{2}} \end{pmatrix}, \quad |\eta_-\rangle = \frac{1}{\sqrt{2}}\begin{pmatrix} e^{-\frac{i\phi}{2}} \\ -e^{\frac{i\phi}{2}} \end{pmatrix}$$

A measurement of the physical quantity spin $(+z, -z)$ is carried out in the quantum state $|\eta_+\rangle$.

(a) What is the *spin state* of the system after measurement?
(b) Find the *probability* of observing spin $(+z, -z)$ in the state of $|\eta_+\rangle$.
(c) Determine the *expectation* of spin $(+z, -z)$ in the state of $|\eta_+\rangle$.

Explain how each of the above can be achieved.

Note: "spin $(+z, -z)$" is an English short form we use to refer to spin along $+z$ or spin along $-z$.

Solution

(a) The process of measurement is given by

$$P_{+z}|\eta_+\rangle = |+z\rangle\langle +z\,|\eta_+\rangle = \begin{pmatrix} 1 & 0 \\ 0 & 0 \end{pmatrix} \frac{1}{\sqrt{2}} \begin{pmatrix} e^{-\frac{i\phi}{2}} \\ e^{\frac{i\phi}{2}} \end{pmatrix}$$

$$= \frac{1}{\sqrt{2}} e^{-\frac{i\phi}{2}} \begin{pmatrix} 1 \\ 0 \end{pmatrix} = \frac{1}{\sqrt{2}} e^{-\frac{i\phi}{2}} |+z\rangle$$

$$P_{-z}|\eta_+\rangle = |-z\rangle\langle -z\,|\eta_+\rangle = \begin{pmatrix} 0 & 0 \\ 0 & 1 \end{pmatrix} \frac{1}{\sqrt{2}} \begin{pmatrix} e^{-\frac{i\phi}{2}} \\ e^{\frac{i\phi}{2}} \end{pmatrix}$$

$$= \frac{1}{\sqrt{2}} e^{-\frac{i\phi}{2}} \begin{pmatrix} 0 \\ 1 \end{pmatrix} = \frac{1}{\sqrt{2}} e^{-\frac{i\phi}{2}} |-z\rangle$$

The final results on the RHS give the final spin states after measurement, i.e.,

$$\begin{pmatrix} 1 \\ 0 \end{pmatrix} \quad \text{and} \quad \begin{pmatrix} 0 \\ 1 \end{pmatrix}$$

(b) To find the *probability* of observing spin $(+z, -z)$ in the state of $|\eta_+\rangle$,

$$\langle \eta_+|P_{+z}|\eta_+\rangle = \langle \eta_+\,|+z\rangle\langle +z\,|\eta_+\rangle = \left(\frac{1}{\sqrt{2}} e^{-\frac{i\phi}{2}}\right)^\dagger \left(\frac{1}{\sqrt{2}} e^{-\frac{i\phi}{2}}\right) = \frac{1}{2}$$

$$\langle \eta_+|P_{-z}|\eta_+\rangle = \langle \eta_+\,|-z\rangle\langle -z\,|\eta_+\rangle = \left(\frac{1}{\sqrt{2}} e^{-\frac{i\phi}{2}}\right)^\dagger \left(\frac{1}{\sqrt{2}} e^{-\frac{i\phi}{2}}\right) = \frac{1}{2}$$

(c) To find the *expectation* of spin $(+z, -z)$ in the state of $|\eta_+\rangle$,

$$\langle \eta_+|S_z|\eta_+\rangle = \langle \eta_+|\sum_n P_n E_n|\eta_+\rangle = \langle \eta_+|(P_{+z}\alpha_{+z} + P_{-z}\alpha_{-z})|\eta_+\rangle$$

Note that E_n is the eigenvalue associated with basis state $|n\rangle$. The eigenvalues of S_z are

$$\alpha_{+z} = \frac{1}{2}\hbar, \quad \alpha_{-z} = -\frac{1}{2}\hbar$$

Therefore, the expectation of S_z given an arbitrary 2D in-plane spin state $|\eta_+\rangle = \frac{1}{\sqrt{2}}\begin{pmatrix} e^{-\frac{i\phi}{2}} \\ e^{\frac{i\phi}{2}} \end{pmatrix}$ is ZERO, as shown in the following:

$$\langle \eta_+|S_z|\eta_+\rangle = \frac{1}{2}\alpha_{+z} + \frac{1}{2}\alpha_{-z} = 0$$

Caution that the projection operator in this form $\sum_n P_n$ plays the role of the resolution identity. However, projection operator in this form $\sum_n P_n O_n$ plays the role of the operator $\hat{O}_n$ itself.

Remarks and Reflections

Now, what if the quantum state is $|+z\rangle$ itself instead of $|\eta_+\rangle$. The results would be

$$\langle +z|P_{+z}|+z\rangle = \langle +z\,|+z\rangle\langle +z\,|+z\rangle = 1$$

$$\langle +z|P_{-z}|+z\rangle = \langle +z\,|-z\rangle\langle -z\,|+z\rangle = 0$$

It makes perfect sense to say the "chance" that a horse is observed to be a horse is 100%. The chance that a horse is to be observed as a non-horse is 0%.

Points to ponder

One could observe in the example above that when a spin state is measured, it shows up in the state it is measured. Therefore, in a general setting, when state 0 (denoted by $|0\rangle$) of the following superposition state is measured, state $|1\rangle$ would disappear,

$$|\psi\rangle = a|0\rangle + b|1\rangle$$

and $|\psi\rangle$ becomes state 0, i.e., $|\psi\rangle = |0\rangle$. Mathematically, measurement is performed by a measurement operator. Let's say we want to measure state 0. The measurement operator is

$$P_0 = |0\rangle\langle 0|$$

The process of a quantum measurement is represented by $P_0|\psi\rangle = a|0\rangle$. But note that the state after measurement is

$$\frac{P_0|\psi\rangle}{|a|} = \frac{a}{|a|}|0\rangle$$

For example, take $|\psi\rangle = \frac{1}{\sqrt{2}}|0\rangle + \frac{1}{\sqrt{2}}|1\rangle$, the state after measurement is $|0\rangle$. In a formal expression, the state of the system after measurement is given by

$$\frac{P_0|\psi\rangle}{\sqrt{\langle\psi|P_0|\psi\rangle}} = \frac{a|0\rangle}{|a|}$$

We know that the probability that state 0 appears after measurement of state 0 is 1/2. Let's see how it works out mathematically.

$$P_0 = \langle\psi|P_0|\psi\rangle = \langle\psi \mid 0\rangle\langle 0 \mid \psi\rangle = a^\dagger a = \frac{1}{2}$$

Problem 8.06 Quantum Measurement II

The 2D spin states $|\eta_+\rangle = \frac{1}{\sqrt{2}}\begin{pmatrix} e^{-\frac{i\phi}{2}} \\ e^{\frac{i\phi}{2}} \end{pmatrix}$, $|\eta_-\rangle = \frac{1}{\sqrt{2}}\begin{pmatrix} e^{-\frac{i\phi}{2}} \\ -e^{\frac{i\phi}{2}} \end{pmatrix}$ are the eigenstates of $S_x \cos\phi + S_y \sin\phi$. Referring to Figure 1, one notes that these spin states describe spin angular momentum along the magnetic field $\boldsymbol{e_n}$ that give energies of, respectively, $\frac{\hbar}{2}\omega_0, -\frac{\hbar}{2}\omega_0$. The energy Hamiltonian is given by $H = -\boldsymbol{S}.\boldsymbol{B} = B(S_x \cos\phi + S_y \sin\phi)$.

(a) Determine the probability of observing a spin (arbitrary ϕ) along $+X$ and $-X$.

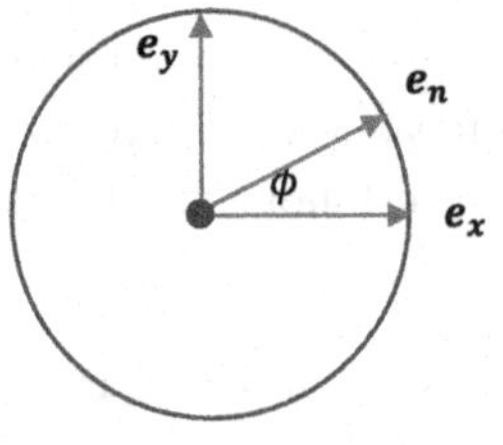

Fig. 1. A 2D system in which the magnetic field lies in the X–Y plane at an arbitrary azimuthal angle.

(b) When the spin is pointing along $+Y$, what is the chance of seeing the spin along $(+X, -X)$? Determine the expectation of a spin (arbitrary ϕ) along X.
(c) Determine the probability of observing a spin (arbitrary ϕ) along $+Y$ and $-Y$.
(d) When the spin is pointing along $+X$, what is the chance of seeing the spin along $(+Y, -Y)$? Determine the expectation of a spin (arbitrary ϕ) along Y.

Solution

The in-plane spin eigenstates are given as follows:

$$|+x\rangle = \frac{1}{\sqrt{2}}\begin{pmatrix}1\\1\end{pmatrix}, |-x\rangle = \frac{1}{\sqrt{2}}\begin{pmatrix}1\\-1\end{pmatrix}, |+y\rangle = \frac{1}{\sqrt{2}}\begin{pmatrix}1\\i\end{pmatrix}, |-y\rangle = \frac{1}{\sqrt{2}}\begin{pmatrix}1\\-i\end{pmatrix}$$

(a) Measure the spin along $+X$,

$$P_{+x}|\eta_+\rangle = |+x\rangle\langle +x \mid \eta_+\rangle = \frac{1}{2\sqrt{2}}\begin{pmatrix}1 & 1\\1 & 1\end{pmatrix}\begin{pmatrix}e^{-\frac{i\phi}{2}}\\e^{\frac{i\phi}{2}}\end{pmatrix}$$

$$= \frac{1}{2\sqrt{2}}\begin{pmatrix}e^{-\frac{i\phi}{2}} + e^{\frac{i\phi}{2}}\\e^{-\frac{i\phi}{2}} + e^{\frac{i\phi}{2}}\end{pmatrix}$$

$$= \frac{1}{2\sqrt{2}}(e^{-\frac{i\phi}{2}} + e^{\frac{i\phi}{2}})\begin{pmatrix}1\\1\end{pmatrix} = \cos\frac{\phi}{2}|+x\rangle$$

Likewise, measure spin angular momentum along $-X$, one has

$$P_{-x}|\eta_+\rangle = |-x\rangle\langle -x \mid \eta_+\rangle = \frac{1}{2\sqrt{2}}\begin{pmatrix}1 & -1\\-1 & 1\end{pmatrix}\begin{pmatrix}e^{-\frac{i\phi}{2}}\\e^{\frac{i\phi}{2}}\end{pmatrix}$$

$$= \frac{1}{2\sqrt{2}}\begin{pmatrix}e^{-\frac{i\phi}{2}} - e^{\frac{i\phi}{2}}\\-e^{-\frac{i\phi}{2}} + e^{\frac{i\phi}{2}}\end{pmatrix}$$

$$= \frac{1}{2\sqrt{2}}(e^{-\frac{i\phi}{2}} - e^{\frac{i\phi}{2}})\begin{pmatrix}1\\-1\end{pmatrix} = -i\sin\frac{\phi}{2}|-x\rangle$$

Probability of measuring spin angular momentum along $+X$ is given by

$$\langle \eta_+ \mid +x\rangle\langle +x \mid \eta_+\rangle = \left|\cos\frac{\phi}{2}\right|^2 = \cos^2\frac{\phi}{2}$$

Likewise, measure along $-X$,

$$\langle \eta_+ \mid -x\rangle\langle -x \mid \eta_+\rangle = \left|-i\sin\frac{\phi}{2}\right|^2 = \sin^2\frac{\phi}{2}$$

(b) When spin state is pointing along $+Y$, angle $\phi = \frac{\pi}{2}$.

$$\langle \eta_+ \mid +x\rangle\langle +x \mid \eta_+\rangle = \cos^2\frac{\phi}{2}$$

$$\langle \eta_+ \mid -x\rangle\langle -x \mid \eta_+\rangle = \sin^2\frac{\phi}{2}$$

The chance of seeing a $+Y$ spin along $+X$ is

$$\cos^2\frac{\pi}{4} = \frac{1}{2}$$

The chance of seeing a $+Y$ spin along $-X$ is

$$\sin^2\frac{\pi}{4} = \frac{1}{2}$$

Expectation of spin (arbitrary ϕ) along X is given by

$$\begin{aligned}\langle \eta_+ | S_x | \eta_+\rangle &= \langle \eta_+ \mid +x\rangle\langle +x \mid \eta_+\rangle\frac{\hbar}{2} - \langle \eta_+ \mid -x\rangle\langle -x \mid \eta_+\rangle\frac{\hbar}{2} \\ &= \frac{\hbar}{2}\left(\cos^2\frac{\phi}{2} - \sin^2\frac{\phi}{2}\right) = \frac{\hbar}{2}\cos\phi\end{aligned}$$

Therefore, the chance of seeing a $+Y$ spin along X is ZERO as $\frac{\hbar}{2}\cos\frac{\pi}{2} = 0$.

(c) Measure the spin along $+Y$

$$
\begin{aligned}
P_{+y}|\eta_+\rangle &= |+y\rangle\langle +y \mid \eta_+\rangle = \frac{1}{2\sqrt{2}}\begin{pmatrix} 1 & -i \\ i & 1 \end{pmatrix}\begin{pmatrix} e^{-\frac{i\phi}{2}} \\ e^{\frac{i\phi}{2}} \end{pmatrix} \\
&= \frac{1}{2\sqrt{2}}\begin{pmatrix} e^{-\frac{i\phi}{2}} - ie^{\frac{i\phi}{2}} \\ ie^{-\frac{i\phi}{2}} + e^{\frac{i\phi}{2}} \end{pmatrix} = \frac{1}{2\sqrt{2}}(e^{-\frac{i\phi}{2}} - ie^{\frac{i\phi}{2}})\begin{pmatrix} 1 \\ i \end{pmatrix} \\
&= \frac{1}{2}\left(\cos\frac{\phi}{2} + \sin\frac{\phi}{2}\right)(1-i)|+y\rangle
\end{aligned}
$$

Likewise, measure along $-Y$

$$
\begin{aligned}
P_{-y}|\eta_+\rangle &= |-y\rangle\langle -y \mid \eta_+\rangle = \frac{1}{2\sqrt{2}}\begin{pmatrix} 1 & i \\ -i & 1 \end{pmatrix}\begin{pmatrix} e^{-\frac{i\phi}{2}} \\ e^{\frac{i\phi}{2}} \end{pmatrix} \\
&= \frac{1}{2\sqrt{2}}\begin{pmatrix} e^{-\frac{i\phi}{2}} + ie^{\frac{i\phi}{2}} \\ -ie^{-\frac{i\phi}{2}} + e^{\frac{i\phi}{2}} \end{pmatrix} = \frac{1}{2\sqrt{2}}(e^{-\frac{i\phi}{2}} + ie^{\frac{i\phi}{2}})\begin{pmatrix} 1 \\ -i \end{pmatrix} \\
&= \frac{1}{2}\left(\cos\frac{\phi}{2} - \sin\frac{\phi}{2}\right)(1-i)|-y\rangle
\end{aligned}
$$

Probability of measuring spin angular momentum along $+Y$ is given by

$$
\begin{aligned}
\langle \eta_+ \mid +y\rangle\langle +y \mid \eta_+\rangle &= \left|\frac{1}{2}\left(\cos\frac{\phi}{2} + \sin\frac{\phi}{2}\right)(1-i)\right|^2 \\
&= \frac{1}{2}\left(\cos\frac{\phi}{2} + \sin\frac{\phi}{2}\right)^2
\end{aligned}
$$

and along $-Y$ is

$$
\begin{aligned}
\langle \eta_+ \mid -y\rangle\langle -y \mid \eta_+\rangle &= \left|\frac{1}{2}\left(\cos\frac{\phi}{2} - \sin\frac{\phi}{2}\right)(1-i)\right|^2 \\
&= \frac{1}{2}\left(\cos\frac{\phi}{2} - \sin\frac{\phi}{2}\right)^2
\end{aligned}
$$

(d) When spin state is pointing along $+X$, angle $\phi = 0$.

$$\langle \eta_+ \mid +y\rangle\langle +y \mid \eta_+\rangle = \frac{1}{2}\left(\cos\frac{\phi}{2} + \sin\frac{\phi}{2}\right)^2$$

$$\langle \eta_+ \mid -y\rangle\langle -y \mid \eta_+\rangle = \frac{1}{2}\left(\cos\frac{\phi}{2} - \sin\frac{\phi}{2}\right)^2$$

The chance of seeing a $+X$ along $+Y$ is

$$\frac{1}{2}(\cos 0 + \sin 0)^2 = \frac{1}{2}$$

The chance of seeing a $+X$ along $-Y$ is

$$\frac{1}{2}(\cos 0 - \sin 0)^2 = \frac{1}{2}$$

Expectation of spin (arbitrary ϕ) along Y is given by

$$\langle \eta_+|S_y|\eta_+\rangle = \langle \eta_+ \mid +y\rangle\langle +y \mid \eta_+\rangle\frac{\hbar}{2} - \langle \eta_+ \mid -y\rangle\langle -y \mid \eta_+\rangle\frac{\hbar}{2}$$
$$= \frac{\hbar}{4}\left(\cos\frac{\phi}{2} + \sin\frac{\phi}{2}\right)^2 - \frac{\hbar}{4}\left(\cos\frac{\phi}{2} - \sin\frac{\phi}{2}\right)^2 = \frac{\hbar}{2}\sin\phi$$

Therefore, the chance of seeing a $+X$ spin along Y is ZERO as $\frac{\hbar}{2}\sin 0 = 0$.

Remarks and Reflections

Given as derived above that for an arbitrary $|\eta_+\rangle$, the probability of measuring spin angular momentum along $+Y$ is given by

$$\langle \eta_+ \mid +y\rangle\langle +y \mid \eta_+\rangle = \left|\frac{1}{2}\left(\cos\frac{\phi}{2} + \sin\frac{\phi}{2}\right)(1-i)\right|^2 = \frac{1}{2}\left(\cos\frac{\phi}{2} + \sin\frac{\phi}{2}\right)^2$$

and that along $-Y$ is

$$\langle \eta_+ \mid -y\rangle\langle -y \mid \eta_+\rangle = \left|\frac{1}{2}\left(\cos\frac{\phi}{2} - \sin\frac{\phi}{2}\right)(1-i)\right|^2 = \frac{1}{2}\left(\cos\frac{\phi}{2} - \sin\frac{\phi}{2}\right)^2$$

Check that the following table makes sense:

Probability of Measured State	**Pre-existing States of $\vert\eta_+\rangle$**	**Probability of Measured State**
$\langle\eta_+ \vert +y\rangle\langle +y \vert \eta_+\rangle = 1$	When $\vert\eta_+\rangle = \vert +y\rangle$	$\phi = 90 \to \frac{1}{2}\left(\cos\frac{\phi}{2} + \sin\frac{\phi}{2}\right)^2 = 1$
$\langle\eta_+ \vert +y\rangle\langle +y \vert \eta_+\rangle = 0$	When $\vert\eta_+\rangle = \vert -y\rangle$	$\phi = 270 \to \frac{1}{2}\left(\cos\frac{\phi}{2} + \sin\frac{\phi}{2}\right)^2 = 0$
$\langle\eta_+ \vert -y\rangle\langle -y \vert \eta_+\rangle = 0$	When $\vert\eta_+\rangle = \vert +y\rangle$	$\phi = 90 \to \frac{1}{2}\left(\cos\frac{\phi}{2} - \sin\frac{\phi}{2}\right)^2 = 0$
$\langle\eta_+ \vert -y\rangle\langle -y \vert \eta_+\rangle = 1$	When $\vert\eta_+\rangle = \vert -y\rangle$	$\phi = 270 \to \frac{1}{2}\left(\cos\frac{\phi}{2} - \sin\frac{\phi}{2}\right)^2 = 1$

Problem 8.07 Projection Operator

The projection operator is given by $m_n = |n\rangle\langle n|$, where $|n\rangle$ is the eigenstate of the physical operator it intends to measure.

(a) Show that m_n can be used alongside the eigenvalues of a physical operator to represent the operator itself (e.g., operator is Hamiltonian (H)).
(b) Show that $H(m_n)$ acts on eigenstates to produce the eigenvalues.
(c) Note its equivalence to the second-quantized Hamiltonian.

Solution

(a) Let's examine the expectation of the Hamiltonian as follows:

$$\langle\psi|H|\psi\rangle = \sum_{n,m} \langle\psi \mid n\rangle\langle n|H|m\rangle\langle m \mid \psi\rangle$$
$$= \sum_{n,m} \langle\psi \mid n\rangle E_m \delta_{mn} \langle m \mid \psi\rangle = \sum_n \langle\psi \mid n\rangle E_n \langle n \mid \psi\rangle$$

Inspecting RHS and LHS,

$$H = \sum_n |n\rangle E_n \langle n| = \sum_n m_n E_n$$

(b) Let H act on an eigenstate

$$H|n''\rangle = \sum_n |n\rangle\langle n|E_n|n''\rangle$$

It follows that

$$H|n''\rangle = \sum_n |n\rangle E_n \delta_{nn''} = E_{n''}|n''\rangle$$

(c) Analogy

Note that $H = \sum_n |n\rangle E_n \langle n|$ has an equivalence in the second-quantized form if one carries out a mapping as follows:

$$|n\rangle \to a_n^\dagger, \quad \langle n| \to a_n$$

It follows that

$$H = \sum_n |n\rangle E_n \langle n| \leftrightarrows \sum_n a_n^\dagger a_n E_n$$

Remarks and Reflections

In summary, the projection operator performs measurement, i.e.,

$$m_n = |n\rangle\langle n|$$

This operator acts on a state $|\psi\rangle$ and projects it into state $|n\rangle$. This action is consistent with the physical description of quantum measurement, which says measuring the physical quantity of an arbitrary quantum state collapses it to the state of the measured quantity. For example, measuring the spin x (a physical quantity) of an arbitrary spin state collapses it to the state of spin x. Recall that

$$|\psi\rangle = \sum_n a_n|n\rangle = \sum_n |n\rangle\langle n \mid \psi\rangle$$

Therefore, the sum of the projection operators over the complete eigenstates of the system produces a ONE, and this is the resolution of identity.

$$m = \sum_n |n\rangle\langle n| = 1$$

It follows that

$$m_n m_m = |n\rangle\langle n \mid m\rangle\langle m| = \delta_{mn}|n\rangle\langle m|$$

The above can be written in any of the following:

$$m_n m_m = m_m = \delta_{mn} m_n = m_n = \delta_{mn} m_m$$

It thus follows that

$$m_n m_n = m_n$$

Quantum Gates and Circuits

Problem 8.08 Quantum Logic Gates: NOT Gate

Quantum computation can be achieved with a set of logic gates that mimic the conventional electronic gates. One example of this is the NOT gate. The NOT gate delivers an outcome that inverts the input. In electronic, the input is a binary (0,1) realized by a 5V or a 0 V. In quantum computation, the input is to be realized by quantum states $(|0\rangle, |1\rangle)$. One physical example is the quantum state of the spin angular momentum (Figure 2).

Mathematically, it is given by $U_{NOT} = (|0\rangle\langle 1| + |1\rangle\langle 0|)$ for use with the standard basis.

(a) Show that the NOT gate inverts the input of

$$(|0\rangle, |1\rangle)$$

(b) Derive the matrix representation for the NOT gate for the standard computational basis of

$$|0\rangle = \begin{pmatrix} 1 \\ 0 \end{pmatrix}, |1\rangle = \begin{pmatrix} 0 \\ 1 \end{pmatrix}$$

(c) Derive the matrix representation for the NOT gate for the following basis:

$$|+\rangle = \frac{1}{\sqrt{2}}\begin{pmatrix} 1 \\ 1 \end{pmatrix}, |-\rangle = \frac{1}{\sqrt{2}}\begin{pmatrix} 1 \\ -1 \end{pmatrix}$$

Note: All the basis states used in the above are orthonormal.

Solution

(a) Examine the effect of the NOT gate on the input state

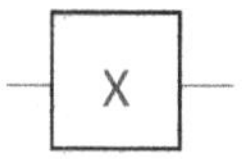

Fig. 2. A graphical NOT gate.

Inverting input state $|0\rangle$:

$$(|0\rangle\langle 1| + |1\rangle\langle 0|)|0\rangle = |0\rangle\langle 1 \mid 0\rangle + |1\rangle\langle 0 \mid 0\rangle = |1\rangle$$

Inverting input state $|1\rangle$:

$$(|0\rangle\langle 1| + |1\rangle\langle 0|)|1\rangle = |0\rangle\langle 1 \mid 1\rangle + |1\rangle\langle 0 \mid 1\rangle = |0\rangle$$

In the above, use is made of the orthogonality of basis states, i.e., $\langle a \mid b\rangle = \delta_{ab}$.

(b) For the standard basis of $|0\rangle = \begin{pmatrix} 1 \\ 0 \end{pmatrix}, |1\rangle = \begin{pmatrix} 0 \\ 1 \end{pmatrix}$,

$$|0\rangle\langle 1| + |1\rangle\langle 0| = \begin{pmatrix} 1 \\ 0 \end{pmatrix} (0\ 1) + \begin{pmatrix} 0 \\ 1 \end{pmatrix} (1\ 0) = \begin{pmatrix} 0 & 1 \\ 1 & 0 \end{pmatrix}$$

(c) For basis $|+\rangle = \frac{1}{\sqrt{2}}\begin{pmatrix} 1 \\ 1 \end{pmatrix}, |-\rangle = \frac{1}{\sqrt{2}}\begin{pmatrix} 1 \\ -1 \end{pmatrix}$,

$$|+\rangle\langle -| + |-\rangle\langle +| = \frac{1}{2}\left(\begin{pmatrix} 1 \\ 1 \end{pmatrix} (1\quad -1) + \begin{pmatrix} 1 \\ -1 \end{pmatrix} (1\quad 1)\right) = \begin{pmatrix} 1 & 0 \\ 0 & -1 \end{pmatrix}$$

Remarks and Reflections

Note that in the standard basis, the NOT gate is the Pauli-X matrix. Therefore, it can be realized with a physical system that performs the action of Pauli-X. The following table is a summary of the single-input quantum gates with their matrices.

Likewise, in the standard basis, the Z gate and the Y gate are, respectively, the Pauli-Z and the Pauli-Y matrix.

$$U_Z = (|0\rangle\langle 0| - |1\rangle\langle 1|) = \begin{pmatrix} 1 & 0 \\ 0 & -1 \end{pmatrix}$$

$$U_Y = i(|1\rangle\langle 0| - |0\rangle\langle 1|) = \begin{pmatrix} 0 & -i \\ i & 0 \end{pmatrix}$$

Problem 8.09 Quantum Logic Gates: Hadamard Gate

While the NOT gate delivers an outcome that inverts the input, the Hadamard gate produces an even superposition of the inputs (Figure 3).

(a) In Bra-Ket form, the Hadamard gate is given by

$$\frac{1}{\sqrt{2}}(|0\rangle\langle 0| + |1\rangle 0| + |0\rangle\langle 1| - |1\rangle\langle 1|)$$

Show that the Hadamard gate can be used to generate an even superposition of a quantum state.

(b) Derive the matrix representation for the Hadamard gate for the standard computation basis of

$$\left(|0\rangle = \begin{pmatrix}1\\0\end{pmatrix}, \quad |1\rangle = \begin{pmatrix}0\\1\end{pmatrix}\right)$$

Note: All the basis states used in the above are orthonormal.

Solution

(a) For input state of $|0\rangle$,

$$\frac{1}{\sqrt{2}}(|0\rangle\langle 0| + |1\rangle\langle 0| + |0\rangle\langle 1| - |1\rangle\langle 1|)|0\rangle$$

$$= \frac{1}{\sqrt{2}}(|0\rangle\langle 0 \mid 0\rangle + |1\rangle\langle 0 \mid 0\rangle) = \frac{1}{\sqrt{2}}(|0\rangle + |1\rangle)$$

For input state of $|1\rangle$,

$$\frac{1}{\sqrt{2}}(|0\rangle\langle 0| + |1\rangle\langle 0| + |0\rangle\langle 1| - |1\rangle\langle 1|)|1\rangle$$

$$= \frac{1}{\sqrt{2}}(|0\rangle\langle 1 \mid 1\rangle - |1\rangle\langle 1 \mid 1\rangle) = \frac{1}{\sqrt{2}}(|0\rangle - |1\rangle)$$

Fig. 3. A graphical NOT gate (left) and a Hadamard gate (right).

(b) Matrix representation for the Hadamard gate is

$$\frac{1}{\sqrt{2}}(|0\rangle\langle 0| + |1\rangle\langle 0| + |0\rangle\langle 1| - |1\rangle\langle 1|)$$

$$= \frac{1}{\sqrt{2}}\left(\begin{pmatrix} 1 \\ 0 \end{pmatrix}\begin{pmatrix} 1 & 0 \end{pmatrix} + \begin{pmatrix} 0 \\ 1 \end{pmatrix}\begin{pmatrix} 1 & 0 \end{pmatrix}\right.$$

$$\left. + \begin{pmatrix} 1 \\ 0 \end{pmatrix}\begin{pmatrix} 0 & 1 \end{pmatrix} - \begin{pmatrix} 0 \\ 1 \end{pmatrix}\begin{pmatrix} 0 & 1 \end{pmatrix}\right)$$

$$= \frac{1}{\sqrt{2}}\begin{pmatrix} 1 & 0 \\ 0 & 0 \end{pmatrix} + \frac{1}{\sqrt{2}}\begin{pmatrix} 0 & 0 \\ 0 & 1 \end{pmatrix} + \frac{1}{\sqrt{2}}\begin{pmatrix} 0 & 1 \\ 0 & 0 \end{pmatrix} + \frac{1}{\sqrt{2}}\begin{pmatrix} 0 & 0 \\ 0 & -1 \end{pmatrix}$$

$$= \frac{1}{\sqrt{2}}\begin{pmatrix} 1 & 1 \\ 1 & -1 \end{pmatrix}$$

Remarks and Reflections

Single-input gates are summarized in the following table for ease of reference.

Graphics	**Matrix Representation**	**Quantum Gate**
X	$\begin{pmatrix} 0 & 1 \\ 1 & 0 \end{pmatrix}$	Pauli-X
Y	$\begin{pmatrix} 0 & -i \\ i & 0 \end{pmatrix}$	Pauli-Y
Z	$\begin{pmatrix} 1 & 0 \\ 0 & -1 \end{pmatrix}$	Pauli-Z
H	$\frac{1}{\sqrt{2}}\begin{pmatrix} 1 & 1 \\ 1 & -1 \end{pmatrix}$	Hadamard
S	$\frac{1}{\sqrt{2}}\begin{pmatrix} 1 & 0 \\ 0 & i \end{pmatrix}$	Phase
T	$\begin{pmatrix} 1 & 0 \\ 0 & e^{\frac{i\pi}{4}} \end{pmatrix}$	$\pi/8$

Problem 8.10 CNOT Gate

Single-input gates have been discussed and tabulated in previous problems. Here, we will pay attention to a very popular and useful gate known as the CNOT. The CNOT is a dual-input gate and a specialized form of the CONTROL gate family (Figure 4).

The CNOT gate takes the first bit as control, and the second bit as target to be acted upon based on the instruction of the control bit. It operates on the standard basis, flips the second bit on a 1 from the control, and retains the second bit (does nothing) on a 0 from the control.

(a) Mathematically, the CNOT gate is given by

$$|0\rangle\langle 0| \otimes I + |1\rangle\langle 1| \otimes \sigma^x$$

Show that the CNOT gate performs the function of flip-on-1 and stay-on-0 for all input combinations.

(b) Derive the matrix representation for the CNOT gate for the standard basis of

$$\left(|0\rangle = \begin{pmatrix} 1 \\ 0 \end{pmatrix}, \quad |1\rangle = \begin{pmatrix} 0 \\ 1 \end{pmatrix} \right)$$

Note: All the basis states used in the above are orthonormal.

Solution

(a) CNOT operation

$$\begin{aligned} &|0\rangle\langle 0| \otimes I + |1\rangle\langle 1| \otimes \sigma^x \\ &\quad = |0\rangle\langle 0| \otimes (|0\rangle\langle 0| + |1\rangle\langle 1|) + |1\rangle\langle 1| \otimes (|1\rangle\langle 0| + |0\rangle\langle 1|) \\ &\quad = |00\rangle\langle 00| + |01\rangle\langle 01| + |11\rangle\langle 10| + |10\rangle\langle 11| \end{aligned}$$

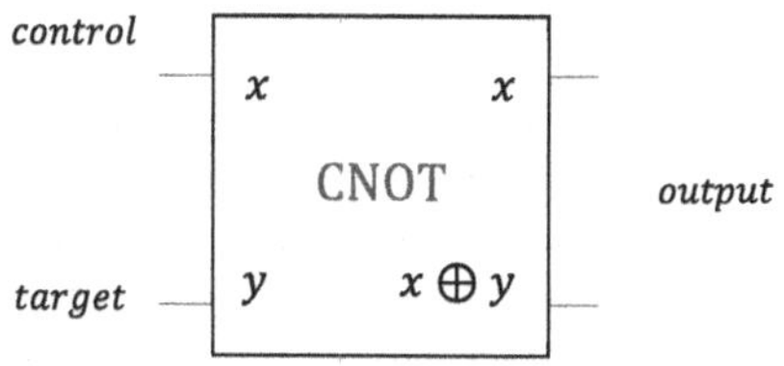

Fig. 4. Schematic of a CNOT gate.

Let it act on all four combinations of the input states as follows:

$$(|00\rangle\langle 00| + |01\rangle\langle 01| + |11\rangle\langle 10| + |10\rangle\langle 11|)|00\rangle = |00\rangle$$
$$(|00\rangle\langle 00| + |01\rangle\langle 01| + |11\rangle\langle 10| + |10\rangle\langle 11|)|01\rangle = |01\rangle$$
$$(|00\rangle\langle 00| + |01\rangle\langle 01| + |11\rangle\langle 10| + |10\rangle\langle 11|)|10\rangle = |11\rangle$$
$$(|00\rangle\langle 00| + |01\rangle\langle 01| + |11\rangle\langle 10| + |10\rangle\langle 11|)|11\rangle = |10\rangle$$

Inspecting the output states, one observes that the CNOT gate performs the function of flip-on-1 and stay-on-0 for all input combinations.

(b) Matrix representation for the CNOT gate

$$\begin{aligned}
&|0\rangle\langle 0| \otimes I + |1\rangle\langle 1| \otimes \sigma^x \\
&= \begin{pmatrix} 1 \\ 0 \end{pmatrix} (1 \quad 0) \otimes \begin{pmatrix} 1 & 0 \\ 0 & 1 \end{pmatrix} + \begin{pmatrix} 0 \\ 1 \end{pmatrix} (0 \quad 1) \otimes \begin{pmatrix} 0 & 1 \\ 1 & 0 \end{pmatrix} \\
&= \begin{pmatrix} 1 & 0 \\ 0 & 0 \end{pmatrix} \otimes \begin{pmatrix} 1 & 0 \\ 0 & 1 \end{pmatrix} + \begin{pmatrix} 0 & 0 \\ 0 & 1 \end{pmatrix} \otimes \begin{pmatrix} 0 & 1 \\ 1 & 0 \end{pmatrix} \\
&= \begin{pmatrix} 1 & 0 & 0 & 0 \\ 0 & 1 & 0 & 0 \\ 0 & 0 & 0 & 0 \\ 0 & 0 & 0 & 0 \end{pmatrix} + \begin{pmatrix} 0 & 0 & 0 & 0 \\ 0 & 0 & 0 & 0 \\ 0 & 0 & 0 & 1 \\ 0 & 0 & 1 & 0 \end{pmatrix} = \begin{pmatrix} 1 & 0 & 0 & 0 \\ 0 & 1 & 0 & 0 \\ 0 & 0 & 0 & 1 \\ 0 & 0 & 1 & 0 \end{pmatrix}
\end{aligned}$$

Remarks and Reflections

The CNOT gate is a useful device to generate an entangled state. The following is an illustration of such a process.

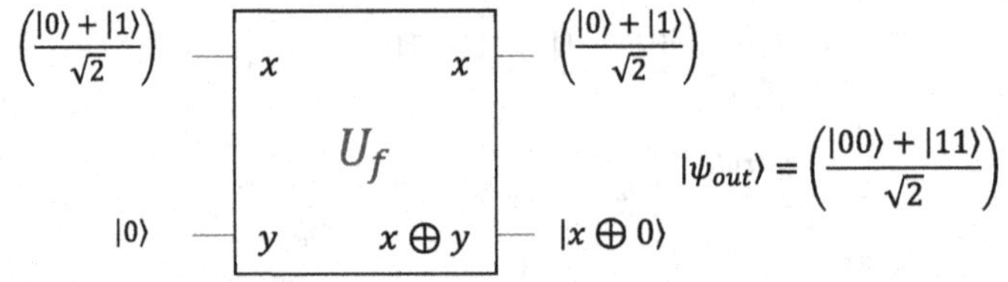

The CNOT acts on the total input and produces a total output as follows:

$$C_{NOT}\left(\frac{|00\rangle + |10\rangle}{\sqrt{2}}\right) = \left(\frac{|00\rangle + |11\rangle}{\sqrt{2}}\right)$$

Likewise, the CNOT gate can perform the inverse

$$C_{NOT}\left(\frac{|00\rangle + |11\rangle}{\sqrt{2}}\right) = \left(\frac{|00\rangle + |10\rangle}{\sqrt{2}}\right)$$

Therefore, the CNOT gate is its own inverse. It is a unitary operator.

Problem 8.11 Toffoli Gate

(a) Referring to Figure 5 for the quantum logic gates, fill out the truth tables for the Fredkin gate, Toffoli gate, Feynman double gate, and Peres gate.

(b) The U matrix for the Toffoli gate can be constructed from the truth table by considering its action on the input that generates a corresponding output. However, the input and output information has to be first converted into their quantum representations as follows:

$$|000\rangle = \begin{pmatrix}1\\0\end{pmatrix} \otimes \begin{pmatrix}1\\0\end{pmatrix} \otimes \begin{pmatrix}1\\0\end{pmatrix} = (1\ 0\ 0\ 0\ 0\ 0\ 0\ 0)^T$$

$$|001\rangle = \begin{pmatrix}1\\0\end{pmatrix} \otimes \begin{pmatrix}1\\0\end{pmatrix} \otimes \begin{pmatrix}0\\1\end{pmatrix} = (0\ 1\ 0\ 0\ 0\ 0\ 0\ 0)^T$$

Show how the other six inputs are derived:

(c) Show the U matrix for the Toffoli gate.

(d) Show (with the aid of Boolean algebra) that the Toffoli gates when connected in series can retrieve themselves.

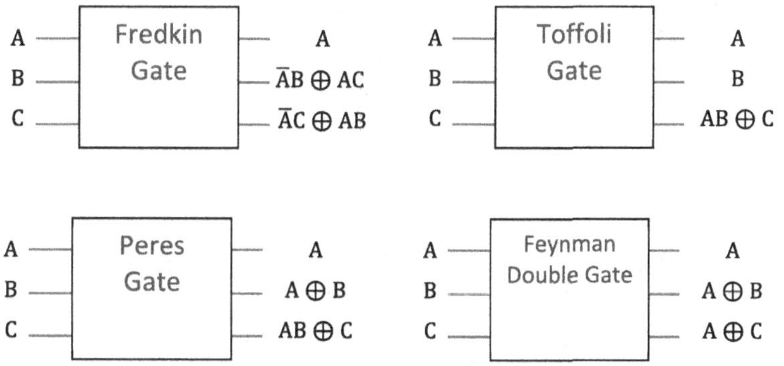

Fig. 5. Schematics of various multiple-input quantum gates as indicated in their labels.

(e) Show (with the aid of Boolean algebra) that the Toffoli gate can simulate the classical NAND gate.
(f) Show (with the aid of Boolean algebra) that the Toffoli gate can simulate the classical FANOUT function.

Solution

(a)

Fredkin Gate

A	B	C	O1	O2	O3
0	0	0	0	0	0
0	0	1	0	0	1
0	1	0	0	1	0
0	1	1	0	1	1
1	0	0	1	0	0
1	0	1	1	1	0
1	1	0	1	0	1
1	1	1	1	1	1

Toffoli Gate

A	B	C	O1	O2	O3
0	0	0	0	0	0
0	0	1	0	0	1
0	1	0	0	1	0
0	1	1	0	1	1
1	0	0	1	0	0
1	0	1	1	0	1
1	1	0	1	1	1
1	1	1	1	1	0

Feynman Double Gate

A	B	C	O1	O2	O3
0	0	0	0	0	0
0	0	1	0	0	1
0	1	0	0	1	0
0	1	1	0	1	1
1	0	0	1	1	1
1	0	1	1	1	0
1	1	0	1	0	1
1	1	1	1	0	0

Peres Gate

A	B	C	O1	O2	O3
0	0	0	0	0	0
0	0	1	0	0	1
0	1	0	0	1	0
0	1	1	0	1	1
1	0	0	1	1	0
1	0	1	1	1	1
1	1	0	1	0	1
1	1	1	1	0	0

(b) The inputs are

$$|000\rangle = \begin{pmatrix} 1 \\ 0 \end{pmatrix} \otimes \begin{pmatrix} 1 \\ 0 \end{pmatrix} \otimes \begin{pmatrix} 1 \\ 0 \end{pmatrix} = (1\ 0\ 0\ 0\ 0\ 0\ 0\ 0)^T$$

$$|001\rangle = \begin{pmatrix} 1 \\ 0 \end{pmatrix} \otimes \begin{pmatrix} 1 \\ 0 \end{pmatrix} \otimes \begin{pmatrix} 0 \\ 1 \end{pmatrix} = (0\ 1\ 0\ 0\ 0\ 0\ 0\ 0)^T$$

$$|010\rangle = \begin{pmatrix}1\\0\end{pmatrix} \otimes \begin{pmatrix}0\\1\end{pmatrix} \otimes \begin{pmatrix}1\\0\end{pmatrix} = (0\;0\;1\;0\;0\;0\;0\;0)^T$$

$$|011\rangle = \begin{pmatrix}1\\0\end{pmatrix} \otimes \begin{pmatrix}0\\1\end{pmatrix} \otimes \begin{pmatrix}0\\1\end{pmatrix} = (0\;0\;0\;1\;0\;0\;0\;0)^T$$

$$|100\rangle = \begin{pmatrix}0\\1\end{pmatrix} \otimes \begin{pmatrix}1\\0\end{pmatrix} \otimes \begin{pmatrix}1\\0\end{pmatrix} = (0\;0\;0\;0\;1\;0\;0\;0)^T$$

$$|101\rangle = \begin{pmatrix}0\\1\end{pmatrix} \otimes \begin{pmatrix}1\\0\end{pmatrix} \otimes \begin{pmatrix}0\\1\end{pmatrix} = (0\;0\;0\;0\;0\;1\;0\;0)^T$$

$$|110\rangle = \begin{pmatrix}0\\1\end{pmatrix} \otimes \begin{pmatrix}0\\1\end{pmatrix} \otimes \begin{pmatrix}1\\0\end{pmatrix} = (0\;0\;0\;0\;0\;0\;1\;0)^T$$

$$|111\rangle = \begin{pmatrix}0\\1\end{pmatrix} \otimes \begin{pmatrix}0\\1\end{pmatrix} \otimes \begin{pmatrix}0\\1\end{pmatrix} = (0\;0\;0\;0\;0\;0\;0\;1)^T$$

(c) The U matrix for the Toffoli gate is

$$\begin{pmatrix}1&0&0&0&0&0&0&0\\0&1&0&0&0&0&0&0\\0&0&1&0&0&0&0&0\\0&0&0&1&0&0&0&0\\0&0&0&0&1&0&0&0\\0&0&0&0&0&1&0&0\\0&0&0&0&0&0&0&1\\0&0&0&0&0&0&1&0\end{pmatrix}\begin{pmatrix}\square\\\square\\\square\\\square\\\square\\\square\\\square\\\square\end{pmatrix} = \begin{pmatrix}\square\\\square\\\square\\\square\\\square\\\square\\\square\\\square\end{pmatrix}$$

(d) The Toffoli gates can retrieve themselves in the following manner:

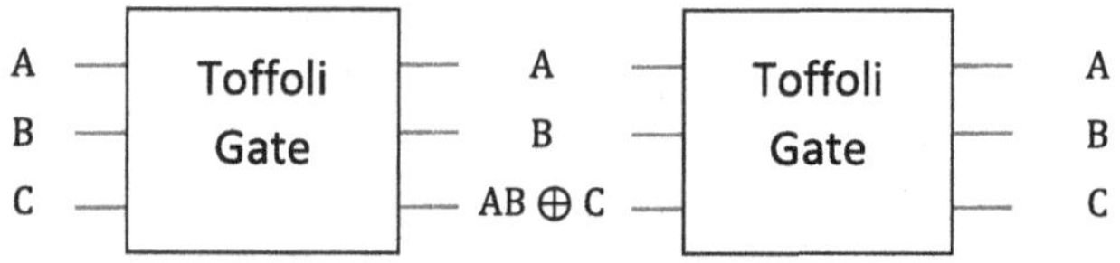

Boolean algebra:

$$(\mathrm{A}, \mathrm{B}, \mathrm{AB} \oplus \mathrm{C}) = (\mathrm{A}, \mathrm{B}, \mathrm{AB} \oplus (\mathrm{AB} \oplus \mathrm{C}) = (\mathrm{A}, \mathrm{B}, (\mathrm{AB} \oplus \mathrm{AB}) \oplus \mathrm{C}$$

But $(\mathrm{AB} \oplus \mathrm{AB}) = 0$

$$(\mathrm{A}, \mathrm{B}, (\mathrm{AB} \oplus \mathrm{AB}) \oplus \mathrm{C} = (\mathrm{A}, \mathrm{B}, 0 \oplus \mathrm{C}) = (\mathrm{A}, \mathrm{B}, \mathrm{C})$$

(e) The Toffoli gate simulates the classical NAND gate as follows:

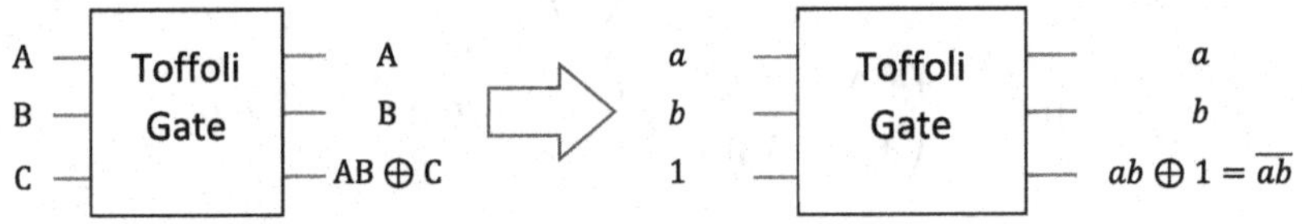

The third input bit is set to 1, and the first two input bits act as the input of the NAND gate. The third output bit would simulate the output of the NAND gate as shown by the Boolean algebra.

(f) The Toffoli gate simulates the classical FANOUT gate as follows:

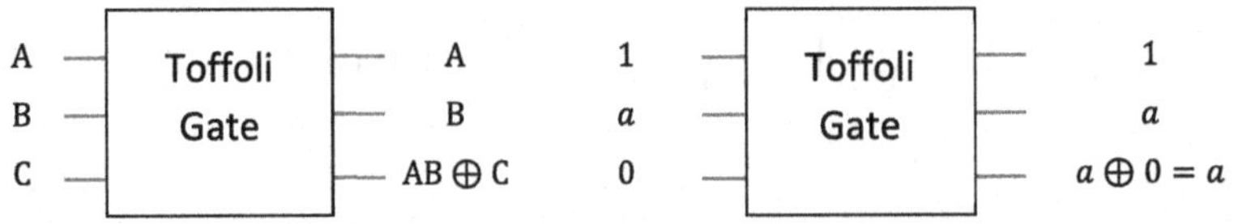

The third input bit is set to 0, and the first input bit is set to 1. The second input bit would act as the input to the FANOUT. The second and third output bits would be the output of the FANOUT.

Problem 8.12 Parallelism: Concept

The chief difference between a classical electronic logic gate and a quantum gate lies in the quantum gate's uncanny ability to take in multiple bits at the same time. This is the concept of parallelism. The Hadamard gate performs an operation as follows:

$$H|0\rangle = \frac{1}{\sqrt{2}}|0\rangle + \frac{1}{\sqrt{2}}|1\rangle$$

$$H|1\rangle = \frac{1}{\sqrt{2}}|0\rangle - \frac{1}{\sqrt{2}}|1\rangle$$

(a) Find $(H \otimes H)(|0\rangle \otimes |0\rangle)$.

(b) Show that the above can be extended to generate an output for multi-qubit parallel processing.

Solution

(a) The operation

$$
\begin{aligned}
(H \otimes H)(|0\rangle \otimes |0\rangle) &= (H|0\rangle) \otimes (H|0\rangle) \\
&= \left(\frac{1}{\sqrt{2}}|0\rangle \frac{1}{\sqrt{2}}|1\rangle\right) \otimes \left(\frac{1}{\sqrt{2}}|0\rangle \frac{1}{\sqrt{2}}|1\rangle\right) \\
&= \frac{1}{2}(|0\rangle \otimes |0\rangle + |0\rangle \otimes |1\rangle + |1\rangle \otimes |0\rangle + |1\rangle \otimes |1\rangle) \\
&= \frac{1}{2}(|00\rangle + |01\rangle + |10\rangle + |11\rangle)
\end{aligned}
$$

(b) Define

$$
H_n = (H \otimes H \otimes \ldots H^{nth}); \quad |0\rangle_n = (|0\rangle \otimes |0\rangle \otimes \ldots |0\rangle^{nth})
$$

Let H_n operate on $|0\rangle_n$

$$
\begin{aligned}
H_n|0\rangle_n &= (H \otimes H \otimes \ldots H^{nth})(|0\rangle \otimes |0\rangle \otimes \ldots |0\rangle^{nth}) \\
&= ((H|0\rangle)_1 \otimes (H|0\rangle)_2 \ldots \otimes (H|0\rangle)_n) \\
&= \left(\frac{1}{\sqrt{2}}(|0\rangle + |1\rangle)\right)_1 \otimes \left(\frac{1}{\sqrt{2}}(|0\rangle + |1\rangle)\right)_2 \\
&\quad \ldots \otimes \left(\frac{1}{\sqrt{2}}(|0\rangle + |1\rangle)\right)_n
\end{aligned}
$$

Subscript n denotes the nth term. It's worth noting that in the above, use has been made of the following:

$$
([A] \otimes [B]) \circ ([C] \otimes [D]) = ([A] \circ [C]) \otimes ([B] \circ [D])
$$

where $\otimes$ is the direct product multiplication and $\circ$ denotes the ordinary matrix multiplication.
Take the example of $n = 3$

$$
\begin{aligned}
H_3|0\rangle_3 = \left(\frac{1}{\sqrt{2}}\right)^3 &(|000\rangle + |001\rangle + |010\rangle + |011\rangle \\
&+ |100\rangle + |101\rangle + |110\rangle + |111\rangle)
\end{aligned}
$$

Write the above as follows:

$$H_3|0\rangle_3 = \left(\frac{1}{\sqrt{2}}\right)^3 (|1\rangle + |2\rangle + |3 + |4\rangle + |5\rangle + |6\rangle + |7\rangle + |8\rangle)$$

There are eight basis states. In the case of n, there will be 2^n basis states. All of the basis states will be processed by the quantum gate at the same time.

Problem 8.13 Parallelism: Quantum Circuit and Bell States

Bell states are some entangled states. The concept of parallelism is fully manifest in the process of creating the Bell states.

(a) Design a simple quantum circuit using standard quantum logic gates to demonstrate the notion of parallelism.
(b) Show that the circuit comprising a Hadamard and a CNOT gate can be used to create the Bell states.

Solution

(a) Let's examine the logic schematic shown in the following figure and focus on input X. If the circuit represents a classical gate, input X takes on just 0 or 1. If it is a quantum gate, it can take 0 and 1 at the same time. How so? A quantum state can be represented as a superposition of some basis states of $|0\rangle$ and $|1\rangle$, i.e., $|\psi_x\rangle = \frac{1}{\sqrt{2}}|0\rangle + \frac{1}{\sqrt{2}}|1\rangle$. This can be prepared by letting a Hadamard gate act on qubit $|0\rangle$. This new quantum state is no longer the computation basis of the quantum chip. To this quantum chip, $|0\rangle$ and $|1\rangle$ are the computation basis.

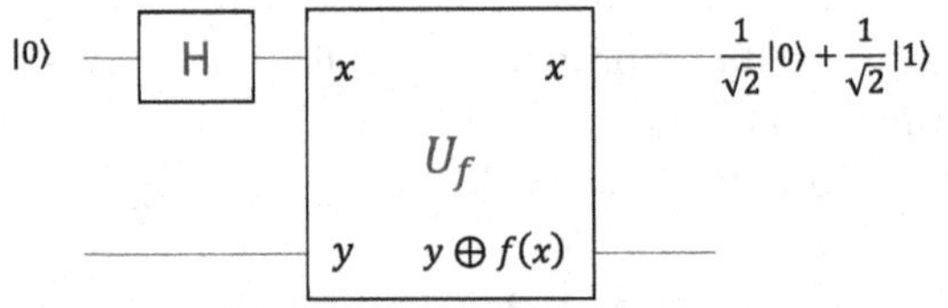

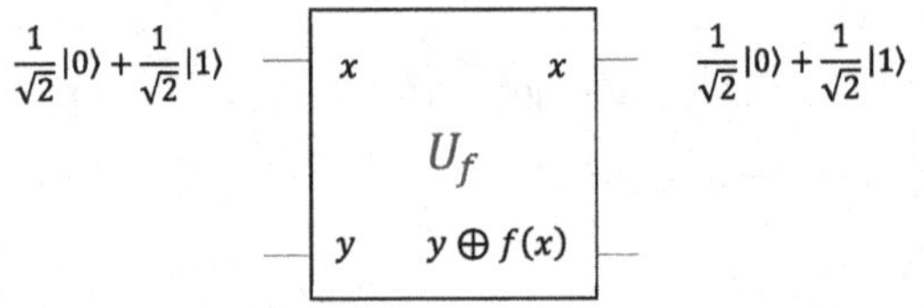

When the quantum state above is fed into X, the 0 and 1 states will be processed simultaneously. Note that the 0 and 1 states are not electrical voltage. They are quantum states associated with the probability of measuring angular momentum values at that location. In general, the input state on the left of the quantum chip is $|x, y\rangle$. The output state on the right of the chip is $|x, y \oplus f(x)\rangle$.

The following shows how the quantum chip processes information in a parallel manner when the input state on the X channel is $|\psi_x\rangle = \left(\frac{|0\rangle + |1\rangle}{\sqrt{2}}\right)$ and the Y channel is $|0\rangle$.

$$\left(\frac{|0\rangle + |1\rangle}{\sqrt{2}}\right) \quad \boxed{\begin{array}{c} x \qquad\qquad x \\[1em] U_f \\[1em] y \quad y \oplus f(x) \end{array}} \quad \left(\frac{|0\rangle + |1\rangle}{\sqrt{2}}\right)$$

$$|0\rangle \qquad\qquad |\psi_{out}\rangle = \left(\frac{|0, f(0)\rangle + |1, f(1)\rangle}{\sqrt{2}}\right)$$

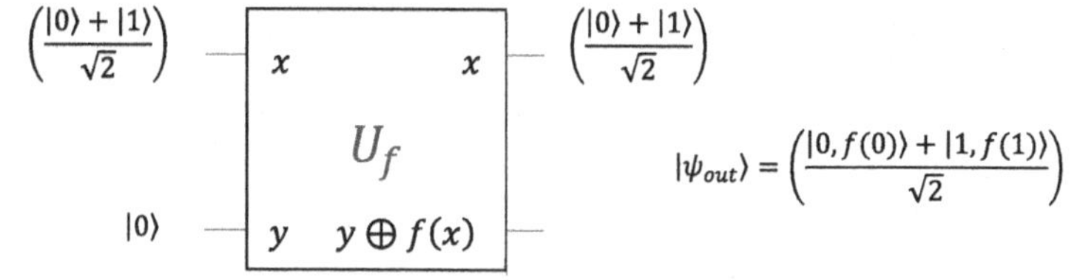

The input of X consists of bits 0 and 1 at the same time. Inside the chip, when input is bit 0, the output of $y \oplus f(x)$ is $0 \oplus f(0) = f(0)$. Likewise, for bit 1, $0 \oplus f(1) = f(1)$. Therefore,

$$|x, y \oplus f(x)\rangle \rightarrow |0, f(0)\rangle \quad and \quad |1, f(1)\rangle$$

The total output state is

$$|\psi_{out}\rangle = \left(\frac{|0, f(0)\rangle + |1, f(1)\rangle}{\sqrt{2}}\right)$$

(b) The Bell states are also known as the EPR states or the EPR pairs. In compact form, one can write the Bell states as

$$|\beta_{xy}\rangle = \left(\frac{|0, y\rangle + (-1)^x |1, \bar{y}\rangle}{\sqrt{2}}\right)$$

To create the Bell states, there are four possible inputs to consider: $|00\rangle, |01\rangle, |10\rangle, |11\rangle$. In the following figure, we will illustrate the process of creating a Bell state for input state of $|00\rangle$.

$$\left(\frac{|0\rangle + |1\rangle}{\sqrt{2}}\right) \quad \boxed{\begin{array}{c} x \qquad\qquad x \\[1em] U_f \\[1em] y \qquad x \oplus y \end{array}} \quad \left(\frac{|0\rangle + |1\rangle}{\sqrt{2}}\right)$$

$$|0\rangle \qquad\qquad |x \oplus 0\rangle \qquad |\psi_{out}\rangle = \left(\frac{|00\rangle + |11\rangle}{\sqrt{2}}\right)$$

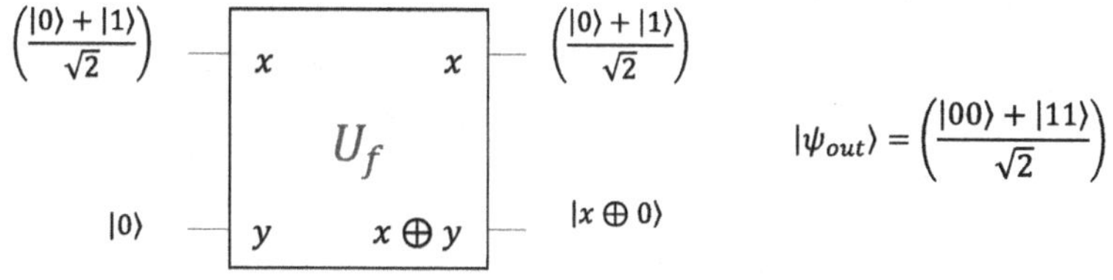

In the above, $|00\rangle$ generates a total input of

$$|\psi_{in}\rangle = \left(\frac{|00\rangle + |10\rangle}{\sqrt{2}}\right)$$

The CNOT gate flips on 1, therefore,

$$\left(\frac{|00\rangle + |10\rangle}{\sqrt{2}}\right) \underset{CNOT}{\rightarrow} \left(\frac{|00\rangle + |11\rangle}{\sqrt{2}}\right)$$

The same principle works for all four inputs.
For input $|00\rangle$, one has

$$\left(\frac{|00\rangle + |10\rangle}{\sqrt{2}}\right) \underset{CNOT}{\rightarrow} \left(\frac{|00\rangle + |11\rangle}{\sqrt{2}}\right)$$

For input $|01\rangle$, one has

$$\left(\frac{|01\rangle + |11\rangle}{\sqrt{2}}\right) \underset{CNOT}{\rightarrow} \left(\frac{|01\rangle + |10\rangle}{\sqrt{2}}\right)$$

For input $|10\rangle$, one has

$$\left(\frac{|00\rangle - |10\rangle}{\sqrt{2}}\right) \underset{CNOT}{\rightarrow} \left(\frac{|00\rangle - |11\rangle}{\sqrt{2}}\right)$$

For input $|11\rangle$, one has

$$\left(\frac{|01\rangle - |11\rangle}{\sqrt{2}}\right) \underset{CNOT}{\rightarrow} \left(\frac{|01\rangle - |10\rangle}{\sqrt{2}}\right)$$

Remarks and Reflections

Referring to (a), if one wants to have more than two bits being processed in parallel, one can perform the following:

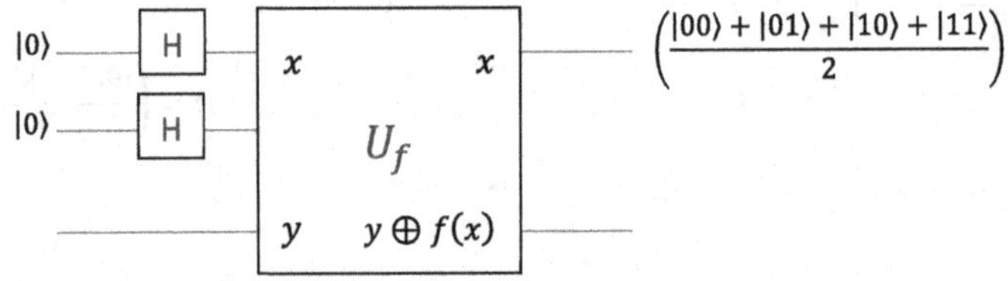

The two Hadamard gates produce a $\frac{1}{\sqrt{2}}(|0\rangle + |1\rangle)$ each. And they combine to form

$$\left(\frac{|0\rangle + |1\rangle}{\sqrt{2}}\right) \otimes \left(\frac{|0\rangle + |1\rangle}{\sqrt{2}}\right) = \left(\frac{|00\rangle + |01\rangle + |10\rangle + |11\rangle}{2}\right)$$

Problem 8.14 No-Cloning Principle

Cloning refers to a quantum operation (C) that copies and pastes a quantum state in the following manner:

$$C|\phi 0\rangle = |\phi\phi\rangle$$

Show that this is an impossible action as far as quantum mechanics is concerned. In other words, the unitary function that performs the quantum copy-and-paste as described above does not exist.

Solution

Consider a quantum state

$$|\psi\rangle = \frac{1}{\sqrt{2}}(|\phi_1\rangle + |\phi_2\rangle)$$

where $|\phi_1\rangle, |\phi_2\rangle$ are two orthonormal states. Imagine a cloning operator C that clones $|\psi\rangle$, and one would expect the following operation to exist without contradiction:

$$C|\psi 0\rangle = |\psi\psi\rangle = \frac{1}{2}(|\phi_1\phi_1\rangle + |\phi_1\phi_2\rangle + |\phi_2\phi_1\rangle + |\phi_2\phi_2\rangle) \quad \text{(A)}$$

Let's examine the following:

$$C|\psi 0\rangle = C\left(\frac{1}{\sqrt{2}}(|\phi_1 0\rangle + |\phi_2 0\rangle)\right)$$

As the C operator is linear,

$$C\left(\frac{1}{\sqrt{2}}(|\phi_1 0\rangle + |\phi_2 0\rangle)\right) = \left(\frac{1}{\sqrt{2}}(C|\phi_1 0\rangle + C|\phi_2 0\rangle)\right)$$
$$= \left(\frac{1}{\sqrt{2}}(|\phi_1\phi_1\rangle + |\phi_2\phi_2\rangle)\right) \quad \text{(B)}$$

It is apparent that Eq. (A) is NOT identical to Eq. (B). Therefore, the cloning operation cannot be performed. In other words, the unitary operator C does not exist.

Problem 8.15 Quantum Circuit: No-Cloning

The principle of no-cloning was elucidated in the previous problem. Design a quantum circuit based on the CNOT to illustrate a practical example of the no-cloning principle.

Solution

For clarity, we consider quantum circuit that comprises a CNOT gate as shown in the following. Enter a quantum state $|\psi\rangle = a|0\rangle + b|1\rangle$ on the control bit of a CNOT gate as follows:

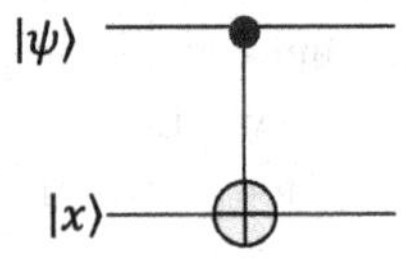

The input state would be

$$|\psi\rangle_{in} = a|0x\rangle + b|1x\rangle$$

As the CNOT flips the target input on control state $|1\rangle$, the output would be

$$|\psi\rangle_{out} = a|0x\rangle + b|1\bar{x}\rangle$$

If the target input is a state $|x\rangle = |0\rangle$, the output is

$$|\psi\rangle_{out} = a|00\rangle + b|11\rangle$$

The desired output would be

$$\begin{aligned} |\psi\rangle_{out} = |\psi\rangle_{in} \otimes |\psi\rangle_{in} &= (a|0\rangle + b|1\rangle) \otimes (a|0\rangle + b|1\rangle) \\ &= a^2|00\rangle + ab|01\rangle + ab|10\rangle + b^2|11\rangle \qquad \text{(A)} \end{aligned}$$

But the actual output is

$$|\psi\rangle_{out} = a|00\rangle + b|11\rangle \qquad \text{(B)}$$

No-cloning principle is illustrated in this simple circuit when one sees that Eq. (A) does not match Eq. (B).

Remarks and Reflections

Inspecting Eqs. (C) and (D), one could note that the desired output matches the actual when $(a,b) = (0,1)$ or $(1,0)$, i.e., $|\psi\rangle_{in} = |00\rangle$ or $|11\rangle$. In other words, the circuit is not copying the quantum state unless $ab = 0$. This is known as the no-cloning theorem.

Problem 8.16 Quantum Circuit: Swapping

In quantum computation, quantum logic gates are used to perform gate operation in analogy with classical logic gates like the NOT, OR, AND, XOR, and so on. The CNOT gate is the quantum version of the classical XOR gate. The quantum circuit in the following (Figure 6) comprises three CNOT gates that perform the simple task of swapping the input states of the two qubits. The CNOT gate and its classical equivalence are shown together with their respective truth tables on the right (Table 1).

(a) Let's begin with the Boolean algebra where $a \oplus b = a.\bar{b} + b.\bar{a}$, and it follows that

$$(a \oplus b) \oplus c = (a \oplus b).\bar{c} + (\overline{a \oplus b}).c = (a.\bar{b} + b.\bar{a}).\bar{c} + (\overline{a.\bar{b} + b.\bar{a}}).c$$

Making use of the relation above, and referring to the standard Boolean algebra identities, prove the associative law for the following Boolean algebra

$$(a \oplus b) \oplus c = a \oplus (b \oplus c)$$

(b) Show using the truth table for the CNOT gate above that the input states will be swapped on the output upon going through all three CNOT gates as shown in the following:

$$|a, b\rangle \rightrightarrows |b, a\rangle$$

(c) Show how the quantum states change at every stage of the circuit as shown above.

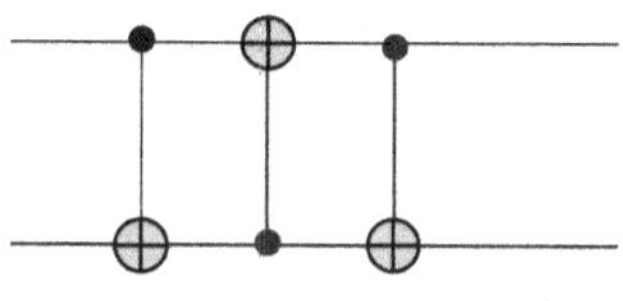

Fig. 6. A swap quantum circuit comprising three CNOT gates connected as shown.

Table 1. Comparing the truth tables of the CNOT and the classical XOR gate.

CNOT Quantum Truth Table

	x	y	x	$y \oplus x$
1.	$\vert 0\rangle$	$\vert 0\rangle$	$\vert 0\rangle$	$\vert 0\rangle$
2.	$\vert 0\rangle$	$\vert 1\rangle$	$\vert 0\rangle$	$\vert 1\rangle$
3.	$\vert 1\rangle$	$\vert 0\rangle$	$\vert 1\rangle$	$\vert 1\rangle$
4.	$\vert 1\rangle$	$\vert 1\rangle$	$\vert 1\rangle$	$\vert 0\rangle$

XOR Classical Truth Table

	x	y	x	$y \oplus x$
1.	0	0	0	0
2.	0	1	0	1
3.	1	0	1	1
4.	1	1	1	0

Note: $\oplus$ *is addition in the sense of the Boolean algebra, aka addition modulo 2.*
Note: ($^\circ$) *denotes the AND operation.* (+) *denotes the OR operation.*

Solution

(a) Let's begin with $a \oplus b = a.\bar{b} + b.\bar{a}$, it follows that

$$\begin{aligned}(a \oplus b) \oplus c &= (a \oplus b).\bar{c} + (\overline{a \oplus b}).c \\ &= (a.\bar{b} + b.\bar{a}).\bar{c} + (\overline{a.\bar{b} + b.\bar{a}}).c \qquad \text{(A)}\end{aligned}$$

Now, let's work on the second term on the RHS of (A): $(\overline{a.\bar{b} + b.\bar{a}}).c$. De Morgan's law leads to

$$(\overline{a \oplus b}).c = (\overline{a.\bar{b} + b.\bar{a}}).c = (\overline{a.\bar{b}}).(\overline{b.\bar{a}}).c = (\bar{a} + b).(\bar{b} + a).c \qquad \text{(B)}$$

Expanding (B) leads to

$$(\overline{a \oplus b}).c = \bar{a}.\bar{b}.c + b.a.c \qquad \text{(C)}$$

Now, reverting to the RHS of (A), and taking note of (C), the following is found:

$$\begin{aligned}(a \oplus b) \oplus c &= (a \oplus b).\bar{c} + (\overline{a \oplus b}).c = (a.\bar{b} + b.\bar{a}).\bar{c} + \bar{a}.\bar{b}.c + b.a.c \\ &= \bar{a}(b.\bar{c} + \bar{b}.c) + a.(\bar{b}.\bar{c} + b.c) = \bar{a}.(b \oplus c) + a.(\bar{b}.\bar{c} + b.c) \qquad \text{(D)}\end{aligned}$$

Last is to address the term $a.(\bar{b}.\bar{c} + b.c)$ of D. Taking note of C once again, i.e., since $(\overline{a \oplus b}).c = (\bar{a}.\bar{b} + b.a).c$, replacing the terms in the following manner

$$a \to b, \quad b \to c, \quad c \to a$$

results in

$$(\bar{a}.\bar{b} + b.a).c \to (\bar{b}.\bar{c} + c.b).a$$
$$(\overline{a \oplus b}).c \to (\overline{b \oplus c}).a$$

Finally,

$$\begin{aligned}(a \oplus b) \oplus c &= \bar{a}.(b \oplus c) + a.(\bar{b}.\bar{c} + b.c)\\ &= \bar{a}.(b \oplus c) + a.(\overline{b \oplus c})\\ &= a \oplus (b \oplus c)\end{aligned}$$

Therefore,

$$(a \oplus b) \oplus c = a \oplus (b \oplus c)$$

(b) Finally, consolidating all side derivations, the swap gate operates as follows:

$$|a, b\rangle \to |a, a \oplus b\rangle \to |a \oplus (a \oplus b), a \oplus b\rangle$$

By the associative law: $a \oplus (a \oplus b) = (a \oplus a) \oplus b = 0 \oplus b = b$. It follows that

$$|a \oplus (a \oplus b), a \oplus b\rangle = |b, b \oplus (a \oplus b)\rangle$$

By the commutative law proven in (a), $b \oplus (a \oplus b) = (a \oplus b) \oplus b$. Moving on with the associative law,

$$(a \oplus b) \oplus b = a \oplus (b \oplus b) = a \oplus 0 = a$$

It thus follows that $|b, b \oplus (a \oplus b)\rangle = |b, a\rangle$.

(c) Quantum states at every stage are illustrated in the swap quantum circuit (see the following figure).

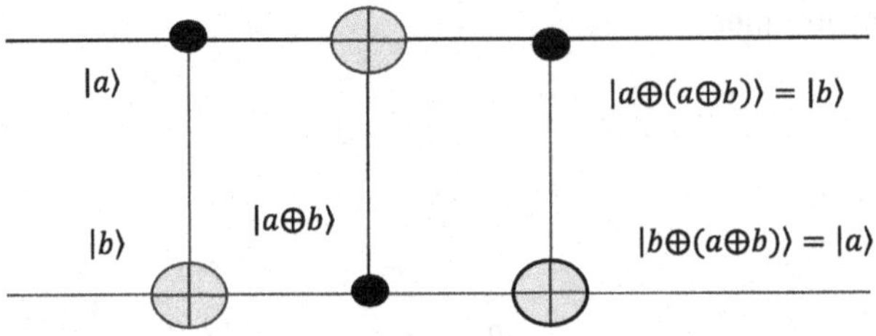

Problem 8.17 Mathematical Methods: Boolean Algebra

Boolean algebra is the language of digital logics commonly used in the design of electronic as well as quantum circuits. Refer to the following:

AND (°)	OR (+)
$1^\circ A = A$	$0 + A = A$
$0^\circ A = 0$	$1 + A = 1$
$A^\circ A = A$	$A + A = A$

	AND (°)	OR (+)
Commutative Law	$A^\circ B = B^\circ A$	$A + B = B + A$
Associative Law	$(A^\circ B)^\circ C = A^\circ(B^\circ C)$	$(A + B) + C = A + (B + C)$
Distributive Law		$A^\circ(B + C) = A^\circ B + A^\circ C$

Prove the following:

(a) Inverse law: $A + \bar{A} = 1$
(b) Inverse law: $A^\circ \bar{A} = 0$
(c) $A^\circ(A + B) = A$
(d) $A + (A^\circ B) = A$
(e) Distributive law for AND: $A + B^\circ C = (A + B)^\circ(A + C)$

Solution

(a) Inverse law: $A + \bar{A} = 1$

$$(A + A) + \bar{A} = A + (A + \bar{A})$$

Since $A + A = A$,

$$A + \bar{A} = A + (A + \bar{A})$$

Denote $A + \bar{A}$ with R,

$$A + R = R \to R \text{ must be } 1$$

(b) Inverse law: $A^\circ \bar{A} = 0$

$$(A^\circ A)^\circ \bar{A} = A^\circ (A^\circ \bar{A})$$

Since $A^\circ A = A$,

$$A^\circ \bar{A} = A^\circ (A^\circ \bar{A})$$

Denote $A^\circ \bar{A}$ with R

$$R = A^\circ R \to R \text{ must be } 0$$

(c) $A^\circ (A + B) = A$
Use is made of the OR distributive law

$$A^\circ (A + B) = A^\circ A + A^\circ B = A + A^\circ B$$

Since $A + 1 = 1$,

$$A + A^\circ B = A^\circ (1 + B) = A$$

(d) $A + (A^\circ B) = A$
Use is made of the OR distributive law

$$A + (A^\circ B) = A^\circ (1 + B) = A$$

(e) Distributive law for AND: $A + B^\circ C = (A + B)^\circ (A + C)$
Use is made of the OR distributive law

$$(A + B)^\circ (A + C) = A^\circ A + A^\circ C + B^\circ A + B^\circ C$$

Denote $(A + B)^\circ (A + C)$ with R. With $A^\circ A = A$,

$$\begin{aligned} R &= A + A^\circ C + B^\circ A + B^\circ C \\ &= A^\circ(1 + C) + B^\circ A + B^\circ C \\ &= A + B^\circ A + B^\circ C \\ &= (1 + B)^\circ A + B^\circ C \\ &= A + B^\circ C \end{aligned}$$

Remarks and Reflections

Boolean Algebra

	AND ($^\circ$)	OR ($+$)	Remark
	$1^\circ A = A$	$0 + A = A$	By inspection
	$0^\circ A = 0$	$1 + A = 1$	By inspection
	$A^\circ A = A$	$A + A = A$	By inspection
Inverse Law	$A^\circ \bar{A} = 0$	$A + \bar{A} = 1$	
Commutative Law	$A^\circ B = B^\circ A$	$A + B = B + A$	By inspection
Associative Law	$(A^\circ B)^\circ C = A^\circ(B^\circ C)$	$(A + B) + C = A + (B + C)$	
Distributive Law	$A + B^\circ C = (A + B)^\circ(A + C)$	$A^\circ(B + C) = A^\circ B + A^\circ C$	
De Morgan's Law	$\overline{A^\circ B} = \bar{A} + \bar{B}$	$\overline{A + B} = \bar{A}^\circ \bar{B}$	

Problem 8.18 Quantum Communication: Teleportation

The following schematic circuit (Figure 7) shows a quantum-classical hybrid circuit that can be employed to teleport a quantum state to a recipient party. Qubits are pushed through lines 1, 2, and 3 from the left of the circuit to the right, and the message encoded in qubit 1 would appear in a jumbled-up form on qubit 3.

The quantum state that comprises sender (Alice) and recipient (Bob)'s states at stage 0 of the information path is given by

$$|\psi\rangle|\psi\rangle_{00} = \frac{1}{\sqrt{2}}[\alpha(|000\rangle + |011\rangle) + \beta(|100\rangle + |111\rangle)]$$

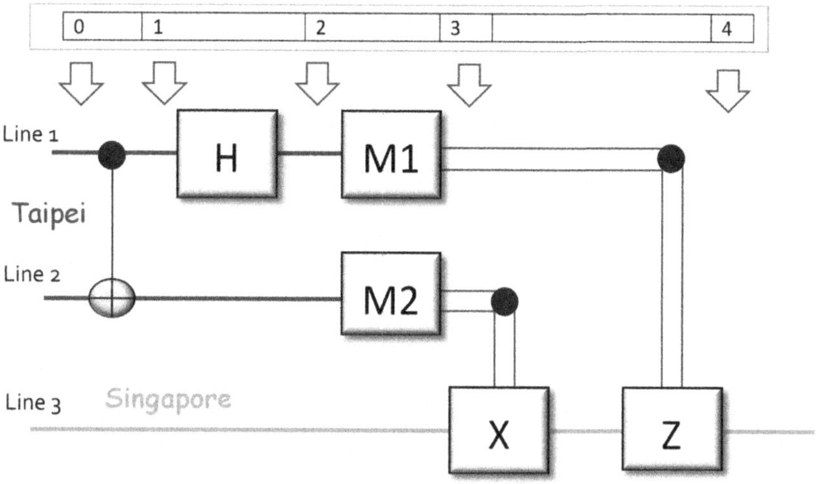

Fig. 7. A teleportation circuit that illustrates the stages of communication between sender and recipient. The circuit is shown with all its component quantum gates.

(a) Show that the quantum state at stage 1 of the information path is as shown in the following. Comment on Alice and Bob's quantum states at this stage.

$$|\psi\rangle_1 = \frac{1}{\sqrt{2}}[\alpha(|000\rangle + |011\rangle) + \beta(|110\rangle + |101\rangle)]$$

(b) Next, qubit 1 is being sent through a Hadamard gate. Show that the quantum state at stage 2 of the information path is as shown in the following:

$$|\psi\rangle_2 = \frac{1}{2}[|00\rangle(\alpha|0\rangle + \beta|1\rangle) + |01\rangle(\alpha|1\rangle + \beta|0\rangle) + |10\rangle(\alpha|0\rangle - \beta|1\rangle) + |11\rangle(\alpha|1\rangle - \beta|0\rangle)]$$

(c) Alice could now measure her own two bits. Explain why it is OK for her to perform measurement on her qubits at stage 2 while it was not advisable for her to do the same at state 0.

(d) There are four possible states that Alice may obtain from her measurement — each of which corresponds to a state on Bob's qubit 3 as shown in the following table. Alice needs to communicate the results of her measurement to Bob. Explain what Bob needs to do upon fixing up his qubit 3 according to the classical information he might obtain from Alice

Alice	Bob
$\|00\rangle$	$\alpha\|0\rangle + \beta\|1\rangle$
$\|01\rangle$	$\alpha\|1\rangle + \beta\|0\rangle$
$\|10\rangle$	$\alpha\|0\rangle - \beta\|1\rangle$
$\|11\rangle$	$\alpha\|1\rangle - \beta\|0\rangle$

Solution

(a) CNOT gate executes a flip-on-1 and stay-on-0. Careful inspection of Alice's qubits (1 and 2) in each of the four terms indeed confirms such operation. Alice's qubit 2 has changed according to the operation of the CNOT gate. Bob's qubit 3 remains unchanged throughout.

(b) Hadamard operation on the first bit leads to

$$|\psi\rangle_2 = \frac{1}{2}[\alpha(|0\rangle + |1\rangle)(|00\rangle + |11\rangle) + \beta(|0\rangle - |1\rangle)(|10\rangle + |01\rangle)]$$

Rearrangement leads to

$$|\psi\rangle_2 = \frac{1}{2}[|00\rangle(\alpha|0\rangle + \beta|1\rangle) + |01\rangle(\alpha|1\rangle + \beta|0\rangle) + |10\rangle(\alpha|0\rangle - \beta|1\rangle) + |11\rangle(\alpha|1\rangle - \beta|0\rangle)]$$

(c) This is because of the following:

(1) At stage 0, the quantum message will be destroyed the moment it is measured. Not to forget that α and β are probability amplitudes. The only way to obtain their accurate values is to measure the quantum message many times. But each time it is measured, it is destroyed and has to be prepared again. This is very impractical.

(2) At stage 0, even if repeating measurements are feasible, the values of α and β are far too elaborate in detail (decimal places). It will take forever for Alice to communicate their values to Bob.

(3) On the other hand, at stage 2, Alice doesn't need to do it many times as all she needs is the quantum state of her two qubits, not their probability amplitude. In other words, it is fine to have her

quantum state destroyed upon measurement because she only needs to measure once.

(d) Bob's states, except for the $|00\rangle$ correspondence, are all jumble-up of the original message on qubit 1. Bob would need to do the following:

Alice	**Bob**	**Gate Operation**	
$\|00\rangle$	$\alpha\|0\rangle + \beta\|1\rangle$	NIL	$\alpha\|0\rangle + \beta\|1\rangle$
$\|01\rangle$	$\alpha\|1\rangle + \beta\|0\rangle$	X gate	$\alpha\|0\rangle + \beta\|1\rangle$
$\|10\rangle$	$\alpha\|0\rangle - \beta\|1\rangle$	Z gate	$\alpha\|0\rangle + \beta\|1\rangle$
$\|11\rangle$	$\alpha\|1\rangle - \beta\|0\rangle$	X gate, then Z gate	$\alpha\|0\rangle + \beta\|1\rangle$

Problem 8.19 Deutsch Algorithm

In Deutsch algorithm, the concepts of superposition and parallelism are exploited. A function $f(x)$ takes a 0 or 1 to produce either 0 a or 1, with an unknown correlation. There are, henceforth, fourpossible outputs (0,0), (0,1), (1,0), (1,1) as shown in the following table. One needs to test the function two times to tabulate all possibilities.

Shown in the following are four combinations of the functions in which two of them correspond to $f(0) = f(1)$ and the other two correspond to $f(0) \neq f(1)$.

$(f(0), f(1)) = (0,0)$	$f(0) = f(1)$
$(f(0), f(1)) = (1,1)$	$f(0) = f(1)$
$(f(0), f(1)) = (0,1)$	$f(0) \neq f(1)$
$(f(0), f(1)) = (1,0)$	$f(0) \neq f(1)$

Using a quantum circuit as shown in the following, we show that one needs to run the circuit just one time to determine if $f(0) = f(1)$ or $f(0) \neq f(1)$. This quantum circuit is designed with three Hadamard logic gates connected to the Oracle chip of the said function as shown in Figure 8. The principle of operation behind the design of this circuit is the algorithm known as the Deutsch algorithm.

Explain how the quantum circuit above is used to perform a fast determination of functions: $f(0), f(1)$.

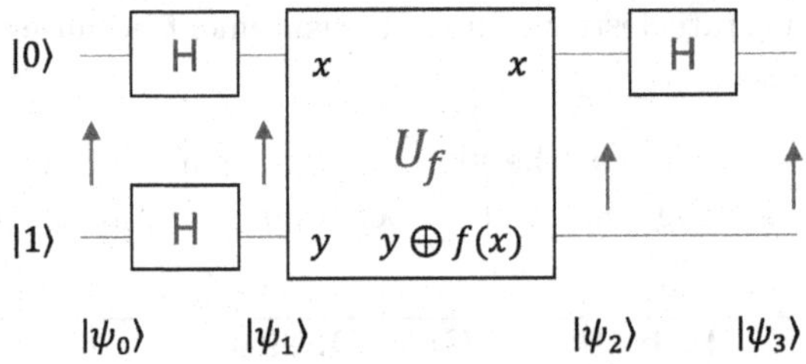

Fig. 8. A quantum circuit designed to realize the Deutsch algorithm.

Solution

First of all, an input state $|\psi_0\rangle = |01\rangle$ is sent through two Hadamard gates as shown above. The first and second Hadamard gates act on $|0\rangle$ and $|1\rangle$, respectively. The results are

$$H|0\rangle = \left(\frac{|0\rangle + |1\rangle}{\sqrt{2}}\right), \quad H|1\rangle = \left(\frac{|0\rangle - |1\rangle}{\sqrt{2}}\right)$$

Therefore, at stage 1, a product state is generated as follows:

$$|\psi_1\rangle = \left(\frac{|0\rangle + |1\rangle}{\sqrt{2}}\right) \otimes \left(\frac{|0\rangle - |1\rangle}{\sqrt{2}}\right)$$

Next is feeding the output of the Hadamard gates into the quantum chip (with a built-in oracle). One would thus obtain $|\psi_2\rangle$ on the output of the quantum chip.

$$\begin{aligned} |\psi_2\rangle = |x, y \oplus f(x)\rangle &= |x\rangle \left| \left(\frac{|0\rangle - |1\rangle}{\sqrt{2}}\right) \oplus f(x) \right\rangle \\ &= |x\rangle \left| \left(\frac{|0\rangle \oplus f(x) - |1\rangle \oplus f(x)}{\sqrt{2}}\right) \right\rangle \end{aligned}$$

Recalling the operations in Boolean algebra in the quantum sense,

$$\begin{aligned} |0 \oplus f(x)\rangle &= |f(x)\rangle \\ |1 \oplus f(x)\rangle &= |\overline{f(x)}\rangle \end{aligned}$$

The following is obtained in compact form:

$$|\psi_2\rangle = |x\rangle \left| \left(\frac{|f(x)\rangle - |\overline{f(x)}\rangle}{\sqrt{2}}\right) \right\rangle = |x\rangle \left(\frac{|0\rangle - |1\rangle}{\sqrt{2}}\right) (-1)^{f(x)}$$

For visual clarity, substitute $f(x) = 1, 0$ into the above. The input on the x-terminal is $|x\rangle = \left(\frac{|0\rangle+|1\rangle}{\sqrt{2}}\right)$, the above can thus be expanded as follows:

$$|\psi_2\rangle = \frac{|0\rangle}{\sqrt{2}}\left(\frac{|0\rangle - |1\rangle}{\sqrt{2}}\right)(-1)^{f(0)} + \frac{|1\rangle}{\sqrt{2}}\left(\frac{|0\rangle - |1\rangle}{\sqrt{2}}\right)(-1)^{f(1)} \qquad \text{(A)}$$

The state of $|\psi_2\rangle$ contains information of all function combinations. A summary is provided in the following table. The RHS of (A) shows that operations of $f(0)$ and $f(1)$ are performed by the quantum chip at the same time. This is parallel processing. As said, $(f(0), f(1))$ is unknown and there are four possibilities. We list the outcome on $\psi_2\rangle$ for all four possibilities as shown in the following:

Function Combinations	$\boldsymbol{|\psi_2\rangle}$
$(f(0), f(1)) = (0, 0)$	$\left(\frac{|0\rangle + |1\rangle}{\sqrt{2}}\right)\left(\frac{|0\rangle - |1\rangle}{\sqrt{2}}\right)$
$(f(0), f(1)) = (0, 1)$	$\left(\frac{|0\rangle - |1\rangle}{\sqrt{2}}\right)\left(\frac{|0\rangle - |1\rangle}{\sqrt{2}}\right)$
$(f(0), f(1)) = (1, 0)$	$-\left(\frac{|0\rangle - |1\rangle}{\sqrt{2}}\right)\left(\frac{|0\rangle - |1\rangle}{\sqrt{2}}\right)$
$(f(0), f(1)) = (\mathbf{1}, \mathbf{1})$	$-\left(\frac{|0\rangle + |1\rangle}{\sqrt{2}}\right)\left(\frac{|0\rangle - |1\rangle}{\sqrt{2}}\right)$

From the above, one can reorganize the four combinations into two sets as follows:

Function Properties	$\boldsymbol{|\psi_2\rangle}$
$f(0) = f(1)$	$\pm\left(\frac{|0\rangle + |1\rangle}{\sqrt{2}}\right)\left(\frac{|0\rangle - |1\rangle}{\sqrt{2}}\right)$
$f(0) \neq f(1)$	$\pm\left(\frac{|0\rangle - |1\rangle}{\sqrt{2}}\right)\left(\frac{|0\rangle - |1\rangle}{\sqrt{2}}\right)$

On $|\psi_3\rangle$, the X-terminal on the right is fed through a Hadamard gate once again (see Figure). One would thus have

Function Properties	$|\psi_3\rangle$
$f(0) = f(1)$	$\pm(|0\rangle)\left(\frac{|0\rangle - |1\rangle}{\sqrt{2}}\right)$
$f(0) \neq f(1)$	$\pm(|1\rangle)\left(\frac{|0\rangle - |1\rangle}{\sqrt{2}}\right)$

Finally, one works out the following, since $|0\rangle$ on the X output corresponds to $f(0) = f(1)$ and $|1\rangle$ on X corresponds to $f(0) \neq f(1)$, one works out the following:

$$|\psi_3\rangle = \pm(f(0) \oplus f(1))\left(\frac{|0\rangle - |1\rangle}{\sqrt{2}}\right)$$